THE PLACE OF SCIENCE IN MODERN CIVILIZATION

THE PLACE OF SCIENCE IN MODERN CIVILIZATION
and Other Essays

Thorstein Veblen

with a New Introduction by
Warren J. Samuels

Transaction Publishers
New Brunswick (U.S.A.) and London (U.K.)

New material copyright © 1990
by Transaction Publishers,
New Brunswick, New Jersey 08903
Originally published in 1919/ Viking Press

Library of Congress Catalog Number: 89-30458
ISBN: 0-88738-808-6
Printed in the United States of America

Library of Congress Cataloging-in-Publication Data

Veblen, Thorstein, 1857–1929.
    [Place of sciene in modern civilisation]
    The place of science in modern civilization /
Thorstein Veblen ; with a new introduction by
Warren J. Samuels.
        p. cm.
    Reprint. Originally published: The place of science in modern civilisation. New York : Viking Press, 1919.
        ISBN 0-88738-808-6
        1. Economics.   2. Science.   I. Title.
HB34.V4   1989
330—dc20   89-30458
                                                    CIP

CONTENTS

INTRODUCTION TO THE
TRANSACTION EDITION

Warren J. Samuels

Thorstein Veblen's *The Place of Science In Modern Civilisation,* published in 1919 as a collection of previously published essays, is a major contribution by a writer who continues to command attention and respect almost three quarters of a century later.

The first three decades of the twentieth century witnessed the work of three writers who, each in their own way, produced comprehensive theories of society. The theoretical systems of Max Weber, Vilfredo Pareto, and Veblen made substantive contributions to the fields of economics, sociology, political science, and psychology, as well as sociolinguistics, broadly defined. The list of their principal modern precursors is arguably small but distinguished: Giambattista Vico, Adam Smith, Henri de Saint-Simon, Auguste Comte, and Karl Marx. Besides their comprehensiveness, Weber, Pareto, and Marx emphasized the interdependence of society. Neither polity nor economy, nor general culture nor language, was self-contained, each had profound impact on the other; society is a set of interacting subsystems. Also, many of their substantive formulations and contributions were parallel. For example, Weber placed emphasis on modern rationalism, Pareto on logico-experimental knowledge, and Veblen on mat-

ter-of-fact habits of thought, each with modern science
(and technology) as a corollary. They also stressed the
fundamental, albeit partly derivative, importance of the
modern state and, *inter alia,* what may be comprehended as the common interactive tripartite elements
of their overall systems, power, knowledge, and psychology.

Veblen was heterodox, iconoclastic, sardonic,
caustic, and satiric. He also was brilliant, penetrating,
original, courageous, literarily dramatic, and unique, as
well as intellectually distant, if not alienated from the
world around him—a distance that may have been both
a help and a hindrance for his thinking. Some of these
characteristics he shared with Weber, others with
Pareto, some with both.

The continuing relevance of the positive analytical
work of Veblen, Weber, and Pareto, derives in part
from the depth and breadth of their analysis, especially
their work on fundamental matters, which makes it in
many respects timeless. The continuing relevance of
Veblen's critical work, directed with often devastating
implications for both existing capitalist society and
mainstream economic theory, as well as for Marxism
and socialism in general, derives from both the penetra-
tion and range of the positive analysis on which it rests,
and the arguable failure of both capitalist society and
neoclassical theory to change in material respects in
the years since Veblen wrote. The continuing relevance
of Veblen's topics, and of his specific ideas, is manifest
in such diverse contemporary areas as the continuing
critique of mainstream economics, much of which is a
series of footnotes to Veblen; to controversies over the
relations of deduction and induction (and the status of

efforts to produce truth, with or without a capital T) to belief systems and language; to disputes about the significance of business mergers and acquisitions, considered as portfolio investment (and not direct contributions to income in the GNP sense), as substitutes for real investment as a principal money-making activity; and *inter alia,* to questions about the historic meaning and status of socialism.

The first six essays of this volume comprise fundamental, systemic, positive contributions to the study of the preconceptions that drive thought and modern science, and the origins of those preconceptions.

The next nine essays represent applications of Veblen's thinking to the critique of both the work of other economists and capitalism in general. These essays also make independently valuable contributions of their own. Three essays, on industrial and pecuniary employments and on the nature of capital, present a fundamental component of Veblen's view of capitalism and its problems, and are of lasting interpretive and analytical value. These nine essays both express Veblen's point of view and identify the limits of the work of such writers as John Bates Clark, Gustav Schmoller, and Karl Marx considered from a positivist's point of view. That Clark was a leading mainstream economist, Schmoller a major historicist, and Marx, well, Marx, indicates something of the breadth and power of Veblen's theoretical position. The arguments of these critical essays are also manifestly applicable to all sorts of comparable present-day theorizing.

The final three essays, especially the last two, reflect, alas, the invidious eugenicism, yes, racism, common among social scientists of Veblen's era; except as examples of that genre, and of the associated conjectural history utilizing natural selection, they have no other permanent value.

The arguments of the first six chapters may be considered as a whole. Veblen argued that mankind pursued its intellectual efforts, at the deepest structural levels, along certain habits of thought and systems of preconceptions, or prepossessions. These habitualized and typically unrecognized and unchallenged preconceptions were derived from the practices of ordinary life, especially how people made a living and the institutional arrangements under which people lived. He argued there have been two principal habits of thought or systems of preconceptions, the animistic or teleological, and the matter-of-fact.

The *animistic or teleological* preconception typically tends to project a personalized conceptualization of ultimate design, reality, and purpose. It involved, as Veblen portrayed it, a combination of projection and ceremonial rationalization, and served both as a psychic balm and as social control functions. At its least animistic it is a taxonomic venture, but more typically it is teleological in its imputation of final causes and inevitability of results, both centering on some notion of the substantiality of reality preeminent to man, the latter even when, as in more modern times, creation is understood to be particularly dedicated to the welfare of mankind. Veblen wrote specifically about classical economics, but in terms applicable to all forms of teleological reasoning.

> The standpoint of the classical economists, in their higher or definitive syntheses and generalisations, may not inaptly be called the standpoint of ceremonial adequacy. The ultimate laws and principles which they formulated were laws of the normal or the natural, according to a preconception regarding the ends to

which, in the nature of things, all things tend. In effect, this preconception imputes to things a tendency to work out what the instructed common sense of the time accepts as the adequate or worthy end of human effort. It is a projection of the accepted ideal of conduct. This ideal of conduct is made to serve as a canon of truth, to the extent that the investigator contents himself with an appeal to its legitimation for premises that run back of the facts with which he is immediately dealing, for the "controlling principles" that are conceived intangibly to underlie the process discussed, and for the "tendencies" that run beyond the situation as it lies before him.[1]

In this context, even the notion of a trend in events imputes purpose to the sequence of events, investing the sequence with a discretionary, teleological character asserting itself as a constraint over all the steps in the sequences by which the supposed objective result is reached.[2] Not surprisingly, in works (such as Clark's) giving effect to this preconception, nuances of a beneficent end are present, as are also "provocations to homiletic discourse."[3] While overtly taxonomic, such analysis subsumes "its data under a rational scheme of categories which are presumed to make up the Order of Nature"[4] giving effect to a range of preconceptions having one metaphysical ground, that there is one right and beautiful definitive scheme of economic life, "to which the whole creation tends."[5] Earlier economists "were believers in a Providential order, or an order of Nature... conceived to work in an effective and just way toward the end which it tends; and in the economic field this objective end is the material welfare of

---

[1] Thorstein Veblen, *The Place of Science in Modern Civilisation* (New York: B. W. Huebsch, 1919)
[2] Ibid., 157
[3] Ibid., 188
[4] Ibid., 191
[5] Ibid., 230

mankind."[6] Even Adam Smith's economics, for all its matter-of-fact elements, is driven by a "preconception of a normal teleological order of procedure in the natural course" which "affects not only those features of theory where he is avowedly concerned with building up a normal scheme of the economic process. Through his normalising the chief causal factor engaged in the process, it affects also his arguments from cause to effect."[7]

What is true of mainstream economics, according to Veblen, is also true of historicist economics, whether Marxian or non-Marxian. Thus Wilhelm Roscher's metaphysical postulates were "the common-sense, commonplace metaphysics afloat in educated German circles in the time of"[8] his youth, a Hegelian metaphysics "of a self-realising life process . . . of a spiritual nature . . . essentially active, self-determining, and unfold[ing] by inner necessity," for which "the laws of the cultural development with which the social sciences, in the Heglian view, have to do are at one with the laws of the processes of the universe at large. . . ."[9]

Moreover, "the universe at large is itself a self-unfolding life process, substantially of a spiritual character, of which the economic life process is but a phase and an aspect."[10] Veblen saw Marx as combining materialistic Hegelianism and English natural rights, reaching polemical conclusions which "run wholly on the ground afforded by the premises of that school" and whose "ideals of . . . propaganda are natural-rights

[6] Ibid., 280
[7] Ibid., 129
[8] Ibid., 258
[9] Ibid., 259
[10] Ibid., 260

ideals" especially that of the right of labor to the whole product of labor.[11]

The *matter-of-fact* preconception is concerned with observable phenomena and material cause and effect studied in an impersonal and dispassionate way. The matter-of-fact preconception focuses on process rather than predetermined outcomes[12] which narrows "the range of discretionary, teleological action to the human agent alone"[13] and is concerned with "the questions of what men do and how and why they do it"[14] and comprehends causation "in an unbroken sequence of cumulative change."[15] It is these characteristics that have made for the primacy of science.[16]

A critical difference between the two types of pre-conceptions centers on the nature of social change. For Veblen the crux of the matter was that matter-of-factness had enabled a specifically Darwinian conception of change as an unfolding sequence without necessary ultimate meaning rather than one or another animistic or teleological conceptions. The Darwinian conception is of a "run of causation unfold[ing] itself in an unbroken sequence of cumulative change"[17] involving a process of natural or artificial human selection brought about, in part, through the exercise of human purpose and choice, one driven not by some transcendental predetermined end, but by the interaction of

[11] Ibid., 411
[12] Ibid., 158
[13] Ibid., 179
[14] Ibid., 312
[15] Ibid., 16
[16] Ibid., 24
[17] Ibid., 16

actors and forces under changing natural and social circumstances.

Post-Darwinian science, Veblen said, focused on the process of causation rather than "that consummation in which causal effect was once presumed to come to rest." It is "substantially a theory of the process of consecutive change, which is taken as a sequence of cumulative change, realized to be self-continuing or self-propagating and to have no final term."[18] The "questions of the class which occupy the modern sciences" are "questions of genesis, growth, variation, process (in short, questions of a dynamic import)"[19] understood in terms of open-ended cumulative causation. For economics to be an "evolutionary science," it "must be the theory of a process of cultural growth as determined by the economic interest, a theory of a cumulative sequence of economic institutions stated in terms of the process itself."[20]

Societies historically have been blends of both habits and thought, though increasingly the latter, which is the hallmark of modernity. Moreover, pragmatic considerations have always been present, even in societies that were substantially animistic. Modern science and technology exist in a mutually reinforcing relationship with the matter-of-fact habit of thought, though there are residues of teleological preconceptions in modern society. The history of economic theory is a history, accordingly, of both the continuing co-existence of teleology and matter-of-fact theorizing, and the gradual eclipse but not total elimination of the former. The

[18] Ibid., 37
[19] Ibid., 192
[20] Ibid., 77

"history of the science," he said, "shows a long and devious course of disintegrating animism."[21]

Preconceptions, the animistic-teleological, or the matter-of-fact doctrines, come into being as part of what amounts to Veblen's large-scale model of society which, unfortunately, was nowhere completely spelled out, though he insisted upon both the key roles of culture and the state of the industrial arts (technology), and the general interdependent nature of the system. The elements of this model include human nature, the material environment, institutions, technology, general culture, and belief systems (predicated upon preconceptions). Varying formulations of this general model are given in different contexts. Preconceptions are the product of both the state of the industrial arts, and the habitualized practices to which they give rise, and the institutional structure under which the community lives.[22] But the "state of the industrial arts is dependent on the traits of human nature, physical, intellectual, and spiritual, and on the character of the material environment."[23] Change in preconceptions "is closely correlated with an analogous change in institutions and habits of life, particularly with the changes which the modern era brings in industry and in the economic organisation of society."[24] In the modern era, "science and technology play into one another's hands."[25]

More generally,

the canons of validity under whose guidance [the scientist] works are those imposed by the modern technology,

[21] Ibid., 64
[22] Ibid., 10
[23] Ibid., 349
[24] Ibid., 12
[25] Ibid., 17

through habituation to its requirements; and therefore his results are available for the technological purpose. His canons of validity are made for him by the cultural situation; they have habits of thought imposed on him by the scheme of life current in the community in which he lives; and under modern conditions this scheme of life is largely machine-made. In the modern culture, industry, industrial processes, and industrial products . . . have become the chief force in shaping men's habits of thought. Hence men have learned to think in the terms in which the technological processes act. This is particularly true of those men who by virtue of a peculiarly strong susceptibility in this direction become addicted to that habit of matter-of-fact inquiry that constitutes scientific research.[26]

A two-word summary is cumulative causation. "And so long as the machine process continues to hold its dominant place as a disciplinary factor in modern culture, so long must the spiritual and intellectual life of this cultural era maintain the character which the machine process gives it"[27]—though the predominance of the machine process is itself a dependent variable. Moreover, the "fabric of institutions intervenes between the material exigencies of life and the speculative scheme of things," in such a way that "habits of thought . . . reflect the habits of life embodied in the institutional structure of society" that "is a matter of law and custom, politics and religion, taste and morals . . . " and with respect to which "the speculative generalisations, the institutions of the realm of knowledge, are created in the image of . . . social institutions of status and personal force . . . ."[28] In addition, practice is always pragmatic, but it is not always matter-of-fact; economic "change is always in the last

[26] Ibid., 17
[27] Ibid., 30
[28] Ibid., 44–45

resort a change in habits of thought."[29] The habits of thought formed in any one line of experience affect thought in any other;[30] varying combinations of disciplines of the mind produce different social results.[31] There is a strong technological determinism here, because it must be comprehended in terms of a much larger model, perhaps better expressed as conditionism ". . . under the Darwinian norm the question of whether and how far material exigencies control human conduct and cultural growth becomes a question of the share which these material exigencies have in shaping men's habits of thought. . . . "[32]

In emphasizing the role of fundamental structuring preconceptions or habits of thought, Veblen emphasized the difference between the terms (language and discourse) in which people express their understanding of nature, society, and mankind, and the actual facts of the matter. As our preconceptions change, even conceptions of the Deity change.[33] What we perceive to be "the facts" are preconception laden. Phenomena do not define themselves, rather, meaning is imputed by the observer.[34] What happens cannot appear to us except on the ground or through the mediating and defining influence of some preconception or prepossession.[35]

The single most encompassing theme of this volume is the presence and discourse-forming role of preconceptions in the history of economic and other thought: The "ultimate term or ground of knowledge is always of

[29] Ibid., 75
[30] Ibid., 79–80, 105
[31] Ibid., 105
[32] Ibid., 438
[33] Ibid., 14
[34] Ibid., 15
[35] Ibid., 76

a metaphysical character. It is something of a preconception, accepted uncritically, but applied in criticism
and demonstration of all else . . . "[36] Veblen was thus
one of the earliest writers to concern himself with the
social construction of meaning, rather than with the
absolute category of truth, and with the formation of
knowledge or belief or both as a product of group life in
particular institutional and cultural contexts.

Veblen was rather adroit with words, especially in
his ability to foment tone and attitude. Consider the
evocative power of the sentence, "The gallantries, the
genteel inanities and devout imbecilities of mediaeval
highlife would be insufferable even to the meanest and
most romatic modern intelligence"[37]—evocative, to be
sure, but at once both descriptive and judgmental.
Consider, too, his use of dramatically unusual words to
elicit a sense of the archaic, "If we are getting restless
under the taxonomy of a monocotyledonous wage doctrine and a cryptogamic theory of interest, with involute, loculicidal, tomentous and moniliform variants,
what is the cytoplasm, centrosome, or karyokinetic
process to which we may turn, and in which we may
find surcease from the metaphysics of normality and
controlling principles?" The message was, however, not
left implicit: "There is the economic life process still in
great measure awaiting theoretical formulation."[38]

One literary or rhetorical device Veblen used
throughout the volume was that of asserting what he
took to be both true and desirable as if it were beyond
dispute at the highest levels. When he wrote that
"economics is helplessly behind the times, and unable

[36] Ibid., 149
[37] Ibid., 23
[38] Ibid., 70

to handle its subject-matter in a way to entitle it to standing as a modern science,"[39] in the opening paragraph of the essay on "Why is economics not an evolutionary science?," he was presuming as the basis of comparison a Darwinian conception of social process and an economic science modelled thereon as beyond cavil, as a given—whereas his own argument establishes the problematic character of such an eventuality. Still, he insisted, when deriding the economics of "normal cases," that "it is only a question of time when that (substantially animistic) habit of thought shall be displaced in the field of economic inquiry by that (substantially materialistic) habit of mind which seeks a comprehension of facts in terms of a cumulative sequence,"[40] to close the penultimate paragraph of the essay.

Veblen argued that people, for the most part, uncritically accept and think in terms of preconceptions and habits of thought that have become a part of them through socialization or enculturation. Beliefs and facts are then system-specific. Veblen unquestionably affirmed and indeed lauded modern Darwinian or evolutionary science and the matter-of-fact preconception or habit of thought on which it rested. Thus, for example:

> ... In a general way, the higher the culture, the greater the share of the mechanical preconception in shaping human thought and knowledge, since, in a general way, the stage of culture attained depends on the efficiency of industry.[41]

But he clearly and unequivocally also affirmed the preconceptional nature of matter-of-factness and cumulative causation (in which metaphysical respects

[39] Ibid., 56
[40] Ibid., 81
[41] Ibid., 103–104

teleological habits of thought are substantively, but not discursively different), the preconception or system-specificity of belief, the consequent normative status of belief, and the applicability of these ideas to his own thinking. All these points, especially the self-referential nature of his argument, are evident in following the extracts:

> [Modern science) will bring under inquiry such questions of knowledge as lie within its particular range of interest, and will seek answers to these questions only in terms that are consonant with the habits of thought current at the time.[42]
> The prime postulate of evolutionary science, the preconception constantly underlying the inquiry, is the notion of a cumulative causal sequence .... [43]
> [The postulate of consecutive, cumulative change] is an unproven and unprovable postulate—that is to say, it is a metaphysical preconception.... [44]
> ... The notion of causal continuity, as a premise of scientific generalisation, is an essentially metaphysical postulate.... Before anything can be said as to the orderliness of the sequence, a point of view must be chosen by the speculator, with respect to which the sequence in question does or does not fulfill this condition of orderliness; that is to say, with respect to which it is a sequence. The endeavor to avoid all metaphysical premises fails here as everywhere.[45]

It will surprise no one that Veblen's several critiques of mainstream economic theory and of capitalism are both derivative and expansive of the arguments so far summarized. Veblen's critique of orthodox theory included the following interrelated arguments or themes.

1. Economic theory has become increasingly matter-of-fact but inevitably continues to have teleological

[42] Ibid., 38
[43] Ibid., 176
[44] Ibid., 33
[45] Ibid., 161–162

elements. In this respect, the economist is like other people: "He is a creature of habits and propensities given through the antecedents, hereditary and cultural, of which he is an outcome; and the habits of thought formed in any one line of experience affect his thinking in any other."[46]

2. Economic theory, as we have already seen, serves the function of ceremonial adequacy, that is, as a social control and also as a psychic balm.[47] As such it represents and replicates "in large part the point of view of the enlightened common sense of [the] time."[48] It is too ready to accept the ground of sufficient reason rather than insist on the ground of efficient cause.[49] It is also too useful for casting luster and reenforcing vested interests.

3. The defect of economic theory lies not in its lack of factual realism, but in its failure to be evolutionary in the Darwinian manner, focusing on development in a self-generating process of cumulative causation and unfolding sequence encompassing all the variables found in Veblen's general model.

4. The foregoing defect is especially clearly manifest in orthodox theory's conception of man as an isolated, passive responder to external stimuli, and not as an effectively purposive economic agent participating in larger processes. Veblen's description of this conception has been, appropriately, widely quoted:

> The hedonistic conception of man is that of a lightning calculator of pleasures and pains, who oscillates like a homogeneous globule of desire of happiness under the impulse of stimuli that shift him about the area, but

[46] Ibid., 79–80
[47] Ibid., 65–66
[48] Ibid., 86
[49] Ibid., 237

leave him intact. He has neither antecedent nor consequent. He is an isolated, definitive human datum, in stable equilibrium except for the buffets of the impinging forces that displace him in one direction or another. Self-imposed in elemental space, he spins symmetrically about his own spiritual axis until the parallelogram of forces bears down upon him, whereupon he follows the line of the resultant. When the force of the impact is spent, he comes to rest, a self-contained globule of desire as before. Spiritually, the hedonistic man is not a prime mover. He is not the seat of a process of living, except in the sense that he is subject to a series of permutations enforced upon him by circumstances external and alien to him.[50]

As a corollary, Veblen argued that orthodox theory failed to consider the respects and means whereby culture and institutions, such as that of private property affect human action.

5. Orthodox theory maintains the fundamental presumption of capitalist society, that to have an income means that one has been productive, disregarding what Veblen considered to be fundamental, namely, the distinction between industrial and pecuniary pursuits, or between making goods and making money.[51] In this respect Veblen was not prepared to assume, along with orthodox theorists, that any business adjustment or any action acquisitive of profit is presumptively instrumental of the welfare of makind, or in his language to the serviceability of the community. As Veblen put it, "the classical theory of production is in good part a doctrine of investment in which the identity of production and pecuniary gain is taken for granted,"[52] an assumption he vehemently rejected.

[50] Ibid., 73–74
[51] Ibid., 122
[52] Ibid., 140

In the essays that comprise this volume Veblen was not principally interested in analyzing capitalism. The argument of the three essays on industrial and pecuniary employments and on the nature of capital, however, present the core of what subsequently became a key thesis for Veblen and for those who follow in his analytical footsteps, that is, the just-mentioned distinction between industrial and pecuniary, between making goods and making money. The production of goods is only incidental and by no means necessary to the acquistion of income. To Veblenians, the distinction is amply manifest in corporate mergers and acquisitions, hostile or friendly, in which real capital assets merely change ownership, and activity is directed toward reaping money rewards without adding to the production of serviceable goods; portfolio manipulations do not constitute real investment.

In Veblen's view, returns on the ownership of capital from both tangible and intangible assets were due to the differential advantages accruing to the owners, advantages typically arising from their class position, or with the assistance of government, as through the law of property, advantages that enable the owners to engross, as he put it, part of the flow of income. This approach, it should be noted, rejected both the orthodox productivity and the Marxian exploitation paradigms in favor of what may be called the Veblen-Weber paradigm of appropriation, in which income (and wealth) is distributed on the basis of a complexly grounded meeting of forces in the political economy, without normative status, except insofar as ideology and general social control impute such status.

For Veblen, the capitalized value of assets is a function of their income-yielding capacity to their

owner. In the case of tangible assets there is a presumption that the objects of wealth involved have at
least some potential serviceability at large, since they
serve materially productive work. In the case of intangible assets, which largely derive from the creation of
pecuniary arrangements, there is no presumption that
the objects of wealth involved have any serviceability
at large, since they serve no materially productive
work, but are only a differential advantage to the owner
in the distribution of the industrial product. Serviceability and claims to income are two different
matters.[53]

One final set of principal topics warrants attention.
Veblen believed the economist must consider: (1) purposive and habitual action by individuals; (2) methodological individualist and methodological collectivist
methodologies, and (3) deliberative and nondeliberative
social control (the collective correlative to the first).
Veblen is appropriately well known for his criticisms of
hedonistic rationality (as above), but the reader will
notice the several elements of this criticism. First, the
individual is said to be heavily motivated by habit,
custom, institutions, and status emulation. Second, the
individual is said to have teleology or purposes of his or
her own, however much these are socially generated or
influenced. Veblen did not insist on either habit, purpose, or socialization alone, but acting in concert. His
comprehension of the inevitability of the combination of
methodological individualism and methodological col-
lectivism is brilliantly expressed in the following ex-
cerpts:

> The growth and mutations of the institutional fabric
> are an outcome of the conduct of the individual members

[53] Ibid., 364–366, 372

of the group, since it is out of the experience of the individuals, through the habituation of individuals, that institutions arise; and it is in this same experience that these institutions act to direct and define the aims and end of conduct. It is, of course, on individuals that the system of institutions imposes those conventional standards, ideals, and canons of conduct that make up the community's scheme of life. Scientific inquiry in this field, therefore, must deal with individual conduct and must formulate its theoretical results in terms of individual conduct. But such an inquiry can serve the purposes of a genetic theory only if and in so far as this individual conduct is attended to in those respects in which it counts toward habituation, and so toward change (or stability) of the institutional fabric, on the one hand, and in those respects in which it is prompted and guided by the received institutional conceptions and ideals on the other hand. The postulates of marginal utility, and the hedonistic preconceptions generally, fail at this point in that they confine the attention to such bearings of economic conduct as are conceived not to be conditioned by habitual standards and ideals and to have no effect in the way of habituation.[54]

Nor is it conceived [by marginal utility economics] that the presence of this institutional element in men's economic relations in any degree affects or disguises the hedonistic calculus, or that its pecuniary conceptions and standards in any degree standardize, color, mitigate, or divert the hedonistic calculator from the direct and unhampered quest of the net sensuous gain. While the institution of property is included... it is allowed to have no force in shaping economic conduct, which is conceived to run its course to its hedonistic outcome as if no such institutional factor intervened between the impulse and its realization... presumed to give rise to no habitual or conventional canons of conduct or standards of valuation, no proximate ends, ideals, or aspirations.[55]

There are also several other matters that will be of interest to many readers. One is Veblen's treatment of pragmatism, where his views were not yet fully worked out or at least not fully stated. This is particularly

[54] Ibid., 243
[55] Ibid., 244–245

noteworthy in light of his affirmation of the ubiquity of pragmatic behavior. Another was his participation in early controversies over the nature of capital, preludes to the modern capital controversy, extending to the deep question of the fundamental coherence of mainstream economic theory, a matter already raised by Veblen.

It should be obvious that Veblen's ideas and theories are subject to criticism. For all his incisive brilliance he may be wrong. Even where he is correct, the Paretian doctrine of the social utility of falsity (or the Marxian doctrine of false consciousness viewed in a non-Marxian manner) may pertain. A society may mislead itself, say, through animistic or teleological reasoning, but this may be instrumental to the operation of that society and its transformation into a more modern society. A few more or less representative criticisms follow.

First, when Veblen said that because of the pressure of modern industrial or technological exigencies "it is only a question of time when that (substantially animistic) habit of mind which proceeds on the notion of a definitive normality shall be displaced in the field of economic inquiry by that (substantially materialistic) habit of mind which seeks a comprehension of facts in terms of a cumulative sequence"[56] he was surely ignoring the psychic balm function of economics as well as the social control function that puts psychic balm solutions to additional effective use. Moreover, he has been empirically wrong: while the tendency which he lauds is present, it remains swamped by another set of preconceptions. Teleological ideology seems inevitable.

Second, Veblen argued that the emulative system with its struggle for economic respectability, on which

[56] Ibid., 81

capitalism rests, breeds discontent with the institution of private property and "is one of the causes, if not the chief cause, of the existing unrest and dissatisfaction with things as they are," indeed, "necessarily adverse to the existing industrial system of free competition."[57] Now Veblen was certainly correct that: "The outcome of modern industrial development has been, so far as concerns the present purpose, to intensify emulation and the jealousy that goes with emulation, and to focus the emulation and the jealousy on the possession and enjoyment of material goods."[58] But to one writing near the end of the twentieth century, rather than in 1892, when those words were published, it seems that while Veblen may be empirically correct in his interpretation of social discontent (notice that it is a very different interpretation from that of Sigmund Freud in his *Civilization and Its Discontents*), certainly there has been no diminution of status emulation, "the struggle to keep up appearances"[59] and its accompaniments in American society, nor does discontent with private property as an institution seem to have grown. This, of course, does not mean that the contradiction between emulation and private property that Veblen noted has not manifested itself in institutional change.

Third, Veblen distinguished between valuation, production, and distribution. He said that valuation was a pecuniary matter fundamentally related to acquisition, and thus distribution, and in this he was surely correct. He said also that "Ownership, no doubt, has its effect upon productive industry, and, indeed, its effect upon industry is very large, . . . but ownership is not itself primarily or immediately a contrivance for production.

[57] Ibid., 398
[58] Ibid., 397
[59] Ibid., 399

Ownership directly touches the results of industry, and only indirectly the methods and processes of industry."[60] A conservative economist would emphasize that the security required for production is provided, willy nilly, by the institution of ownership, but my point is a different one. Consider Veblen's related argument that "Pecuniary capital is a matter of market values, while industrial capital is, in the last analysis, a matter of mechanical efficiency, or rather of mechanical effects not reducible to a common measure or a collective magnitude,"[61] and that "Capital pecuniarily considered rests on a basis of subjective value; capital industrially considered rests on material circumstances reducible to objective terms of mechanical, chemical and physiological effect."[62] The problem is not with the distinction but with its force: Industrial capital produces goods, but which goods? Somehow society has to determine whose interests—whose subjective valuations— are to count in determining the allocation of resources and the production of real goods and services. (For Veblen, technological change and resource allocation were influenced by pecuniary factors.)

This consideration is underscored by Veblen's very candid recognition, given his stress upon the state of industrial arts, that "technological proficiency is not of itself and intrinsically serviceable or disserviceable to mankind—it is only a means of efficiency for good or ill."[63] The question is, serviceable for whom? The owner or the community? And if the community, how are different and conflicting serviceabilities to be rec-

---

[60] Ibid., 296
[61] Ibid., 310
[62] Ibid., 311
[63] Ibid., 359

onciled? Matter-of-factness, science, the machine process, and culture may be interpreted or evaluated differently. "Seen in certain lights, tested by certain standards, it is doubtless better; by other standards, worse."[64] The results of technology are not conclusively good.[65] Subjective, normative inputs are necessary.

Fourth, Veblen wrote in an essay published in 1906, that "It is not the Marxism of Marx, but the materialism of Darwin, which the socialists of to-day have adopted."[66] Granted the place of pragmatic adjustments in Marxian theory (abetted by its dialecticism) as well as practice, it does not seem that Marxists have become Darwinists. There continues to this day controversy within Marxism between the deterministic and the conditionistic interpretation of Marx's dialectical materialism, but neither the latter nor, certainly, the former are evidently Darwinian in the manner understood by Veblen.

Finally, we are entitled by Veblen's own argument to ask of him the same question which he in effect posed to other schools of thought: what is reality (and in what sense of "reality") and what is preconception? We must take seriously the self-referential character of his argument. Of the "cumulative process of development, and its complex and unstable outcome, that are to be the economist's subject-matter," he said[67] What is objective and what is subjective? What is a matter of preconception and what is independent of preconception? What is it, precisely, which produces that which is matter-of-fact? What putative status are we to ascribe to

[64] Ibid., 29–30
[65] Ibid., 31
[66] Ibid., 432
[67] Ibid., 267

"knowledge"? These are serious questions; the answers are not self-evident, even after reading Veblen and especially after such writers as Richard Rorty, Mary Douglas, Peter Berger, Michel Foucault, and Jerome Brunner. Veblen's argument about preconceptions is truly for the ages; belief, even in Veblen, must be held with some sense of diffidence. But if Veblen's Darwinism is correct, then what of a society in which belief is in fact held with a sense of diffidence? After all, it is not too much to say that Veblenian ideas are part of the Darwinian process of cumulation variation and selection. That fact may be the greatest monument to Veblen—who, of course, did not have good things to say about such ceremonial fetishes as monuments.

THE PLACE OF SCIENCE IN MODERN CIVILISATION [1]

It is commonly held that modern Christendom is superior to any and all other systems of civilised life. Other ages and other cultural regions are by contrast spoken of as lower, or more archaic, or less mature. The claim is that the modern culture is superior on the whole, not that it is the best or highest in all respects and at every point. It has, in fact, not an all-around superiority, but a superiority within a closely limited range of intellectual activities, while outside this range many other civilisations surpass that of the modern occidental peoples. But the peculiar excellence of the modern culture is of such a nature as to give it a decisive practical advantage over all other cultural schemes that have gone before or that have come into competition with it. It has proved itself fit to survive in a struggle for existence as against those civilisations which differ from it in respect of its distinctive traits.

Modern civilisation is peculiarly matter-of-fact. It contains many elements that are not of this character, but these other elements do not belong exclusively or characteristically to it. The modern civilised peoples are in a peculiar degree capable of an impersonal, dispassionate insight into the material facts with which mankind has to deal. The apex of cultural growth is at this point. Compared with this trait the rest of what is comprised in the cultural scheme is adventitious, or at the best it is a

[1] Reprinted by permission from *The American Journal of Sociology,* Vol. XI, March, 1906.

by-product of this hard-headed apprehension of facts. This quality may be a matter of habit or of racial endowment, or it may be an outcome of both; but whatever be the explanation of its prevalence, the immediate consequence is much the same for the growth of civilisation. A civilisation which is dominated by this matter-of-fact insight must prevail against any cultural scheme that lacks this element. This characteristic of western civilisation comes to a head in modern science, and it finds its highest material expression in the technology of the machine industry. In these things modern culture is creative and self-sufficient; and these being given, the rest of what may seem characteristic in western civilisation follows by easy consequence. The cultural structure clusters about this body of matter-of-fact knowledge as its substantial core. Whatever is not consonant with these opaque creations of science is an intrusive feature in the modern scheme, borrowed or standing over from the barbarian past.

Other ages and other peoples excel in other things and are known by other virtues. In creative art, as well as in critical taste, the faltering talent of Christendom can at the best follow the lead of the ancient Greeks and the Chinese. In deft workmanship the handicraftsmen of the middle Orient, as well as of the Far East, stand on a level securely above the highest European achievement, old or new. In myth-making, folklore, and occult symbolism many of the lower barbarians have achieved things beyond what the latter-day priests and poets know how to propose. In metaphysical insight and dialectical versatility many orientals, as well as the Schoolmen of the Middle Ages, easily surpass the highest reaches of the New Thought and the Higher Criticism. In a shrewd sense of the religious verities, as well as in an unsparing faith in devout observances, the people of India or Thibet, or even the mediæval Christians, are past-masters in comparison even with the select of the faith of modern times. In political finesse, as well as in unreasoning, brute loyalty, more than one of the ancient peoples give evidence of a capacity to which no modern civilised nation may aspire. In warlike malevolence and abandon, the hosts of Islam, the Sioux Indian, and the " heathen of the northern sea " have set the mark above the reach of the most strenuous civilised warlord.

To modern civilised men, especially in their intervals of sober reflection, all these things that distinguish the barbarian civilisations seem of dubious value and are required to show cause why they should not be slighted. It is not so with the knowledge of facts. The making of states and dynasties, the founding of families, the prosecution of feuds, the propagation of creeds and the creation of sects, the accumulation of fortunes, the consumption of superfluities — these have all in their time been felt to justify themselves as an end of endeavor; but in the eyes of modern civilised men all these things seem futile in comparison with the achievements of science. They dwindle in men's esteem as time passes, while the achievements of science are held higher as time passes. This is the one secure holding-ground of latter-day conviction, that " the increase and diffusion of knowledge among men " is indefeasibly right and good. When seen in such perspective as will clear it of the trivial perplexities of workday life, this proposition is not questioned within the horizon of the western culture, and no other cultural ideal holds a similar unquestioned place in the convictions of civilised mankind.

On any large question which is to be disposed of for good and all the final appeal is by common consent taken to the scientist. The solution offered in the name of science is decisive so long as it is not set aside by a still more searching scientific inquiry. This state of things may not be altogether fortunate, but such is the fact. There are other, older grounds of finality that may conceivably be better, nobler, worthier, more profound, more beautiful. It might conceivably be preferable, as a matter of cultural ideals, to leave the last word with the lawyer, the duelist, the priest, the moralist, or the college of heraldry. In past times people have been content to leave their weightiest questions to the decision of some one or other of these tribunals, and, it cannot be denied, with very happy results in those respects that were then looked to with the greatest solicitude. But whatever the common-sense of earlier generations may have held in this respect, modern common-sense holds that the scientist's answer is the only ultimately true one. In the last resort enlightened common-sense sticks by the opaque truth and refuses to go behind the returns given by the tangible facts.

Quasi lignum vitae in paradiso Dei, et quasi lucerna fulgoris in domo Domini, such is the place of science in modern civilisation. This latterday faith in matter-of-fact knowledge may be well grounded or it may not. It has come about that men assign it this high place, perhaps idolatrously, perhaps to the detriment of the best and most intimate interests of the race. There is room for much more than a vague doubt that this cult of science is not altogether a wholesome growth — that the unmitigated quest of knowledge, of this matter-of-fact kind, makes for race-deterioration and discomfort on the whole, both in its immediate effects upon the spiritual life of mankind, and in the material consequences that follow from a great advance in matter-of-fact knowledge.

But we are not here concerned with the merits of the case. The question here is : How has this cult of science arisen? What are its cultural antecedents? How far is it in consonance with hereditary human nature? and, What is the nature of its hold on the convictions of civilised men?

In dealing with pedagogical problems and the theory of education, current psychology is nearly at one in saying that all learning is of a " pragmatic " character; that knowledge is inchoate action inchoately directed to an end; that all knowledge is " functional "; that it is of the nature of use. This, of course, is only a corollary under the main postulate of the latter-day psychologists, whose catchword is that The Idea is essentially active. There is no need of quarreling with this " pragmatic " school of psychologists. Their aphorism may not contain the whole truth, perhaps, but at least it goes nearer to the heart of the epistemological problem than any earlier formulation. It may confidently be said to do so because, for one thing, its argument meets the requirements of modern science. It is such a concept as matter-of-fact science can make effective use of; it is drawn in terms which are, in the last analysis, of an impersonal, not to say tropismatic, character; such as is demanded by science, with its insistence on opaque cause and effect. While knowledge is construed in teleological terms, in terms of personal interest and attention, this teleological aptitude is itself reducible to a product of unteleological natural selection. The teleological bent of intelligence is an hereditary trait settled upon the race by the selective action of forces that look to no end. The foundations of pragmatic intelligence are not pragmatic, nor even personal or sensible.

This impersonal character of intelligence is, of course, most evident on the lower levels of life. If we follow Mr. Loeb, e. g., in his inquiries into the psychology of

that life that lies below the threshold of intelligence, what
we meet with is an aimless but unwavering motor response
to stimulus.[2]   The response is of the nature of motor impulse, and in so far it is " pragmatic," if that term may
fairly be applied to so rudimentary a phase of sensibility.
The responding organism may be called an " agent " in so
far.   It is only by a figure of speech that these terms are
made to apply to tropismatic reactions.   Higher in the
scale of sensibility and nervous complication instincts
work to a somewhat similar outcome.   On the human
plane, intelligence (the selective effect of inhibitive complication) may throw the response into the form of a reasoned line of conduct looking to an outcome that shall
be expedient for the agent.   This is naïve pragmatism of
the developed kind.   There is no longer a question but
that the responding organism is an " agent " and that
his intelligent response to stimulus is of a teleological
character.   But that is not all.   The inhibitive nervous
complication may also detach another chain of response
to the given stimulus, which does not spend itself in a line
of motor conduct and does not fall into a system of uses.
Pragmatically speaking, this outlying chain of response is
unintended and irrelevant.   Except in urgent cases, such
an idle response seems commonly to be present as a subsidiary phenomenon.   If credence is given to the view
that intelligence is, in its elements, of the nature of an
inhibitive selection, it seems necessary to assume some
such chain of idle and irrelevant response to account for
the further course of the elements eliminated in giving
the motor response the character of a reasoned line of
conduct.   So that associated with the pragmatic attention there is found more or less of an irrelevant atten-

[2] Jacques Loeb, *Heliotropismus der Thiere,* and *Comparative
Psychology and Physiology of the Brain.*

tion, or idle curiosity. This is more particularly the case where a higher range of intelligence is present. This idle curiosity is, perhaps, closely related to the aptitude for play, observed both in man and in the lower animals.[3] The aptitude for play, as well as the functioning of idle curiosity, seems peculiarly lively in the young, whose aptitude for sustained pragmatism is at the same time relatively vague and unreliable.

This idle curiosity formulates its response to stimulus, not in terms of an expedient line of conduct, nor even necessarily in a chain of motor activity, but in terms of the sequence of activities going on in the observed phenomena. The " interpretation " of the facts under the guidance of this idle curiosity may take the form of anthropomorphic or animistic explanations of the " conduct " of the objects observed. The interpretation of the facts takes a dramatic form. The facts are conceived in an animistic way, and a pragmatic animus is imputed to them. Their behavior is construed as a reasoned procedure on their part looking to the advantage of these animistically conceived objects, or looking to the achievement of some end which these objects are conceived to have at heart for reasons of their own.

Among the savage and lower barbarian peoples there is commonly current a large body of knowledge organised in this way into myths and legends, which need have no pragmatic value for the learner of them and no intended bearing on his conduct of practical affairs. They may come to have a practical value imputed to them as a ground of superstitious observances, but they may also not.[4] All students of the lower cultures are aware of

[3] Cf. Gross, *Spiele der Thiere*, chap. 2 (esp. pp. 65–76), and chap. 5; *The Play of Man*, Part III, sec. 3; Spencer, *Principles of Psychology*, secs. 533–35.

[4] The myths and legendary lore of the Eskimo, the Pueblo

the dramatic character of the myths current among these peoples, and they are also aware that, particularly among the peaceable communities, the great body of mythical lore is of an idle kind, as having very little intended bearing on the practical conduct of those who believe in these myth-dramas. The myths on the one hand, and the workday knowledge of uses, materials, appliances, and expedients on the other hand, may be nearly independent of one another. Such is the case in an especial degree among those peoples who are prevailingly of a peaceable habit of life, among whom the myths have not in any great measure been canonised into precedents of divine malevolence.

The lower barbarian's knowledge of the phenomena of nature, in so far as they are made the subject of deliberate speculation and are organised into a consistent body, is of the nature of life-histories. This body of knowledge is in the main organised under the guidance of an idle curiosity. In so far as it is systematised under the canons of curiosity rather than of expediency, the test of truth applied throughout this body of barbarian knowledge is the test of dramatic consistency. In addition to their dramatic cosmology and folk legends, it is needless to say, these peoples have also a considerable body of worldly wisdom in a more or less systematic form. In this the test of validity is usefulness.[5]

Indians, and some tribes of the northwest coast afford good instances of such idle creations. Cf. various *Reports* of the Bureau of American Ethnology; also, e. g., Tylor, *Primitive Culture*, esp. the chapters on " Mythology " and " Animism."

[5] " Pragmatic " is here used in a more restricted sense than the distinctively pragmatic school of modern psychologists would commonly assign the term. " Pragmatic," " teleological," and the like terms have been extended to cover imputation of purpose as well as conversion to use. It is not intended to criticise this ambiguous use of terms, nor to correct it; but the terms are

The pragmatic knowledge of the early days differs scarcely at all in character from that of the maturest phases of culture. Its highest achievements in the direction of systematic formulation consist of didactic exhortations to thrift, prudence, equanimity, and shrewd management — a body of maxims of expedient conduct. In this field there is scarcely a degree of advance from Confucius to Samuel Smiles. Under the guidance of the idle curiosity, on the other hand, there has been a continued advance toward a more and more comprehensive system of knowledge. With the advance in intelligence and experience there come closer observation and more detailed analysis of facts.[6] The dramatisation of the sequence of phenomena may then fall into somewhat less personal, less anthropomorphic formulations of the processes observed; but at no stage of its growth — at least at no stage hitherto reached — does the output of this work of the idle curiosity lose its dramatic character. Comprehensive generalisations are made and cosmologies are built up, but always in dramatic form. General principles of explanation are settled on, which in the earlier days of theoretical speculation seem invariably to run back to the broad vital principle of generation. Procreation, birth, growth, and decay constitute the cycle of postulates within which the dramatised processes of natural phenomena run their course. Creation is procreation in these archaic theoretical systems, and causation is gesta-

here used only in the latter sense, which alone belongs to them by force of early usage and etymology. " Pragmatic " knowledge, therefore, is such as is designed to serve an expedient end for the knower, and is here contrasted with the imputation of expedient conduct to the facts observed. The reason for preserving this distinction is simply the present need of a simple term by which to mark the distinction between worldly wisdom and idle learning.

[6] Cf. Ward, *Pure Sociology,* esp. pp. 437-48.

tion and birth. The archaic cosmological schemes of Greece, India, Japan, China, Polynesia, and America, all run to the same general effect on this head.[7] The like seems true for the Elohistic elements in the Hebrew scriptures.

Throughout this biological speculation there is present, obscurely in the background, the tacit recognition of a material causation, such as conditions the vulgar operations of workday life from hour to hour. But this causal relation between vulgar work and product is vaguely taken for granted and not made a principle for comprehensive generalisations. It is overlooked as a trivial matter of course. The higher generalisations take their color from the broader features of the current scheme of life. The habits of thought that rule in the working-out of a system of knowledge are such as are fostered by the more impressive affairs of life, by the institutional structure under which the community lives. So long as the ruling institutions are those of blood-relationship, descent, and clannish discrimination, so long the canons of knowledge are of the same complexion.

When presently a transformation is made in the scheme of culture from peaceable life with sporadic predation to a settled scheme of predaceous life, involving mastery and servitude, gradations of privilege and honor, coercion and personal dependence, then the scheme of knowledge undergoes an analogous change. The predaceous, or higher barbarian, culture is, for the present purpose, peculiar in that it is ruled by an accentuated pragmatism. The institutions of this cultural phase are conventionalised relations of force and fraud. The questions of life are questions of expedient conduct as carried on under the current relations of mastery and subservience. The

[7] Cf., e. g., Tylor, *Primitive Culture,* chap. 8.

habitual distinctions are distinctions of personal force, advantage, precedence, and authority. A shrewd adaptation to this system of graded dignity and servitude becomes a matter of life and death, and men learn to think in these terms as ultimate and definitive. The system of knowledge, even in so far as its motives are of a dispassionate or idle kind, falls into the like terms, because such are the habits of thought and the standards of discrimination enforced by daily life.[8]

The theoretical work of such a cultural era, as, for instance, the Middle Ages, still takes the general shape of dramatisation, but the postulates of the dramaturgic theories and the tests of theoretic validity are no longer the same as before the scheme of graded servitude came to occupy the field. The canons which guide the work of the idle curiosity are no longer those of generation, blood-relationship, and homely life, but rather those of graded dignity, authenticity, and dependence. The higher generalisations take on a new complexion, it may be without formally discarding the older articles of belief. The cosmologies of these higher barbarians are cast in terms of a feudalistic hierarchy of agents and elements, and the causal nexus between phenomena is conceived animistically after the manner of sympathetic magic. The laws that are sought to be discovered in the natural universe are sought in terms of authoritative enactment. The relation in which the deity, or deities, are conceived to stand to facts is no longer the relation of progenitor, so much as that of suzerainty. Natural laws are corollaries under the arbitrary rules of status imposed on the natural universe by an all-powerful Providence with a view to the maintenance of his own prestige. The science that grows in such a spiritual environment is of the class

[8] Cf. James, *Psychology*, chap. 9, esp. sec. 5.

represented by alchemy and astrology, in which the imputed degree of nobility and prepotency of the objects and the symbolic force of their names are looked to for an explanation of what takes place.

The theoretical output of the Schoolmen has necessarily an accentuated pragmatic complexion, since the whole cultural scheme under which they lived and worked was of a strenuously pragmatic character. The current concepts of things were then drawn in terms of expediency, personal force, exploit, prescriptive authority, and the like, and this range of concepts was by force of habit employed in the correlation of facts for purposes of knowledge even where no immediate practical use of the knowledge so gained was had in view. At the same time a very large proportion of the scholastic researches and speculations aimed directly at rules of expedient conduct, whether it took the form of a philosophy of life under temporal law and custom, or of a scheme of salvation under the decrees of an autocratic Providence. A naïve apprehension of the dictum that all knowledge is pragmatic would find more satisfactory corroboration in the intellectual output of scholasticism than in any system of knowledge of an older or a later date.

With the advent of modern times a change comes over the nature of the inquiries and formulations worked out under the guidance of the idle curiosity — which from this epoch is often spoken of as the scientific spirit. The change in question is closely correlated with an analogous change in institutions and habits of life, particularly with the changes which the modern era brings in industry and in the economic organisation of society. It is doubtful whether the characteristic intellectual interests and teachings of the new era can properly be spoken of as less " pragmatic," as that term is sometimes understood, than those of the scholastic times; but they are of another kind, being conditioned by a different cultural and industrial situation.[9]   In the life of the new era conceptions of authentic rank and differential dignity have grown weaker in practical affairs, and notions of preferential reality and authentic tradition similarly count for less in the new science.   The forces at work in the external world are conceived in a less animistic manner, although anthropomorphism still prevails, at least to the degree required in order to give a dramatic interpretation of the sequence of phenomena.

The changes in the cultural situation which seem to have had the most serious consequences for the methods and animus of scientific inquiry are those changes that took place in the field of industry.   Industry in early modern times is a fact of relatively greater preponderance, more of a tone-giving factor, than it was under the régime of feudal status.   It is the characteristic trait of the modern culture, very much as exploit and fealty were the characteristic cultural traits of the earlier time. This early-modern industry is, in an obvious and convincing degree, a matter of workmanship.   The same has not been true in the same degree either before or since. The workman, more or less skilled and with more or less specialised efficiency, was the central figure in the cultural situation of the time; and so the concepts of the scientists came to be drawn in the image of the workman.   The dramatisations of the sequence of external

[9] As currently employed, the term "pragmatic" is made to cover both conduct looking to the agent's preferential advantage, expedient conduct, and workmanship directed to the production of things that may or may not be of advantage to the agent.   If the term be taken in the latter meaning, the culture of modern times is no less "pragmatic" than that of the Middle Ages.   It is here intended to be used in the former sense.

phenomena worked out under the impulse of the idle curiosity were then conceived in terms of workmanship. Workmanship gradually supplanted differential dignity as the authoritative canon of scientific truth, even on the higher levels of speculation and research. This, of course, amounts to saying in other words that the law of cause and effect was given the first place, as contrasted with dialectical consistency and authentic tradition. But this early-modern law of cause and effect — the law of efficient causes — is of an anthropomorphic kind. " Like causes produce like effects," in much the same sense as the skilled workman's product is like the workman; " nothing is found in the effect that was not contained in the cause," in much the same manner.

These dicta are, of course, older than modern science, but it is only in the early days of modern science that they come to rule the field with an unquestioned sway and to push the higher grounds of dialectical validity to one side. They invade even the highest and most recondite fields of speculation, so that at the approach to the transition from the early-modern to the late-modern period, in the eighteenth century, they determine the outcome even in the counsels of the theologians. The deity, from having been in mediæval times primarily a suzerain concerned with the maintenance of his own prestige, becomes primarily a creator engaged in the workmanlike occupation of making things useful for man. His relation to man and the natural universe is no longer primarily that of a progenitor, as it is in the lower barbarian culture, but rather that of a talented mechanic. The " natural laws " which the scientists of that era make so much of are no longer decrees of a preternatural legislative authority, but rather details of the workshop specifications handed down by the master-craftsman for the guidance of handicraftsmen working out his designs. In the eighteenth-century science these natural laws are laws specifying the sequence of cause and effect, and will bear characterisation as a dramatic interpretation of the activity of the causes at work, and these causes are conceived in a quasi-personal manner. In later modern times the formulations of causal sequence grow more impersonal and more objective, more matter-of-fact; but the imputation of activity to the observed objects never ceases, and even in the latest and maturest formulations of scientific research the dramatic tone is not wholly lost. The causes at work are conceived in a highly impersonal way, but hitherto no science (except ostensibly mathematics) has been content to do its theoretical work in terms of inert magnitude alone. Activity continues to be imputed to the phenomena with which science deals; and activity is, of course, not a fact of observation, but is imputed to the phenomena by the observer.[10] This is, also of course, denied by those who insist on a purely mathematical formulation of scientific theories, but the denial is maintained only at the cost of consistency. Those eminent authorities who speak for a colorless mathematical formulation invariably and necessarily fall back on the (essentially metaphysical) preconception of causation as soon as they go into the actual work of scientific inquiry.[11]

Since the machine technology has made great advances, during the nineteenth century, and has become a cultural force of wide-reaching consequence, the formulations of

[10] Epistemologically speaking, activity is imputed to phenomena for the purpose of organising them into a dramatically consistent system.

[11] Cf., e. g., Karl Pearson, *Grammar of Science,* and compare his ideal of inert magnitudes as set forth in his exposition with his actual work as shown in chaps. 9, 10, and 12, and more particularly in his discussions of " Mother Right " and related topics in *The Chances of Death.*

science have made another move in the direction of impersonal matter-of-fact. The machine process has displaced the workman as the archetype in whose image causation is conceived by the scientific investigators. The dramatic interpretation of natural phenomena has thereby become less anthropomorphic; it no longer constructs the life-history of a cause working to produce a given effect — after the manner of a skilled workman producing a piece of wrought goods — but it constructs the life-history of a process in which the distinction between cause and effect need scarcely be observed in an itemised and specific way, but in which the run of causation unfolds itself in an unbroken sequence of cumulative change. By contrast with the pragmatic formulations of worldly wisdom these latter-day theories of the scientists appear highly opaque, impersonal, and matter-of-fact; but taken by themselves they must be admitted still to show the constraint of the dramatic prepossessions that once guided the savage myth-makers.

In so far as touches the aims and the animus of scientific inquiry, as seen from the point of view of the scientist, it is a wholly fortuitous and insubstantial coincidence that much of the knowledge gained under machine-made canons of research can be turned to practical account. Much of this knowledge is useful, or may be made so, by applying it to the control of the processes in which natural forces are engaged. This employment of scientific knowledge for useful ends is technology, in the broad sense in which the term includes, besides the machine industry proper, such branches of practice as engineering, agriculture, medicine, sanitation, and economic reforms. The reason why scientific theories can be turned to account for these practical ends is not that these ends are included in the scope of scientific inquiry.

These useful purposes lie outside the scientist's interest. It is not that he aims, or can aim, at technological improvements. His inquiry is as "idle" as that of the Pueblo myth-maker. But the canons of validity under whose guidance he works are those imposed by the modern technology, through habituation to its requirements; and therefore his results are available for the technological purpose. His canons of validity are made for him by the cultural situation; they are habits of thought imposed on him by the scheme of life current in the community in which he lives; and under modern conditions this scheme of life is largely machine-made. In the modern culture, industry, industrial processes, and industrial products have progressively gained upon humanity, until these creations of man's ingenuity have latterly come to take the dominant place in the cultural scheme; and it is not too much to say that they have become the chief force in shaping men's daily life, and therefore the chief factor in shaping men's habits of thought. Hence men have learned to think in the terms in which the technological processes act. This is particularly true of those men who by virtue of a peculiarly strong susceptibility in this direction become addicted to that habit of matter-of-fact inquiry that constitutes scientific research.

Modern technology makes use of the same range of concepts, thinks in the same terms, and applies the same tests of validity as modern science. In both, the terms of standardisation, validity, and finality are always terms of impersonal sequence, not terms of human nature or of preternatural agencies. Hence the easy copartnership between the two. Science and technology play into one another's hands. The processes of nature with which science deals and which technology turns to account, the sequence of changes in the external world, animate and inanimate, run in terms of brute causation, as do the theories of science. These processes take no thought of human expediency or inexpediency. To make use of them they must be taken as they are, opaque and unsympathetic. Technology, therefore, has come to proceed on an interpretation of these phenomena in mechanical terms, not in terms of imputed personality nor even of workmanship. Modern science, deriving its concepts from the same source, carries on its inquiries and states its conclusions in terms of the same objective character as those employed by the mechanical engineer.

So it has come about, through the progressive change of the ruling habits of thought in the community, that the theories of science have progressively diverged from the formulations of pragmatism, ever since the modern era set in. From an organisation of knowledge on the basis of imputed personal or animistic propensity the theory has changed its base to an imputation of brute activity only, and this latter is conceived in an increasingly matter-of-fact manner; until, latterly, the pragmatic range of knowledge and the scientific are more widely out of touch than ever, differing not only in aim, but in matter as well. In both domains knowledge runs in terms of activity, but it is on the one hand knowledge of what had best be done, and on the other hand knowledge of what takes place; on the one hand knowledge of ways and means, on the other hand knowledge without any ulterior purpose. The latter range of knowledge may serve the ends of the former, but the converse does not hold true.

These two divergent ranges of inquiry are to be found together in all phases of human culture. What distinguishes the present phase is that the discrepancy between the two is now wider than ever before. The present is nowise distinguished above other cultural eras by any exceptional urgency or acumen in the search for pragmatic expedients. Neither is it safe to assert that the present excels all other civilisations in the volume or the workmanship of that body of knowledge that is to be credited to the idle curiosity. What distinguishes the present in these premises is (1) that the primacy in the cultural scheme has passed from pragmatism to a disinterested inquiry whose motive is idle curiosity, and (2) that in the domain of the latter the making of myths and legends in terms of imputed personality, as well as the construction of dialectical systems in terms of differential reality, has yielded the first place to the making of theories in terms of matter-of-fact sequence.[12]

Pragmatism creates nothing but maxims of expedient conduct. Science creates nothing but theories.[13] It knows nothing of policy or utility, of better or worse. None of all that is comprised in what is to-day accounted scientific knowledge. Wisdom and proficiency of the pragmatic sort does not contribute to the advance of a knowledge of fact. It has only an incidental bearing on scientific research, and its bearing is chiefly that of inhibition and misdirection. Wherever canons of expediency are intruded into or are attempted to be incorporated in the inquiry, the consequence is an unhappy one for science, however happy it may be for some other purpose extraneous to science. The mental attitude of worldly wisdom is at cross-purposes with the disinterested scientific spirit, and the pursuit of it induces an intellectual bias that is incompatible with scientific insight. Its intellectual output is a body of shrewd rules of conduct, in great part designed to take advantage of human infirmity.

[12] Cf. James, *Psychology*, Vol. II, chap. 28, pp. 633–71, esp. p. 640 note.
[13] Cf. Ward, *Principles of Psychology*, pp. 439–43.

Its habitual terms of standarisation and validity are terms of human nature, of human preference, prejudice, aspiration, endeavor, and disability, and the habit of mind that goes with it is such as is consonant with these terms. No doubt, the all-pervading pragmatic animus of the older and non-European civilisations has had more than anything else to do with their relatively slight and slow advance in scientific knowledge.   In the modern scheme of knowledge it holds true, in a similar manner and with analogous effect, that training in divinity, in law, and in the related branches of diplomacy, business tactics, military affairs, and political theory, is alien to the skeptical scientific spirit and subversive of it.

The modern scheme of culture comprises a large body of worldly wisdom, as well as of science.   This pragmatic lore stands over against science with something of a jealous reserve.   The pragmatists value themselves somewhat on being useful as well as being efficient for good and evil.   They feel the inherent antagonism between themselves and the scientists, and look with some doubt on the latter as being merely decorative triflers, although they sometimes borrow the prestige of the name of science — as is only good and well, since it is of the essence of worldly wisdom to borrow anything that can be turned to account.   The reasoning in these fields turns about questions of personal advantage of one kind or another, and the merits of the claims canvassed in these discussions are decided on grounds of authenticity.   Personal claims make up the subject of the inquiry, and these claims are construed and decided in terms of precedent and choice, use and wont, prescriptive authority, and the like.   The higher reaches of generalisation in these pragmatic inquiries are of the nature of deductions from authentic tradition, and the training in this class of reasoning gives discrimination in respect of authenticity and expediency. The resulting habit of mind is a bias for substituting dialectical distinctions and decisions *de jure* in the place of explanations *de facto*. The so-called " sciences " associated with these pragmatic disciplines, such as jurisprudence, political science, and the like, are a taxonomy of credenda. Of this character was the greater part of the " science " cultivated by the Schoolmen, and large remnants of the same kind of authentic convictions are, of course, still found among the tenets of the scientists, particularly in the social sciences, and no small solicitude is still given to their cultivation. Substantially the same value as that of the temporal pragmatic inquiries belongs also, of course, to the " science " of divinity. Here the questions to which an answer is sought, as well as the aim and method of inquiry, are of the same pragmatic character, although the argument runs on a higher plane of personality, and seeks a solution in terms of a remoter and more metaphysical expediency.

In the light of what has been said above, the questions recur: How far is the scientific quest of matter-of-fact knowledge consonant with the inherited intellectual aptitudes and propensities of the normal man? and, What foothold has science in the modern culture? The former is a question of the temperamental heritage of civilised mankind, and therefore it is in large part a question of the circumstances which have in the past selectively shaped the human nature of civilised mankind. Under the barbarian culture, as well as on the lower levels of what is currently called civilised life, the dominant note has been that of competitive expediency for the individual or the group, great or small, in an avowed struggle for the means of life. Such is still the ideal of the politician

and business man, as well as of other classes whose habits
of life lead them to cling to the inherited barbarian traditions.   The upper-barbarian and lower-civilised culture,
as has already been indicated, is pragmatic, with a thoroughness that nearly bars out any non-pragmatic ideal of
life or of knowledge.   Where this tradition is strong
there is but a precarious chance for any consistent effort
to formulate knowledge in other terms than those drawn
from the prevalent relations of personal mastery and subservience and the ideals of personal gain.

During the Dark and Middle Ages, for instance, it is
true in the main that any movement of thought not controlled by considerations of expediency and conventions
of status are to be found only in the obscure depths of
vulgar life, among those neglected elements of the population that lived below the reach of the active class struggle.   What there is surviving of this vulgar, non-pragmatic intellectual output takes the form of legends and
folk-tales, often embroidered on the authentic documents
of the Faith.   These are less alien to the latest and highest culture of Christendom than are the dogmatic, dialectical, and chivalric productions that occupied the attention of the upper classes in mediæval times.   It may seem
a curious paradox that the latest and most perfect flower
of the western civilisation is more nearly akin to the
spiritual life of the serfs and villeins than it is to that of
the grange or the abbey.   The courtly life and the chivalric habits of thought of that past phase of culture have
left as nearly no trace in the cultural scheme of later modern times as could well be.   Even the romancers who
ostensibly rehearse the phenomena of chivalry, unavoidably make their knights and ladies speak the language
and the sentiments of the slums of that time, tempered
with certain schematised modern reflections and specu-

lations. The gallantries, the genteel inanities and devout imbecilities of mediæval high-life would be insufferable even to the meanest and most romantic modern intelligence. So that in a later, less barbarian age the precarious remnants of folklore that have come down through that vulgar channel — half savage and more than half pagan — are treasured as containing the largest spiritual gains which the barbarian ages of Europe have to offer.

The sway of barbarian pragmatism has, everywhere in the western world, been relatively brief and relatively light; the only exceptions would be found in certain parts of the Mediterranean seaboard. But wherever the barbarian culture has been sufficiently long-lived and unmitigated to work out a thoroughly selective effect in the human material subjected to it, there the pragmatic animus may be expected to have become supreme and to inhibit all movement in the direction of scientific inquiry and eliminate all effective aptitude for other than worldly wisdom. What the selective consequences of such a protracted régime of pragmatism would be for the temper of the race may be seen in the human flotsam left by the great civilisations of antiquity, such as Egypt, India, and Persia. Science is not at home among these leavings of barbarism. In these instances of its long and unmitigated dominion the barbarian culture has selectively worked out a temperamental bias and a scheme of life from which objective, matter-of-fact knowledge is virtually excluded in favor of pragmatism, secular and religious. But for the greater part of the race, at least for the greater part of civilised mankind, the régime of the mature barbarian culture has been of relatively short duration, and has had a correspondingly superficial and transient selective effect. It has not had force and time to eliminate certain elements of human nature handed down from an earlier phase of life, which are not in full consonance with the barbarian animus or with the demands of the pragmatic scheme of thought. The barbarian-pragmatic habit of mind, therefore, is not properly speaking a temperamental trait of the civilised peoples, except possibly within certain class limits (as, *e.g.,* the German nobility). It is rather a tradition, and it does not constitute so tenacious a bias as to make head against the strongly materialistic drift of modern conditions and set aside that increasingly urgent resort to matter-of-fact conceptions that makes for the primacy of science. Civilised mankind does not in any great measure take back atavistically to the upper-barbarian habit of mind. Barbarism covers too small a segment of the life-history of the race to have given an enduring temperamental result. The unmitigated discipline of the higher barbarism in Europe fell on a relatively small proportion of the population, and in the course of time this select element of the population was crossed and blended with the blood of the lower elements whose life always continued to run in the ruts of savagery rather than in those of the high-strung, finished barbarian culture that gave rise to the chivalric scheme of life.

Of the several phases of human culture the most protracted, and the one which has counted for most in shaping the abiding traits of the race, is unquestionably that of savagery. With savagery, for the purpose in hand, is to be classed that lower, relatively peaceable barbarism that is not characterised by wide and sharp class discrepancies or by an unremitting endeavor of one individual or group to get the better of another. Even under the full-grown barbarian culture — as, for instance, during the Middle Ages — the habits of life and the spiritual interests of the great body of the population continue in large measure to bear the character of savagery. The savage phase of culture accounts for by far the greater portion of the life-history of mankind, particularly if the lower barbarism and the vulgar life of later barbarism be counted in with savagery, as in a measure they properly should. This is particularly true of those racial elements that have entered into the composition of the leading peoples of Christendom.

The savage culture is characterised by the relative absence of pragmatism from the higher generalisations of its knowledge and beliefs. As has been noted above, its theoretical creations are chiefly of the nature of mythology shading off into folklore. This genial spinning of apocryphal yarns is, at its best, an amiably inefficient formulation of experiences and observations in terms of something like a life-history of the phenomena observed. It has, on the one hand, little value, and little purpose, in the way of pragmatic expediency, and so it is not closely akin to the pragmatic-barbarian scheme of life; while, on the other hand, it is also ineffectual as a systematic knowledge of matter-of-fact. It is a quest of knowledge, perhaps of systematic knowledge, and it is carried on under the incentive of the idle curiosity. In this respect it falls in the same class with the civilised man's science; but it seeks knowledge not in terms of opaque matter-of-fact, but in terms of some sort of spiritual life imputed to the facts. It is romantic and Hegelian rather than realistic and Darwinian. The logical necessities of its scheme of thought are necessities of spiritual consistency rather than of quantitative equivalence. It is like science in that it has no ulterior motive beyond the idle craving for a systematic correlation of data; but it is unlike science in that its standardisation and correlation of data run in terms of the free play of imputed personal initiative rather than in terms of the constraint of objective cause and effect.

By force of the protracted selective discipline of this past phase of culture, the human nature of civilised mankind is still substantially the human nature of savage man. The ancient equipment of congenital aptitudes and propensities stands over substantially unchanged, though overlaid with barbarian traditions and conventionalities and readjusted by habituation to the exigencies of civilised life. In a measure, therefore, but by no means altogether, scientific inquiry is native to civilised man with his savage heritage, since scientific inquiry proceeds on the same general motive of idle curiosity as guided the savage myth-makers, though it makes use of concepts and standards in great measure alien to the myth-makers' habit of mind. The ancient human predilection for discovering a dramatic play of passion and intrigue in the phenomena of nature still asserts itself. In the most advanced communities, and even among the adepts of modern science, there comes up persistently the revulsion of the native savage against the inhumanly dispassionate sweep of the scientific quest, as well as against the inhumanly ruthless fabric of technological processes that have come out of this search for matter-of-fact knowledge. Very often the savage need of a spiritual interpretation (dramatisation) of phenomena breaks through the crust of acquired materialistic habits of thought, to find such refuge as may be had in articles of faith seized on and held by sheer force of instinctive conviction. Science and its creations are more or less uncanny, more or less alien, to that fashion of craving for knowledge that by ancient inheritance animates mankind. Furtively or by an overt breach of consistency, men still seek comfort in marvelous articles of savage-born lore, which contradict the truths of that modern science whose dominion they dare not question, but whose findings at the same time go beyond the breaking point of their jungle-fed spiritual sensibilities.

The ancient ruts of savage thought and conviction are smooth and easy; but however sweet and indispensable the archaic ways of thinking may be to the civilised man's peace of mind, yet such is the binding force of matter-of-fact analysis and inference under modern conditions that the findings of science are not questioned on the whole. The name of science is after all a word to conjure with. So much so that the name and the mannerisms, at least, if nothing more of science, have invaded all fields of learning and have even overrun territory that belongs to the enemy. So there are " sciences " of theology, law, and medicine, as has already been noted above. And there are such things as Christian Science, and " scientific " astrology, palmistry, and the like. But within the field of learning proper there is a similar predilection for an air of scientific acumen and precision where science does not belong. So that even that large range of knowledge that has to do with general information rather than with theory — what is loosely termed scholarship — tends strongly to take on the name and forms of theoretical statement. However decided the contrast between these branches of knowledge on the one hand, and science properly so called on the other hand, yet even the classical learning, and the humanities generally, fall in with this predilection more and more with each succeeding generation of students. The students of literature, for instance, are more and more prone to substitute critical analysis and linguistic speculation, as the end of their endeavors, in the place of that discipline of taste and that cultivated sense of literary form and literary feeling that must always remain the chief end of literary training, as distinct from philology and the social sciences. There is, of course, no intention to question the legitimacy of a science of philology or of the analytical study of literature as a fact in cultural history, but these things do not constitute training in literary taste, nor can they take the place of it. The effect of this straining after scientific formulations in a field alien to the scientific spirit is as curious as it is wasteful. Scientifically speaking, these quasi-scientific inquiries necessarily begin nowhere and end in the same place; while in point of cultural gain they commonly come to nothing better than spiritual abnegation. But these blindfold endeavors to conform to the canons of science serve to show how wide and unmitigated the sway of science is in the modern community.

Scholarship — that is to say an intimate and systematic familiarity with past cultural achievements — still holds its place in the scheme of learning, in spite of the unadvised efforts of the short-sighted to blend it with the work of science, for it affords play for the ancient genial propensities that ruled men's quest of knowledge before the coming of science or of the outspoken pragmatic barbarism. Its place may not be so large in proportion to the entire field of learning as it was before the scientific era got fully under way. But there is no intrinsic antagonism between science and scholarship, as there is between pragmatic training and scientific inquiry. Modern scholarship shares with modern science the quality of not being pragmatic in its aim. Like science it has no ulterior end. It may be difficult here and there to draw the line between science and scholarship, and it may even more be unnecessary to draw such a line; yet while the two ranges of discipline belong together in many ways, and while there are many points of contact and sympathy between the two; while the two together make up the modern scheme of learning; yet there is no need of confounding the one with the other, nor can the one do the work of the other. The scheme of learning has changed in such manner as to give science the more commanding place, but the scholar's domain has not thereby been invaded, nor has it suffered contraction at the hands of science, whatever may be said of the weak-kneed abnegation of some whose place, if they have one, is in the field of scholarship rather than of science.

All that has been said above has of course nothing to say as to the intrinsic merits of this quest of matter-of-fact knowledge. In point of fact, science gives its tone to modern culture. One may approve or one may deprecate the fact that this opaque, materialistic interpretation of things pervades modern thinking. That is a question of taste, about which there is no disputing. The prevalence of this matter-of-fact inquiry is a feature of modern culture, and the attitude which critics take toward this phenomenon is chiefly significant as indicating how far their own habit of mind coincides with the enlightened common-sense of civilised mankind. It shows in what degree they are abreast of the advance of culture. Those in whom the savage predilection or the barbarian tradition is stronger than their habituation to civilised life will find that this dominant factor of modern life is perverse, if not calamitous; those whose habits of thought have been fully shaped by the machine process and scientific inquiry are likely to find it good. The modern western culture, with its core of matter-of-fact knowledge, may be better or worse than some other cultural scheme, such as the classic Greek, the mediæval Christian, the Hindu, or the Pueblo Indian. Seen in certain lights, tested by certain standards, it is doubtless better; by other standards, worse. But the fact remains that the current cultural scheme, in its maturest growth, is of that complexion; its characteristic force lies in this matter-of-fact insight; its highest discipline and its maturest aspirations are these.

In point of fact, the sober common-sense of civilised mankind accepts no other end of endeavor as self-sufficient and ultimate. That such is the case seems to be due chiefly to the ubiquitous presence of the machine technology and its creations in the life of modern communities. And so long as the machine process continues to hold its dominant place as a disciplinary factor in modern culture, so long must the spiritual and intellectual life of this cultural era maintain the character which the machine process gives it.

But while the scientist's spirit and his achievements stir an unqualified admiration in modern men, and while his discoveries carry conviction as nothing else does, it does not follow that the manner of man which this quest of knowledge produces or requires comes near answering to the current ideal of manhood, or that his conclusions are felt to be as good and beautiful as they are true. The ideal man, and the ideal of human life, even in the apprehension of those who most rejoice in the advances of science, is neither the finikin skeptic in the laboratory nor the animated slide-rule. The quest of science is relatively new. It is a cultural factor not comprised, in anything like its modern force, among those circumstances whose selective action in the far past has given to the race the human nature which it now has. The race reached the human plane with little of this searching knowledge of facts; and throughout the greater part of its life-history on the human plane it has been accustomed to make its higher generalisations and to formulate its larger principles of life in other terms than those of passionless matter-of-fact. This manner of knowledge has occupied an increasing share of men's attention in the past, since it bears in a decisive way upon the minor affairs of workday life; but it has never until now been put in the first place, as the dominant note of human culture. The normal man, such as his inheritance has made him, has therefore good cause to be restive under its dominion.

THE EVOLUTION OF THE SCIENTIFIC
POINT OF VIEW [1]

A DISCUSSION of the scientific point of view which avowedly proceeds from this point of view itself has necessarily
the appearance of an argument in a circle; and such in
great part is the character of what here follows. It is
in large part an attempt to explain the scientific point of
view in terms of itself, but not altogether. This inquiry
does not presume to deal with the origin or the legitimation of the postulates of science, but only with the growth
of the habitual use of these postulates, and the manner
of using them. The point of inquiry is the changes which
have taken place in the secondary postulates involved in
the scientific point of view — in great part a question of
the progressive redistribution of emphasis among the preconceptions under whose guidance successive generations
of scientists have gone to their work.

The sciences which are in any peculiar sense modern
take as an (unavowed) postulate the fact of consecutive
change. Their inquiry always centers upon some manner
of process. This notion of process about which the researches of modern science cluster, is a notion of a sequence, or complex, of consecutive change in which the
nexus of the sequence, that by virtue of which the change
inquired into is consecutive, is the relation of cause and
effect. The consecution, moreover, runs in terms of
persistence of quantity or of force. In so far as the sci-

[1] Read before the Kosmos Club, at the University of California,
May 4, 1908. Reprinted by permission from the *University of
California Chronicle,* Vol. X, No. 4.

ence is of a modern complexion, in so far as it is not of the nature of taxonomy simply, the inquiry converges upon a matter of process; and it comes to rest, provisionally, when it has disposed of its facts in terms of process. But modern scientific inquiry in any case comes to rest only provisionally; because its prime postulate is that of consecutive change, and consecutive change can, of course, not come to rest except provisionally. By its own nature the inquiry cannot reach a final term in any direction. So it is something of a homiletical commonplace to say that the outcome of any serious research can only be to make two questions grow where one question grew before. Such is necessarily the case because the postulate of the scientist is that things change consecutively. It is an unproven and unprovable postulate — that is to say, it is a metaphysical preconception — but it gives the outcome that every goal of research is necessarily a point of departure; every term is transitional.[2]

[2] It is by no means unusual for modern scientists to deny the truth of this characterization, so far as regards this alleged recourse to the concept of causation. They deny that such a concept — of efficiency, activity, and the like — enters, or can legitimately enter, into their work, whether as an instrument of research or as a means or guide to theoretical formulation. They even deny the substantial continuity of the sequence of changes that excite their scientific attention. This attitude seems particularly to commend itself to those who by preference attend to the mathematical formulations of theory and who are chiefly occupied with proving up and working out details of the system of theory which have previously been left unsettled or uncovered. The concept of causation is recognized to be a metaphysical postulate, a matter of imputation, not of observation; whereas it is claimed that scientific inquiry neither does nor can legitimately, nor, indeed, currently, make use of a postulate more metaphysical than the concept of an idle concomitance of variation, such as is adequately expressed in terms of mathematical function.

The contention seems sound, to the extent that the materials — essentially statistical materials — with which scientific inquiry

A hundred years ago, or even fifty years ago, scientific men were not in the habit of looking at the matter

is occupied are of this non-committal character, and that the
mathematical formulations of theory include no further element
than that of idle variation. Such is necessarily the case because
causation is a fact of imputation, not of observation, and so cannot be included among the data; and because nothing further
than non-committal variation can be expressed in mathematical
terms. A bare notation of quantity can convey nothing further.

If it were the intention to claim only that the conclusions of
the scientists are, or should be, as a matter of conservative caution, overtly stated in terms of function alone, then the contention might well be allowed. Causal sequence, efficiency or
continuity is, of course, a matter of metaphysical imputation.
It is not a fact of observation, and cannot be asserted of the
facts of observation except as a trait imputed to them. It is so
imputed, by scientists and others, as a matter of logical necessity, as a basis of a systematic knowledge of the facts of observation.

Beyond this, in their exercise of scientific initiative, as well as
in the norms which guide the systematisation of scientific results,
the contention will not be made good — at least not for the current phase of scientific knowledge. The claim, indeed, carries
its own refutation. In making such a claim, both in rejecting
the imputation of metaphysical postulates and in defending their
position against their critics, the arguments put forward by the
scientists run in causal terms. For the polemical purposes,
where their antagonists are to be scientifically confuted, the
defenders of the non-committal postulate of concomitance find
that postulate inadequate. They are not content, in this precarious conjuncture, simply to attest a relation of idle quantitative concomitance (mathematical function) between the allegations of their critics, on the one hand, and their own con-
troversial exposition of these matters on the other hand. They
argue that they do not " make use of " such a postulate as " efficiency," whereas they claim to " make use of " the concept of
function. But " make use of " is not a notion of functional variation but of causal efficiency in a somewhat gross and highly
anthropomorphic form. The relation between their own thinking
and the " principles " which they " apply " or the experiments and
calculations which they " institute " in their " search " for facts,
is not held to be of this non-committal kind. It will not be
claimed that the shrewd insight and the bold initiative of a man
eminent in the empirical sciences bear no more efficient or con-

in this way. At least it did not then seem a matter of
course, lying in the nature of things, that scientific inquiry

sequential a relation than that of mathematical function to the
ingenious experiments by which he tests his hypotheses and extends the secure bounds of human knowledge. Least of all is the
masterly experimentalist himself in a position to deny that his
intelligence counts for something more efficient than idle concomitance in such a case. The connection between his premises,
hypotheses, and experiments, on the one hand, and his theoretical results, on the other hand, is not felt to be of the nature
of mathematical function. Consistently adhered to, the principle of "function" or concomitant variation precludes recourse
to experiment, hypotheses or inquiry — indeed, it precludes "recourse" to anything whatever. Its notation does not comprise
anything so anthropomorphic.

The case is illustrated by the latter-day history of theoretical
physics. Of the sciences which affect a non-committal attitude
in respect of the concept of efficiency and which claim to get
along with the notion of mathematical function alone, physics
is the most outspoken and the one in which the claim has the
best *prima facie* validity. At the same time, latter-day physicists,
for a hundred years or more, have been much occupied with
explaining how phenomena which to all appearance involve action
at a distance do not involve action at a distance at all. The
greater theoretical achievements of physics during the past century lie within the sweep of this (metaphysical) principle that
action at a distance does not take place, that apparent action
at a distance must be explained by effective contact, through a
continuum, or by a material transference. But this principle is
nothing better than an unreasoning repugnance on the part of the
physicists to admitting action at a distance. The requirement of
a continuum involves a gross form of the concept of efficient
causation. The "functional" concept, concomitant variation, requires no contact and no continuum. Concomitance at a distance is quite as simple and convincing a notion as concomitance
within contact or by the intervention of a continuum, if not more
so. What stands in the way of its acceptance is the irrepressible anthropomorphism of the physicists. And yet the great
achievements of physics are due to the initiative of men animated
with this anthropomorphic repugnance to the notion of concomitant variation at a distance. All the generalisations on un-
dulatory motion and translation belong here. The latter-day
researches in light, electrical transmission, the theory of ions,
together with what is known of the obscure and late-found

could not reach a final term in any direction. To-day it is a matter of course, and will be so avowed without argument. Stated in the broadest terms, this is the substantial outcome of that nineteenth-century movement in science with which the name of Darwin is associated as a catch-word.

This use of Darwin's name does not imply that this epoch of science is mainly Darwin's work. What merit may belong to Darwin, specifically, in these premises, is a question which need not detain the argument. He may, by way of creative initiative, have had more or less to do with shaping the course of things scientific. Or, if you choose, his voice may even be taken as only one of the noises which the wheels of civilisation make when they go round. But by scientifically colloquial usage we have come to speak of pre-Darwinian and post-Darwinian science, and to appreciate that there is a significant difference in the point of view between the scientific era which preceded and that which followed the epoch to which his name belongs.

Before that epoch the animus of a science was, on the whole, the animus of taxonomy; the consistent end of scientific inquiry was definition and classification,— as it still continues to be in such fields of science as have not been affected by the modern notion of consecutive change. The scientists of that era looked to a final term, a consummation of the changes which provoked their inquiry, as well as to a first beginning of the matters with which their researches were concerned. The questions of science were directed to the problem, essentially classi-

radiations and emanations, are to be credited to the same metaphysical preconception, which is never absent in any " scientific " inquiry in the field of physical science. It is only the " occult " and " Christian " " Sciences " that can dispense with this metaphysical postulate and take recourse to " absent treatment."

ficatory, of how things had been in the presumed primordial stable equilibrium out of which they, putatively, had come, and how they should be in the definitive state of settlement into which things were to fall as the outcome of the play of forces which intervened between this primordial and the definitive stable equilibrium. To the pre-Darwinian taxonomists the center of interest and attention, to which all scientific inquiry must legitimately converge, was the body of natural laws governing phenomena under the rule of causation. These natural laws were of the nature of rules of the game of causation. They formulated the immutable relations in which things "naturally" stood to one another before causal disturbance took place between them, the orderly unfolding of the complement of causes involved in the transition over this interval of transient activity, and the settled relations that would supervene when the disturbance had passed and the transition from cause to effect had been consummated,— the emphasis falling on the consummation.

The characteristic feature by which post-Darwinian science is contrasted with what went before is a new distribution of emphasis, whereby the process of causation, the interval of instability and transition between initial cause and definitive effect, has come to take the first place in the inquiry; instead of that consummation in which causal effect was once presumed to come to rest. This change of the point of view was, of course, not abrupt or catastrophic. But it has latterly gone so far that modern science is becoming substantially a theory of the process of consecutive change, which is taken as a sequence of cumulative change, realized to be self-continuing or self-propagating and to have no final term. Questions of a primordial beginning and a definitive outcome have fallen into abeyance within the modern sciences, and such questions are in a fair way to lose all claim to consideration at the hands of the scientists. Modern science is ceasing to occupy itself with the natural laws — the codified rules of the game of causation —and is concerning itself wholly with what has taken place and what is taking place.

Rightly seen from this ultra-modern point of view, this modern science and this point of view which it affects are, of course, a feature of the current cultural situation, — of the process of life as it runs along under our eyes. So also, when seen from this scientific point of view, it is a matter of course that any marked cultural era will have its own characteristic attitude and animus toward matters of knowledge, will bring under inquiry such questions of knowledge as lie within its peculiar range of interest, and will seek answers to these questions only in terms that are consonant with the habits of thought current at the time. That is to say, science and the scientific point of view will vary characteristically in response to those variations in the prevalent habits of thought which constitute the sequence of cultural development; the current science and the current scientific point of view, the knowledge sought and the manner of seeking it, are a product of the cultural growth. Perhaps it would all be better characterised as a by-product of the cultured growth.

This question of a scientific point of view, of a particular attitude and animus in matters of knowledge, is a question of the formation of habits of thought; and habits of thought are an outcome of habits of life. A scientific point of view is a consensus of habits of thought current in the community, and the scientist is constrained to believe that this consensus is formed in response to a more or less consistent discipline of habituation to which the community is subjected, and that the consensus can extend only so far and maintain its force only so long as the discipline of habituation exercised by the circumstances of life enforces it and backs it up. The scheme of life, within which lies the scheme of knowledge, is a consensus of habits in the individuals which make up the community. The individual subjected to habituation is each a single individual agent, and whatever affects him in any one line of activity, therefore, necessarily affects him in some degree in all his various activities. The cultural scheme of any community is a complex of the habits of life and of thought prevalent among the members of the community. It makes up a more or less congruous and balanced whole, and carries within it a more or less consistent habitual attitude toward matters of knowledge — more or less consistent according as the community's cultural scheme is more or less congruous throughout the body of the population; and this in its turn is in the main a question of how nearly uniform or consonant are the circumstances of experience and tradition to which the several classes and members of the community are subject.

So, then, the change which has come over the scientific point of view between pre-Darwinian and post-Darwinian times is to be explained, at least in great part, by the changing circumstances of life, and therefore of habituation, among the people of Christendom during the life-history of modern science. But the growth of a scientific point of view begins farther back than modern Christendom, and a record of its growth would be a record of the growth of human culture. Modern science demands a genetic account of the phenomena with which it deals, and a genetic inquiry into the scientific point of view necessarily will have to make up its account with the earlier phases of cultural growth. A life-history of human culture is a large topic, not to be attempted here even in the sketchiest outline. The most that can be attempted is a hasty review of certain scattered questions and salient points in this life-history.

In what manner and with what effect the idle curiosity of mankind first began to tame the facts thrown in its way, far back in the night of time, and to break them in under a scheme of habitual interpretation; what may have been the earliest norms of systematic knowledge, such as would serve the curiosity of the earliest generations of men in a way analogous to the service rendered the curiosity of later generations by scientific inquiry — all that is, of course, a matter of long-range conjecture, more or less wild, which cannot be gone into here. But among such peoples of the lower cultures as have been consistently observed, norms of knowledge and schemes for its systematization are always found. These norms and systems of knowledge are naïve and crude, perhaps, but there is fair ground for presuming that out of the like norms and systems in the remoter ages of our own antecedents have grown up the systems of knowledge cultivated by the peoples of history and by their representatives now living.

It is not unusual to say that the primitive systems of knowledge are constructed on animistic lines; that animistic sequence is the rule to which the facts are broken in. This seems to be true, if "animism" be construed in a sufficiently naïve and inchoate sense. But this is not the whole case. In their higher generalisations, in what Powell calls their "sophiology," it appears that the primitive peoples are guided by animistic norms; they make up their cosmological schemes, and the like, in terms of personal or quasi-personal activity, and the whole is thrown into something of a dramatic form. Through the early cosmological lore runs a dramatic consistency which imputes something in the way of initiative and propensity to the phenomena that are to be accounted for. But this dramatisation of the facts, the accounting for phenomena in terms of spiritual or quasi-spiritual initiative, is by no means the whole case of primitive men's systematic knowledge of facts. Their theories are not all of the nature of dramatic legend, myth, or animistic life-history, although the broader and more picturesque generalisations may take that form. There always runs along by the side of these dramaturgic life-histories, and underlying them, an obscure system of generalisations in terms of matter-of-fact. The system of matter-of-fact generalisations, or theories, is obscurer than the dramatic generalisations only in the sense that it is left in the background as being less picturesque and of less vital interest, not in the sense of being less familiar, less adequately apprehended, or less secure. The peoples of the lower cultures "know" that the broad scheme of things is to be explained in terms of creation, perhaps of procreation, gestation, birth, growth, life and initiative; and these matters engross the attention and stimulate speculation. But they know equally well the matter of fact that water will run down hill, that two stones are heavier than one of them, that an edge-tool will cut softer substances, that two things may be tied together with a string, that a pointed stick may be stuck in the ground, and the like. There is no range of knowledge that is held more securely by any people than such matters of fact; and these are generalisations from experience; they are theoretical knowledge, and they are a matter of course. They underlie the dramatical generalisations of the broad scheme of things, and are so employed in the speculations of the myth-makers and the learned.

It may be that the exceptional efficiency of a given edge-tool, *e.g.,* will be accounted for on animistic or quasi-personal grounds,— grounds of magical efficacy; but it is the exceptional behavior of such a tool that calls for explanation on the higher ground of animistic potency, not its work-day performance of common work. So also if an edge-tool should fail to do what is expected of it as a matter of course, its failure may require an explanation in other terms than matter-of-fact. But all that only serves to bring into evidence the fact that a scheme of generalisations in terms of matter-of-fact is securely held and is made use of as a sufficient and ultimate explanation of the more familiar phenomena of experience. These commonplace matter-of-fact generalisations are not questioned and do not clash with the higher scheme of things.

All this may seem like taking pains about trivialities. But the data with which any scientific inquiry has to do are trivialities in some other bearing than that one in which they are of account.

In all succeeding phases of culture, developmentally subsequent to the primitive phase supposed above, there is found a similar or analogous division of knowledge between a higher range of theoretical explanations of phenomena, an ornate scheme of things, on the one hand, and such an obscure range of matter-of-fact generalisations as is here spoken of, on the other hand. And the evolution of the scientific point of view is a matter of the shifting fortunes which have in the course of cultural growth overtaken the one and the other of these two divergent methods of apprehending and systematising the facts of experience.

The historians of human culture have, no doubt justly, commonly dealt with the mutations that have occurred on the higher levels of intellectual enterprise, in the more ambitious, more picturesque, and less secure of these two contrasted ranges of theoretical knowledge; while the lower range of generalisations, which has to do with work-day experience, has in great part been passed over with scant ceremony as lying outside the current of ideas, and as belonging rather among the things which engage the attention than among the modes, expedients and creations of this attention itself. There is good reason for this relative neglect of the work-day matters of fact. It is on the higher levels of speculative generalisation that the impressive mutations in the development of thought have taken place, and that the shifting of points of view and the clashing of convictions have drawn men into controversy and analysis of their ideas and have given rise to schools of thought. The matter-of-fact generalisations have met with relatively few adventures and have afforded little scope for intellectual initiative and profoundly picturesque speculation. On the higher levels speculation is freer, the creative spirit has some scope, because its excursions are not so immediately and harshly checked by material facts.

In these speculative ranges of knowledge it is possible to form and to maintain habits of thought which shall be consistent with themselves and with the habit of mind and run of tradition prevalent in the community at the time, though not thereby consistent with the material actualities of life in the community. Yet this range of speculative generalisation, which makes up the higher learning of the barbarian culture, is also controlled, checked, and guided by the community's habits of life; it, too, is an integral part of the scheme of life and is an outcome of the habituation enforced by experience. But it does not rest immediately on men's dealings with the refractory phenomena of brute creation, nor is it guided, undisguised and directly, by the habitual material (industrial) occupations. The fabric of institutions intervenes between the material exigencies of life and the speculative scheme of things.

The higher theoretical knowledge, that body of tenets which rises to the dignity of a philosophical or scientific system, in the early culture, is a complex of habits of thought which reflect the habits of life embodied in the institutional structure of society; while the lower, matter-of-fact generalisations of work-day efficiency — the trivial matters of course — reflect the workmanlike habits of life enforced by the commonplace material exigencies under which men live. The distinction is analogous, and indeed, closely related, to the distinction between " intangible " and " tangible " assets. And the institutions are more flexible, they involve or admit a larger margin of error, or of tolerance, than the material exigencies. The latter are systematised into what economists have called " the state of the idustrial arts," which enforce a somewhat rigorous standardisation of whatever knowledge falls within their scope; whereas the institutional scheme is a matter of law and custom, politics and religion, taste and morals, on all of which matters men have opinions and convictions, and on which all men " have a right to their own opinions." The scheme of institutions is also not necessarily uniform throughout the several classes of society; and the same institution (as, *e.g.,* slavery, ownership, or royalty) does not impinge with the same effect on all parties touched by it. The discipline of

any institution of servitude, *e.g.,* is not the same for the
master as for the serf, etc. If there is a considerable
institutional discrepancy between an upper and a lower
class in the community, leading to divergent lines of
habitual interest or discipline; if by force of the cultural
scheme the institutions of society are chiefly in the keeping of one class, whose attention is then largely engrossed
with the maintenance of the scheme of law and order;
while the workmanlike activities are chiefly in the hands
of another class, in whose apprehension the maintenance
of law and order is at the best a wearisome tribulation,
there is likely to be a similarly considerable divergence
or discrepancy between the speculative knowledge, cultivated primarily by the upper class, and the work-day
knowledge which is primarily in the keeping of the lower
class. Such, in particular, will be the case if the community is organised on a coercive plan, with well-marked
ruling and subject classes. The important and interesting institutions in such a case, those institutions which
fill a large angle in men's vision and carry a great force
of authenticity, are the institutions of coercive control,
differential authority and subjection, personal dignity and
consequence; and the speculative generalisations, the institutions of the realm of knowledge, are created in
the image of these social institutions of status and personal force, and fall into a scheme drawn after the plan
of the code of honor. The work-day generalisations,
which emerge from the state of the industrial arts, concomitantly fall into a deeper obscurity, answering to the
depth of indignity to which workmanlike efficiency sinks
under such a cultural scheme; and they can touch and
check the current speculative knowledge only remotely
and incidentally. Under such a bifurcate scheme of culture, with its concomitant two-cleft systematisation of

knowledge, "reality" is likely to be widely dissociated from fact — that is to say, the realities and verities which are accepted as authentic and convincing on the plane of speculative generalisation; while science has no show — that is to say, science in that modern sense of the term which implies a close contact, if not a coincidence, of reality with fact.

Whereas, if the institutional fabric, the community's scheme of life, changes in such a manner as to throw the work-day experience into the foreground of attention and to center the habitual interest of the people on the immediate material relations of men to the brute actualities, then the interval between the speculative realm of knowledge, on the one hand, and the work-day generalisations of fact, on the other hand, is likely to lessen, and the two ranges of knowledge are likely to converge more or less effectually upon a common ground. When the growth of culture falls into such lines, these two methods and norms of theoretical formulation may presently come to further and fortify one another, and something in the way of science has at least a chance to arise.

On this view there is a degree of interdependence between the cultural situation and the state of theoretical inquiry. To illustrate this interdependence, or the concomitance between the cultural scheme and the character of theoretical speculation, it may be in place to call to mind certain concomitant variations of a general character which occur in the lower cultures between the scheme of life and the scheme of knowledge. In this tentative and fragmentary presentation of evidence there is nothing novel to be brought forward; still less is there anything to be offered which carries the weight of authority.

On the lower levels of culture, even more decidedly than on the higher, the speculative systematisation of knowledge is prone to take the form of theology (mythology) and cosmology. This theological and cosmological lore serves the savage and barbaric peoples as a theoretical account of the scheme of things, and its characteristic traits vary in response to the variations of the institutional scheme under which the community lives. In a prevailingly peaceable agricultural community, such, *e.g.*, as the more peaceable Pueblo Indians or the more settled Indians of the Middle West, there is little coercive authority, few and slight class distinctions involving superiority and inferiority; property rights are few, slight and unstable; relationship is likely to be counted in the female line. In such a culture the cosmological lore is likely to offer explanations of the scheme of things in terms of generation or germination and growth. Creation by fiat is not obtrusively or characteristically present. The laws of nature bear the character of an habitual behavior of things, rather than that of an authoritative code of ordinances imposed by an overruling providence. The theology is likely to be polytheistic in an extreme degree and in an extremely loose sense of the term, embodying relatively little of the suzerainty of God. The relation of the deities to mankind is likely to be that of consanguinity, and as if to emphasise the peaceable, noncoercive character of the divine order of things, the deities are, in the main, very apt to be females. The matters of interest dealt with in the cosmological theories are chiefly matters of the livelihood of the people, the growth and care of the crops, and the promotion of industrial ways and means.

With these phenomena of the peaceable culture may be contrasted the order of things found among a predatory

pastoral people — and pastoral peoples tend strongly to
take on a predatory cultural scheme. Such a people will
adopt male deities, in the main, and will impute to them a
coercive, imperious, arbitrary animus and a degree of
princely dignity. They will also tend strongly to a monotheistic, patriarchal scheme of divine government; to explain things in terms of creative fiat; and to a belief in
the control of the natural universe by rules imposed by
divine ordinance. The matters of prime consequence in
this theology are matters of the servile relation of man
to God, rather than the details of the quest of a livelihood.
The emphasis falls on the glory of God rather than on
the good of man. The Hebrew scriptures, particularly
the Jahvistic elements, show such a scheme of pastoral
cultural and predatory theoretical generalisations.

The learning cultivated on the lower levels of culture
might be gone into at some length if space and time permitted, but even what has been said may serve to show,
in the most general way, what are the characteristic marks
of this savage and barbarian lore. A similarly summary
characterisation of a cultural situation nearer home will
bear more directly on the immediate topic of inquiry.
The learning of mediæval Christendom shows such a concomitance between the scheme of knowledge and the
scheme of institutions, somewhat analogous to the barbaric Hebrew situation. The mediæval scheme of institutions was of a coercive, authoritative character, essentially a scheme of graded mastery and graded servitude,
in which a code of honor and a bill of differential dignity
held the most important place. The theology of that
time was of a like character. It was a monotheistic, or
rather a monarchical system, and of a despotic complexion. The cosmological scheme was drawn in terms
of fiat; and the natural philosophy was occupied, in the

main and in its most solemn endeavors, with the corollaries to be subsumed under the divine fiat. When the
philosophical speculation dealt with facts it aimed to
interpret them into systematic consistency with the glory
of God and the divine purpose. The " realities " of the
scholastic lore were spiritual, quasi-personal, intangible,
and fell into a scale of differential dignity and prepotency.
Matter-of-fact knowledge and work-day information were
not then fit topics of dignified inquiry. The interval, or
discrepancy, between reality and actuality was fairly wide.
Throughout that era, of course, work-day knowledge also
continually increased in volume and consistency; technological proficiency was gaining; the effective control
of natural processes was growing larger and more secure;
showing that matter-of-fact theories drawn from experience were being extended and were made increasing
use of. But all this went on in the field of industry; the
matter-of-fact theories were accepted as substantial and
ultimate only for the purposes of industry, only as technological maxims, and were beneath the dignity of science.

With the transition to modern times industry comes into
the foreground in the west-European scheme of life, and
the institutions of European civilisation fall into a more
intimate relation with the exigencies of industry and
technology. The technological range of habituation progressively counts for more in the cultural complex, and
the discrepancy between the technological discipline and
the discipline of law and order under the institutions then
in force grows progressively less. The institutions of law
and order take on a more impersonal, less coercive character. Differential dignity and invidious discriminations
between classes gradually lose force.

The industry which so comes into the foreground and
so affects the scheme of institutions is peculiar in that its

most obvious and characteristic trait is the workmanlike initiative and efficiency of the individual handicraftsman and the individual enterprise of the petty trader. The technology which embodies the theoretical substance of this industry is a technology of workmanship, in which the salient factors are personal skill, force and diligence. Such a technology, running as it does in great part on personal initiative, capacity, and application, approaches nearer to the commonplace features of the institutional fabric than many another technological system might; and its disciplinary effects in some considerable measure blend with those of the institutional discipline. The two lines of habituation, in the great era of handicraft and petty trade, even came to coalesce and fortify one another; as in the organisation of the craft gilds and of the industrial towns. Industrial life and usage came to intrude creatively into the cultural scheme on the one hand and into the scheme of authentic knowledge on the other hand. So the body of matter-of-fact knowledge, in modern times, is more and more drawn into the compass of theoretical inquiry; and theoretical inquiry takes on more and more of the animus and method of technological generalisation. But the matter-of-fact elements so drawn in are construed in terms of workmanlike initiative and efficiency, as required by the technological preconceptions of the era of handicraft.

In this way, it may be conceived, modern science comes into the field under the cloak of technology and gradually encroaches on the domain of authentic theory previously held by other, higher, nobler, more profound, more spiritual, more intangible conceptions and systems of knowledge. In this early phase of modern science its central norm and universal solvent is the concept of workmanlike initiative and efficiency. This is the new organon.

Whatever is to be explained must be reduced to this notation and explained in these terms; otherwise the inquiry does not come to rest. But when the requirements of this notation in terms of workmanship have been duly fulfilled the inquiry does come to rest.

By the early decades of the nineteenth century, with a passable degree of thoroughness, other grounds of validity and other interpretations of phenomena, other vouchers for truth and reality, had been eliminated from the quest of authentic knowledge and from the terms in which theoretical results were conceived or expressed. The new organon had made good its pretensions. In this movement to establish the hegemony of workmanlike efficiency — under the style and title of the " law of causation," or of " efficient cause "— in the realm of knowledge, the English-speaking communities took the lead after the earlier scientific onset of the south-European communities had gone up in the smoke of war, politics and religion during the great era of state-making. The ground of this British lead in science is apparently the same as that of the British lead in technology which came to a head in the Industrial Revolution; and these two associated episodes of European civilisation are apparently both traceable to the relatively peaceable run of life, and so of habituation, in the English-speaking communities, as contrasted with the communities of the continent.[3]

Along with the habits of thought peculiar to the tech-

[3] A broad exception may perhaps be taken at this point, to the effect that this sketch of the growth of the scientific animus overlooks the science of the Ancients. The scientific achievements of classical antiquity are a less obscure topic to-day than ever before during modern times, and the more there is known of them the larger is the credit given them. But it is to be noted that, *(a)* the relatively large and free growth of scientific in-

nology of handicraft, modern science also took over and
assimilated much of the institutional preconceptions of
the era of handicraft and petty trade. The "natural
laws," with the formulation of which this early modern
science is occupied, are the rules governing natural " uniformities of sequence "; and they punctiliously formulate
the due procedure of any given cause creatively working
out the achievement of a given effect, very much as the
craft rules sagaciously specified the due routine for turning out a staple article of merchantable goods. But these
" natural laws " of science are also felt to have something of that integrity and prescriptive moral force that
belongs to the principles of the system of "natural
rights " which the era of handicraft has contributed to
the institutional scheme of later times. The natural laws
were not only held to be true to fact, but they were also

quiry in classical antiquity is to be found in the relatively peaceable and industrial Greek communities (with an industrial culture
of unknown pre-Hellenic antiquity), and *(b)* that the sciences
best and chiefly cultivated were those which rest on a mathematical basis, if not mathematical sciences in the simpler sense
of the term. Now, mathematics occupies a singular place among
the sciences, in that it is, in its pure form, a logical discipline
simply; its subject matter being the logic of quantity, and its
researches being of the nature of an analysis of the intellect's
modes of dealing with matters of quantity. Its generalisations
are generalisations of logical procedure, which are tested and
verified by immediate self-observation. Such a science is in a
peculiar degree, but only in a peculiar degree, independent of the
detail-discipline of daily life, whether technological or institutional; and, given the propensity — the intellectual enterprise,
or "idle curiosity"— to go into speculation in such a field, the
results can scarcely vary in a manner to make the variants inconsistent among themselves; nor need the state of institutions
or the state of the industrial arts seriously color or distort such
analytical work in such a field. Mathematics is peculiarly independent of cultural circumstances, since it deals analytically with
mankind's native gifts of logic, not with the ephemeral traits acquired by habituation.

felt to be right and good. They were looked upon as intrinsically meritorious and beneficent, and were held to carry a sanction of their own. This habit of uncritically imputing merit and equity to the " natural laws " of science continued in force through much of the nineteenth century; very much as the habitual acceptance of the principles of " natural rights " has held on by force of tradition long after the exigencies of experience ·out of which these " rights " sprang ceased to shape men's habits of life.[4] This traditional attitude of submissive approval toward the " natural laws " of science has not yet been wholly lost, even among the scientists of the passing generation, many of whom have uncritically invested these " laws " with a prescriptive rectitude and excellence; but so far, at least, has this animus progressed toward disuse that it is now chiefly a matter for expatiation in the pulpit, the accredited vent for the exudation of effete matter from the cultural organism.

The traditions of the handicraft technology lasted over as a commonplace habit of thought in science long after that technology had ceased to be the decisive element in the industrial situation; while a new technology, with its inculcation of new habits of thought, new preconceptions, gradually made its way among the remnants of the old, altering them, blending with them, and little by little

---

[4] " Natural laws," which are held to be not only correct formulations of the sequence of cause and effect in a given situation but also meritoriously right and equitable rules governing the run of events, necessarily impute to the facts and events in question a tendency to a good and equitable, if not beneficent, consummation; since it is necessarily the consummation, the effect considered as an accomplished outcome, that is to be adjudged good and equitable, if anything. Hence these " natural laws," as traditionally conceived, are laws governing the accomplishment of an end — that is to say, laws as to how a sequence of cause and effect comes to rest in a final term.

superseding them. The new technological departure, which made its first great epoch in the so-called industrial revolution, in the technological ascendancy of the machine-process, brought a new and characteristic discipline into the cultural situation. The beginnings of the machine-era lie far back, no doubt; but it is only of late, during the past century at the most, that the machine-process can be said to have come into the dominant place in the technological scheme; and it is only later still that its discipline has, even in great part, remodeled the current preconceptions as to the substantial nature of what goes on in the current of phenomena whose changes excite the scientific curiosity. It is only relatively very lately, whether in technological work or in scientific inquiry, that men have fallen into the habit of thinking in terms of process rather than in terms of the workmanlike efficiency of a given cause working to a given effect.

These machine-made preconceptions of modern science, being habits of thought induced by the machine technology in industry and in daily life, have of course first and most consistently affected the character of those sciences whose subject matter lies nearest to the technological field of the machine-process; and in these material sciences the shifting to the machine-made point of view has been relatively very consistent, giving a highly impersonal interpretation of phenomena in terms of consecutive change, and leaving little of the ancient preconceptions of differential reality or creative causation. In such a science as physics or chemistry, *e.g.,* we are threatened with the disappearance or dissipation of all stable and efficient substances; their place being supplied, or their phenomena being theoretically explained, by appeal to unremitting processes of inconceivably high-pitched consecutive change.

In the sciences which lie farther afield from the technological domain, and which, therefore, in point of habituation, are remoter from the center of disturbance, the effect of the machine discipline may even yet be scarcely appreciable. In such lore as ethics, *e.g.*, or political theory, or even economics, much of the norms of the régime of handicraft still stands over; and very much of the institutional preconceptions of natural rights, associated with the régime of handicraft in point of genesis, growth and content, is not only still intact in this field of inquiry, but it can scarcely even be claimed that there is ground for serious apprehension of its prospective obsolescence. Indeed, something even more ancient than handicraft and natural rights may be found surviving in good vigor in this "moral" field of inquiry, where tests of authenticity and reality are still sought and found by those who cultivate these lines of inquiry that lie beyond the immediate sweep of the machine's discipline. Even the evolutionary process of cumulative causation as conceived by the adepts of these sciences is infused with a preternatural, beneficent trend; so that "evolution" is conceived to mean amelioration or "improvement." The metaphysics of the machine technology has not yet wholly, perhaps not mainly, superseded the metaphysics of the code of honor in those lines of inquiry that have to do with human initiative and aspiration. Whether such a shifting of the point of view in these sciences shall ever be effected is still an open question. Here there still are spiritual verities which transcend the sweep of consecutive change. That is to say, there are still current habits of thought which definitively predispose their bearers to bring their inquiries to rest on grounds of differential reality and invidious merit.

WHY IS ECONOMICS NOT AN EVOLUTIONARY SCIENCE? [1]

M. G. DE LAPOUGE recently said, " Anthropology is destined to revolutionise the political and the social sciences as radically as bacteriology has revolutionised the science of medicine." [2] In so far as he speaks of economics, the eminent anthropologist is not alone in his conviction that the science stands in need of rehabilitation. His words convey a rebuke and an admonition, and in both respects he speaks the sense of many scientists in his own and related lines of inquiry. It may be taken as the consensus of those men who are doing the serious work of modern anthropology, ethnology, and psychology, as well as of those in the biological sciences proper, that economics is helplessly behind the times, and unable to handle its subject-matter in a way to entitle it to standing as a modern science. The other political and social sciences come in for their share of this obloquy, and perhaps on equally cogent grounds. Nor are the economists themselves buoyantly indifferent to the rebuke. Probably no economist to-day has either the hardihood or the inclination to say that the science has now reached a definitive formulation, either in the detail of results or as regards the fundamental features of theory. The nearest recent approach to such a position on the part of an economist of

[1] Reprinted by permission from *The Quarterly Journal of Economics,* vol. xii, July, 1898.

[2] "The Fundamental Laws of Anthropo-sociology," *Journal of Political Economy,* December, 1897, p. 54. The same paper, in substance, appears in the *Rivista Italiana di Sociologia* for November, 1897.

accredited standing is perhaps to be found in Professor Marshall's Cambridge address of a year and a half ago.[3] But these utterances are so far from the jaunty confidence shown by the classical economists of half a century ago that what most forcibly strikes the reader of Professor Marshall's address is the exceeding modesty and the un-called-for humility of the spokesman for the " old genera-tion." With the economists who are most attentively looked to for guidance, uncertainty as to the definitive value of what has been and is being done, and as to what we may, with effect, take to next, is so common as to suggest that indecision is a meritorious work. Even the Historical School, who made their innovation with so much home-grown applause some time back, have been unable to settle down contentedly to the pace which they set themselves.

The men of the sciences that are proud to own them-selves " modern " find fault with the economists for being still content to occupy themselves with repairing a struc-ture and doctrines and maxims resting on natural rights, utilitarianism, and administrative expediency. This as-persion is not altogether merited, but is near enough to the mark to carry a sting. These modern sciences are evolu-tionary sciences, and their adepts contemplate that charac-teristic of their work with some complacency. Economics is not an evolutionary science — by the confession of its spokesmen ; and the economists turn their eyes with some-thing of envy and some sense of baffled emulation to these rivals that make broad their phylacteries with the legend, " Up to date."

Precisely wherein the social and political sciences, in-cluding economics, fall short of being evolutionary sci-

[3] " The Old Generation of Economists and the New," *Quar-terly Journal of Economics,* January, 1897, p. 133.

ences, is not so plain.   At least, it has not been satisfactorily pointed out by their critics.   Their successful rivals in this matter — the sciences that deal with human nature among the rest — claim as their substantial distinction that they are realistic: they deal with facts.   But economics, too, is realistic in this sense: it deals with facts, often in the most painstaking way, and latterly with an increasingly strenuous insistence on the sole efficacy of data.   But this " realism " does not make economics an evolutionary science.   The insistence on data could scarcely be carried to a higher pitch than it was carried by the first generation of the Historical School; and yet no economics is farther from being an evolutionary science than the received economics of the Historical School. The whole broad range of erudition and research that engaged the energies of that school commonly falls short of being science, in that, when consistent, they have contented themselves with an enumeration of data and a narrative account of industrial development, and have not presumed to offer a theory of anything or to elaborate their results into a consistent body of knowledge.

Any evolutionary science, on the other hand, is a close-knit body of theory.   It is a theory of a process, of an unfolding sequence.   But here, again, economics seems to meet the test in a fair measure, without satisfying its critics that its credentials are good.   It must be admitted, *e.g.,* that J. S. Mill's doctrines of production, distribution, and exchange, are a theory of certain economic processes, and that he deals in a consistent and effective fashion with the sequences of fact that make up his subject-matter. So, also, Cairnes's discussion of normal value, of the rate of wages, and of international trade, are excellent instances of a theoretical handling of economic processes of sequence and the orderly unfolding development of fact. But an attempt to cite Mill and Cairnes as exponents of an evolutionary economics will produce no better effect than perplexity, and not a great deal of that. Very much of monetary theory might be cited to the same purpose and with the like effect. Something similar is true even of late writers who have avowed some penchant for the evolutionary point of view; as, *e.g.,* Professor Hadley,— to cite a work of unquestioned merit and unusual reach. Measurably, he keeps the word of promise to the ear; but any one who may cite his *Economics* as having brought political economy into line as an evolutionary science will convince neither himself nor his interlocutor. Something to the like effect may fairly be said of the published work of that later English strain of economists represented by Professors Cunningham and Ashley, and Mr. Cannan, to name but a few of the more eminent figures in the group.

Of the achievements of the classical economists, recent and living, the science may justly be proud; but they fall short of the evolutionist's standard of adequacy, not in failing to offer a theory of a process or of a developmental relation, but through conceiving their theory in terms alien to the evolutionist's habits of thought. The difference between the evolutionary and the pre-evolutionary sciences lies not in the insistence on facts. There was a great and fruitful activity in the natural sciences in collecting and collating facts before these sciences took on the character which marks them as evolutionary. Nor does the difference lie in the absence of efforts to formulate and explain schemes of process, sequence, growth, and development in the pre-evolutionary days. Efforts of this kind abounded, in number and diversity; and many schemes of development, of great subtlety and beauty, gained a vogue both as theories of organic and inorganic development and as schemes of the life history of nations and societies. It will not even hold true that our elders overlooked the presence of cause and effect in formulating their theories and reducing their data to a body of knowledge. But the terms which were accepted as the definitive terms of knowledge were in some degree different in the early days from what they are now. The terms of thought in which the investigators of some two or three generations back definitively formulated their knowledge of facts, in their last analyses, were different in kind from the terms in which the modern evolutionist is content to formulate his results. The analysis does not run back to the same ground, or appeal to the same standard of finality or adequacy, in the one case as in the other.

The difference is a difference of spiritual attitude or point of view in the two contrasted generations of scientists. To put the matter in other words, it is a difference in the basis of valuation of the facts for the scientific purpose, or in the interest from which the facts are appreciated. With the earlier as with the later generation the basis of valuation of the facts handled is, in matters of detail, the causal relation which is apprehended to subsist between them. This is true to the greatest extent for the natural sciences. But in their handling of the more comprehensive schemes of sequence and relation — in their definitive formulation of the results — the two generations differ. The modern scientist is unwilling to depart from the test of causal relation or quantitative sequence. When he asks the question, Why? he insists on an answer in terms of cause and effect. He wants to reduce his solution of all problems to terms of the conservation of energy or the persistence of quantity. This is his last recourse. And this last recourse has in our time been made available for the handling of schemes of develop-

ment and theories of a comprehensive process by the notion of a cumulative causation. The great deserts of the
evolutionist leaders — if they have great deserts as leaders — lie, on the one hand, in their refusal to go back of
the colorless sequence of phenomena and seek higher
ground for their ultimate syntheses, and, on the other
hand, in their having shown how this colorless impersonal
sequence of cause and effect can be made use of for
theory proper, by virtue of its cumulative character.

For the earlier natural scientists, as for the classical
economists, this ground of cause and effect is not definitive. Their sense of truth and substantiality is not satisfied with a formulation of mechanical sequence. The
ultimate term in their systematisation of knowledge is a
" natural law." This natural law is felt to exercise some
sort of a coercive surveillance over the sequence of
events, and to give a spiritual stability and consistence to
the causal relation at any given juncture. To meet the
high classical requirement, a sequence — and a developmental process especially — must be apprehended in terms
of a consistent propensity tending to some spiritually
legitimate end. When facts and events have been reduced
to these terms of fundamental truth and have been made
to square with the requirements of definitive normality,
the investigator rests his case. Any causal sequence
which is apprehended to traverse the imputed propensity
in events is a " disturbing factor." Logical congruity
with the apprehended propensity is, in this view, adequate
ground of procedure in building up a scheme of knowledge or of development. The objective point of the
efforts of the scientists working under the guidance of
this classical tradition, is to formulate knowledge in terms
of absolute truth; and this absolute truth is a spiritual
fact. It means a coincidence of facts with the deliver-

ances of an enlightened and deliberate common sense.

The development and the attenuation of this preconception of normality or of a propensity in events might be traced in detail from primitive animism down through the elaborate discipline of faith and metaphysics, overruling Providence, order of nature, natural rights, natural law, underlying principles. But all that may be necessary here is to point out that, by descent and by psychological content, this constraining normality is of a spiritual kind. It is for the scientific purpose an imputation of spiritual coherence to the facts dealt with. The question of interest is how this preconception of normality has fared at the hands of modern science, and how it has come to be superseded in the intellectual primacy by the latter-day preconception of a non-spiritual sequence. This question is of interest because its answer may throw light on the question as to what chance there is for the indefinite persistence of this archaic habit of thought in the methods of economic science.

Under primitive conditions, men stand in immediate personal contact with the material facts of the environment; and the force and discretion of the individual in shaping the facts of the environment count obviously, and to all appearance solely, in working out the conditions of life. There is little of impersonal or mechanical sequence visible to primitive men in their every-day life; and what there is of this kind in the processes of brute nature about them is in large part inexplicable and passes for inscrutable. It is accepted as malignant or beneficent, and is construed in the terms of personality that are familiar to all men at first hand,— the terms known to all men by first-hand knowledge of their own acts. The inscrutable movements of the seasons and of the natural forces are apprehended as actions guided by discretion, will power, or propensity looking to an end, much as human actions are. The processes of inanimate nature are agencies whose habits of life are to be learned, and who are to be coerced, outwitted, circumvented, and turned to account, much as the beasts are. At the same time the community is small, and the human contact of the individual is not wide. Neither the industrial life nor the non-industrial social life forces upon men's attention the ruthless impersonal sweep of events that no man can withstand or deflect, such as becomes visible in the more complex and comprehensive life process of the larger community of a later day. There is nothing decisive to hinder men's knowledge of facts and events being formulated in terms of personality — in terms of habit and propensity and will power.

As time goes on and as the situation departs from this archaic character,— where it does depart from it,— the circumstances which condition men's systematisation of facts change in such a way as to throw the impersonal character of the sequence of events more and more into the foreground. The penalties for failure to apprehend facts in dispassionate terms fall surer and swifter. The sweep of events is forced home more consistently on men's minds. The guiding hand of a spiritual agency or a propensity in events becomes less readily traceable as men's knowledge of things grows ampler and more searching. In modern times, and particularly in the industrial countries, this coercive guidance of men's habits of thought in the realistic direction has been especially pronounced; and the effect shows itself in a somewhat reluctant but cumulative departure from the archaic point of view. The departure is most visible and has gone farthest in those homely branches of knowledge that have to do immediately with modern mechanical processes, such as engineering designs and technological contrivances generally. Of the sciences, those have wandered farthest on this way (of integration or disintegration, according as one may choose to view it) that have to do with mechanical sequence and process; and those have best and longest retained the archaic point of view intact which — like the moral, social, or spiritual sciences — have to do with process and sequence that is less tangible, less traceable by the use of the senses, and that therefore less immediately forces upon the attention the phenomenon of sequence as contrasted with that of propensity.

There is no abrupt transition from the pre-evolutionary to the post-evolutionary standpoint. Even in those natural sciences which deal with the processes of life and the evolutionary sequence of events the concept of dispassionate cumulative causation has often and effectively been helped out by the notion that there is in all this some sort of a meliorative trend that exercises a constraining guidance over the course of causes and effects. The faith in this meliorative trend as a concept useful to the science has gradually weakened, and it has repeatedly been disavowed; but it can scarcely be said to have yet disappeared from the field.

The process of change in the point of view, or in the terms of definitive formulation of knowledge, is a gradual one; and all the sciences have shared, though in an unequal degree, in the change that is going forward. Economics is not an exception to the rule, but it still shows too many reminiscences of the "natural" and the "normal," of "verities" and "tendencies," of "controlling principles" and "disturbing causes" to be classed as an evolutionary science. This history of the science shows a long and devious course of disintegrating animism,—

from the days of the scholastic writers, who discussed usury from the point of view of its relation to the divine suzerainty, to the Physiocrats, who rested their case on an "*ordre naturel*" and a "*loi naturelle*" that decides what is substantially true and, in a general way, guides the course of events by the constraint of logical congruence. There has been something of a change from Adam Smith, whose recourse in perplexity was to the guidance of "an unseen hand," to Mill and Cairnes, who formulated the laws of "natural" wages and "normal" value, and the former of whom was so well content with his work as to say, "Happily, there is nothing in the laws of Value which remains for the present or any future writer to clear up: the theory of the subject is complete." [4] But the difference between the earlier and the later point of view is a difference of degree rather than of kind.

  The standpoint of the classical economists, in their higher or definitive syntheses and generalisations, may not inaptly be called the standpoint of ceremonial adequacy. The ultimate laws and principles which they formulated were laws of the normal or the natural, according to a preconception regarding the ends to which, in the nature of things, all things tend. In effect, this preconception imputes to things a tendency to work out what the instructed common sense of the time accepts as the adequate or worthy end of human effort. It is a projection of the accepted ideal of conduct. This ideal of conduct is made to serve as a canon of truth, to the extent that the investigator contents himself with an appeal to its legitimation for premises that run back of the facts with which he is immediately dealing, for the "controlling principles" that are conceived intangibly to underlie the process discussed, and for the "tendencies" that run beyond the

[4] *Political Economy*, Book III, chap. i.

situation as it lies before him. As instances of the use
of this ceremonial canon of knowledge may be cited the
" conjectural history " that plays so large a part in the
classical treatment of economic institutions, such as the
normalized accounts of the beginnings of barter in the
transactions of the putative hunter, fisherman, and boat-
builder, or the man with the plane and the two planks,
or the two men with the basket of apples and the basket
of nuts.[5]　Of a similar import is the characterisation of
money as " the great wheel of circulation "[6] or as " the
medium of exchange."　Money is here discussed in terms
of the end which, " in the normal case," it should work
out according to the given writer's ideal of economic life,
rather than in terms of causal relation.

With later writers especially, this terminology is no
doubt to be commonly taken as a convenient use of metaphor, in which the concept of normality and propensity to
an end has reached an extreme attenuation.　But it is
precisely in this use of figurative terms for the formulation of theory that the classical normality still lives its
attenuated life in modern economics ; and it is this facile
recourse to inscrutable figures of speech as the ultimate
terms of theory that has saved the economists from being
dragooned into the ranks of modern science.　The metaphors are effective, both in their homiletical use and as a
labor-saving device,— more effective than their user designs them to be.　By their use the theorist is enabled
serenely to enjoin himself from following out an elusive
train of causal sequence.　He is also enabled, without
misgivings, to construct a theory of such an institution

[5] Marshall, *Principles of Economics* (2d ed.), Book V, chap.
ii, p. 395, note.
[6] Adam Smith, *Wealth of Nations* (Bohn ed.), Book II, chap.
ii, p. 289.

as money or wages or land-ownership without descending to a consideration of the living items concerned, except for convenient corroboration of his normalised scheme of symptoms. By this method the theory of an institution or a phase of life may be stated in conventionalised terms of the apparatus whereby life is carried on, the apparatus being invested with a tendency to an equilibrium at the normal, and the theory being a formulation of the conditions under which this putative equilibrium supervenes. In this way we have come into the usufruct of a cost-of-production theory of value which is pungently reminiscent of the time when Nature abhorred a vacuum. The ways and means and the mechanical structure of industry are formulated in a conventionalised nomenclature, and the observed motions of this mechanical apparatus are then reduced to a normalised scheme of relations. The scheme so arrived at is spiritually binding on the behavior of the phenomena contemplated. With this normalised scheme as a guide, the permutations of a given segment of the apparatus are worked out according to the values assigned the several items and features comprised in the calculation ; and a ceremonially consistent formula is constructed to cover that much of the industrial field. This is the deductive method. The formula is then tested by comparison with observed permutations, by the polariscopic use of the " normal case " ; and the results arrived at are thus authenticated by induction. Features of the process that do not lend themselves to interpretation in the terms of the formula are abnormal cases and are due to disturbing causes. In all this the agencies or forces causally at work in the economic life process are neatly avoided. The outcome of the method, at its best, is a body of logically consistent propositions concerning the normal relations of things — a system of economic taxonomy. At its worst, it is a body of maxims for the conduct of business and a polemical discussion of disputed points of policy.

In all this, economic science is living over again in its turn the experiences which the natural sciences passed through some time back. In the natural sciences the work of the taxonomist was and continues to be of great value, but the scientists grew restless under the régime of symmetry and system-making. They took to asking why, and so shifted their inquiries from the structure of the coral reefs to the structure and habits of life of the polyp that lives in and by them. In the science of plants, systematic botany has not ceased to be of service; but the stress of investigation and discussion among the botanists to-day falls on the biological value of any given feature of structure, function, or tissue rather than on its taxonomic bearing. All the talk about cytoplasm, centrosomes, and karyokinetic process, means that the inquiry now looks consistently to the life process, and aims to explain it in terms of cumulative causation.

What may be done in economic science of the taxonomic kind is shown at its best in Cairnes's work, where the method is well conceived and the results effectively formulated and applied. Cairnes handles the theory of the normal case in economic life with a master hand. In his discussion the metaphysics of propensity and tendencies no longer avowedly rules the formulation of theory, nor is the inscrutable meliorative trend of a harmony of interests confidently appealed to as an engine of definitive use in giving legitimacy to the economic situation at a given time. There is less of an exercise of faith in Cairnes's economic discussions than in those of the writers that went before him. The definitive terms of the formulation are still the terms of normality and natural law, but the metaphysics underlying this appeal to normality is so far removed from the ancient ground of the beneficent "order of nature" as to have become at least nominally impersonal and to proceed without a constant regard to the humanitarian bearing of the "tendencies" which it formulates. The metaphysics has been attenuated to something approaching in colorlessness the naturalist's conception of natural law. It is a natural law which, in the guise of "controlling principles," exercises a constraining surveillance over the trend of things; but it is no longer conceived to exercise its constraint in the interest of certain ulterior human purposes. The element of beneficence has been well-nigh eliminated, and the system is formulated in terms of the system itself. Economics as it left Cairnes's hand, so far as his theoretical work is concerned, comes near being taxonomy for taxonomy's sake.

No equally capable writer has come as near making economics the ideal "dismal" science as Cairnes in his discussion of pure theory. In the days of the early classical writers economics had a vital interest for the laymen of the time, because it formulated the common sense metaphysics of the time in its application to a department of human life. But in the hands of the later classical writers the science lost much of its charm in this regard. It was no longer a definition and authentication of the deliverances of current common sense as to what ought to come to pass; and it, therefore, in large measure lost the support of the people out of doors, who were unable to take an interest in what did not concern them; and it was also out of touch with that realistic or evolutionary habit of mind which got under way about the middle of the century in the natural sciences. It was neither vitally metaphysical nor matter-of-fact, and it found comfort with very few outside of its own ranks. Only for those who by the fortunate accident of birth or education have been able to conserve the taxonomic animus has the science during the last third of a century continued to be of absorbing interest. The result has been that from the time when the taxonomic structure stood forth as a completed whole in its symmetry and stability the economists themselves, beginning with Cairnes, have been growing restive under its discipline of stability, and have made many efforts, more or less sustained, to galvanise it into movement. At the hands of the writers of the classical line these excursions have chiefly aimed at a more complete and comprehensive taxonomic scheme of permutations; while the historical departure threw away the taxonomic ideal without getting rid of the preconceptions on which it is based; and the later Austrian group struck out on a theory of process, but presently came to a full stop because the process about which they busied themselves was not, in their apprehension of it, a cumulative or unfolding sequence.

But what does all this signify? If we are getting restless under the taxonomy of a monocotyledonous wage doctrine and a cryptogamic theory of interest, with involute, loculicidal, tomentous and moniliform variants, what is the cytoplasm, centrosome, or karyokinetic process to which we may turn, and in which we may find surcease from the metaphysics of normality and controlling principles? What are we going to do about it? The question is rather, What are we doing about it? There is the economic life process still in great measure awaiting theoretical formulation. The active material in which the economic process goes on is the human material of the industrial community. For the purpose of economic science the process of cumulative change that is to be accounted for is the sequence of change in the methods of doing things,— the methods of dealing with the material means of life.

What has been done in the way of inquiry into this economic life process? The ways and means of turning material objects and circumstances to account lie before the investigator at any given point of time in the form of mechanical contrivances and arrangements for compassing certain mechanical ends. It has therefore been easy to accept these ways and means as items of inert matter having a given mechanical structure and thereby serving the material ends of man. As such, they have been scheduled and graded by the economists under the head of capital, this capital being conceived as a mass of material objects serviceable for human use. This is well enough for the purposes of taxonomy; but it is not an effective method of conceiving the matter for the purpose of a theory of the developmental process. For the latter purpose, when taken as items in a process of cumulative change or as items in the scheme of life, these productive goods are facts of human knowledge, skill, and predilection; that is to say, they are, substantially, prevalent habits of thought, and it is as such that they enter into the process of industrial development. The physical properties of the materials accessible to man are constants: it is the human agent that changes,— his insight and his appreciation of what these things can be used for is what develops. The accumulation of goods already on hand conditions his handling and utilisation of the materials offered, but even on this side — the " limitation of industry by capital "— the limitation imposed is on what men can do and on the methods of doing it. The changes that take place in the mechanical contrivances are an expression of changes in the human factor. Changes in the material facts breed further change only through the human factor. It is in the human material that the continuity of development is to be looked for; and it is here, therefore, that the motor forces of the process of economic development must be studied if they are to be studied in action at all. Economic action must be the subject-matter of the science if the science is to fall into line as an evolutionary science.

Nothing new has been said in all this. But the fact is all the more significant for being a familiar fact. It is a fact recognised by common consent throughout much of the later economic discussion, and this current recognition of the fact is a long step towards centering discussion and inquiry upon it. If economics is to follow the lead or the analogy of the other sciences that have to do with a life process, the way is plain so far as regards the general direction in which the move will be made.

The economists of the classical trend have made no serious attempt to depart from the standpoint of taxonomy and make their science a genetic account of the economic life process. As has just been said, much the same is true for the Historical School. The latter have attempted an account of developmental sequence, but they have followed the lines of pre-Darwinian speculations on development rather than lines which modern science would recognise as evolutionary. They have given a narrative survey of phenomena, not a genetic account of an unfolding process. In this work they have, no doubt, achieved results of permanent value; but the results achieved are scarcely to be classed as economic theory. On the other hand, the Austrians and their precursors and their co-adjutors in the value discussion have taken up a detached portion of economic theory, and have inquired with great nicety into the process by which the phenomena within their limited field are worked out. The entire discussion of marginal utility and subjective value as the outcome of a valuation process must be taken as a genetic study of this range of facts. But here, again, nothing further has come of the inquiry, so far as regards a rehabilitation of economic theory as a whole. Accepting Menger as their spokesman on this head, it must be said that the Austrians have on the whole showed themselves unable to break with the classical tradition that economics is a taxonomic science.

The reason for the Austrian failure seems to lie in a faulty conception of human nature,— faulty for the present purpose, however adequate it may be for any other. In all the received formulations of economic theory, whether at the hands of English economists or those of the Continent, the human material with which the inquiry is concerned is conceived in hedonistic terms; that is to say, in terms of a passive and substantially inert and immutably given human nature. The psychological and anthropological preconceptions of the economists have been those which were accepted by the psychological and social sciences some generations ago. The hedonistic conception of man is that of a lightning calculator of pleasures and pains, who oscillates like a homogeneous globule of desire of happiness under the impulse of stimuli that shift him about the area, but leave him intact. He has neither antecedent nor consequent. He is an isolated, definitive human datum, in stable equilibrium except for the buffets of the impinging forces that displace him in one direction or another. Self-imposed in elemental space, he spins symmetrically about his own spiritual axis until the parallelogram of forces bears down upon him, whereupon he follows the line of the resultant. When the force of the impact is spent, he comes to rest, a self-contained globule of desire as before. Spiritually, the hedonistic man is not a prime mover. He is not the seat of a process of living, except in the sense that he is subject to a series of permutations enforced upon him by circumstances external and alien to him.

The later psychology, reënforced by modern anthropological research, gives a different conception of human nature. According to this conception, it is the characteristic of man to do something, not simply to suffer pleasures and pains through the impact of suitable forces. He is not simply a bundle of desires that are to be saturated by being placed in the path of the forces of the environment, but rather a coherent structure of propensities and habits which seeks realisation and expression in an unfolding activity. According to this view, human activity, and economic activity among the rest, is not apprehended as something incidental to the process of saturating given desires. The activity is itself the substantial fact of the process, and the desires under whose guidance the action takes place are circumstances of temperament which determine the specific direction in which the activity will unfold itself in the given case. These circumstances of temperament are ultimate and definitive for the individual who acts under them, so far as regards his attitude as agent in the particular action in which he is engaged. But, in the view of the science, they are elements of the existing frame of mind of the agent, and are the outcome of his antecedents and his life up to the point at which he stands. They are the products of his hereditary traits and his past experience, cumulatively wrought out under a given body of traditions, conventionalities, and material circumstances; and they afford the point of departure for the next step in the process. The economic life history of the individual is a cumulative process of adaptation of means to ends that cumulatively change as the process goes on, both the agent and his environment being at any point the outcome of the last process. His methods of life to-day are enforced upon him by his habits of life carried over from yesterday and by the circumstances left as the mechanical residue of the life of yesterday.

What is true of the individual in this respect is true of the group in which he lives. All economic change is a change in the economic community,— a change in the community's methods of turning material things to account. The change is always in the last resort a change in habits of thought. This is true even of changes in the mechanical processes of industry. A given contrivance for effecting certain material ends becomes a circumstance which affects the further growth of habits of thought — habitual methods of procedure — and so becomes a point of departure for further development of the methods of compassing the ends sought and for the further variation of ends that are sought to be compassed. In all this flux there is no definitively adequate method of life and no definitive or absolutely worthy end of action, so far as concerns the science which sets out to formulate a theory of the process of economic life. What remains as a hard and fast residue is the fact of activity directed to an objective end. Economic action is teleological, in the sense that men always and everywhere seek to do something. What, in specific detail, they seek, is not to be answered except by a scrutiny of the details of their activity; but, so long as we have to do with their life as members of the economic community, there remains the generic fact that their life is an unfolding activity of a teleological kind.

It may or may not be a teleological process in the sense

that it tends or should tend to any end that is conceived to be worthy or adequate by the inquirer or by the consensus of inquirers. Whether it is or is not, is a question with which the present inquiry is not concerned; and it is also a question of which an evolutionary economics need take no account. The question of a tendency in events can evidently not come up except on the ground of some preconception or prepossession on the part of the person looking for the tendency. In order to search for a tendency, we must be possessed of some notion of a definitive end to be sought, or some notion as to what is the legitimate trend of events. The notion of a legitimate trend in a course of events is an extra-evolutionary preconception, and lies outside the scope of an inquiry into the causal sequence in any process. The evolutionary point of view, therefore, leaves no place for a formulation of natural laws in terms of definitive normality, whether in economics or in any other branch of inquiry. Neither does it leave room for that other question of normality, What should be the end of the developmental process under discussion?

The economic life history of any community is its life history in so far as it is shaped by men's interest in the material means of life. This economic interest has counted for much in shaping the cultural growth of all communities. Primarily and most obviously, it has guided the formation, the cumulative growth, of that range of conventionalities and methods of life that are currently recognized as economic institutions; but the same interest has also pervaded the community's life and its cultural growth at points where the resulting structural features are not chiefly and most immediately of an economic bearing. The economic interest goes with men through life, and it goes with the race throughout its process of cultural development. It affects the cultural structure at all points, so that all institutions may be said to be in some measure economic institutions. This is necessarily the case, since the base of action — the point of departure — at any step in the process is the entire organic complex of habits of thought that have been shaped by the past process. The economic interest does not act in isolation, for it is but one of several vaguely isolable interests on which the complex of teleological activity carried out by the individual proceeds. The individual is but a single agent in each case; and he enters into each successive action as a whole, although the specific end sought in a given action may be sought avowedly on the basis of a particular interest; as *e.g.,* the economic, æsthetic, sexual, humanitarian, devotional interests. Since each of these passably isolable interests is a propensity of the organic agent man, with his complex of habits of thought, the expression of each is affected by habits of life formed under the guidance of all the rest. There is, therefore, no neatly isolable range of cultural phenomena that can be rigorously set apart under the head of economic institutions, although a category of " economic institutions " may be of service as a convenient caption, comprising those institutions in which the economic interest most immediately and consistently finds expression, and which most immediately and with the least limitation are of an economic bearing.

From what has been said it appears that an evolutionary economics must be the theory of a process of cultural growth as determined by the economic interest, a theory of a cumulative sequence of economic institutions stated in terms of the process itself. Except for the want of space to do here what should be done in some detail if it is done at all, many efforts by the later economists in this direction might be cited to show the trend of economic discussion in this direction. There is not a little evidence to this effect, and much of the work done must be rated as effective work for this purpose. Much of the work of the Historical School, for instance, and that of its later exponents especially, is too noteworthy to be passed over in silence, even with all due regard to the limitations of space.

We are now ready to return to the question why economics is not an evolutionary science. It is necessarily the aim of such an economics to trace the cumulative working-out of the economic interest in the cultural sequence. It must be a theory of the economic life process of the race or the community. The economists have accepted the hedonistic preconceptions concerning human nature and human action, and the conception of the economic interest which a hedonistic psychology gives does not afford material for a theory of the development of human nature. Under hedonism the economic interest is not conceived in terms of action. It is therefore not readily apprehended or appreciated in terms of a cumulative growth of habits of thought, and does not provoke, even if it did lend itself to, treatment by the evolutionary method. At the same time the anthropological preconceptions current in that common-sense apprehension of human nature to which economists have habitually turned has not enforced the formulation of human nature in terms of a cumulative growth of habits of life. These received anthropological preconceptions are such as have made possible the normalized conjectural accounts of primitive barter with which all economic readers are familiar, and the no less normalized conventional derivation of landed property and its rent, or the sociologico-philosophical discussions of the "function" of this or that class in the life of society or of the nation.

The premises and the point of view required for an evolutionary economics have been wanting. The economists have not had the materials for such a science ready to their hand, and the provocation to strike out in such a direction has been absent. Even if it has been possible at any time to turn to the evolutionary line of speculation in economics, the possibility of a departure is not enough to bring it about. So long as the habitual view taken of a given range of facts is of the taxonomic kind and the material lends itself to treatment by that method, the taxonomic method is the easiest, gives the most gratifying immediate results, and bests fits into the accepted body of knowledge of the range of facts in question. This has been the situation in economics. The other sciences of its group have likewise been a body of taxonomic discipline, and departures from the accredited method have lain under the odium of being meretricious innovations. The well-worn paths are easy to follow and lead into good company. Advance along them visibly furthers the accredited work which the science has in hand. Divergence from the paths means tentative work, which is necessarily slow and fragmentary and of uncertain value.

It is only when the methods of the science and the syntheses resulting from their use come to be out of line with habits of thought that prevail in other matters that the scientist grows restive under the guidance of the received methods and standpoints, and seeks a way out. Like other men, the economist is an individual with but one intelligence. He is a creature of habits and propensities given through the antecedents, hereditary and cultural, of which he is an outcome; and the habits of thought formed in any one line of experience affect his thinking in any other.　Methods of observation and of handling facts that are familiar through habitual use in the general range of knowledge, gradually assert themselves in any given special range of knowledge.　They may be accepted slowly and with reluctance where their acceptance involves innovation; but, if they have the continued backing of the general body of experience, it is only a question of time when they shall come into dominance in the special field. The intellectual attitude and the method of correlation enforced upon us in the apprehension and assimilation of facts in the more elementary ranges of knowledge that have to do with brute facts assert themselves also when the attention is directed to those phenomena of the life process with which economics has to do; and the range of facts which are habitually handled by other methods than that in traditional vogue in economics has now become so large and so insistently present at every turn that we are left restless, if the new body of facts cannot be handled according to the method of mental procedure which is in this way becoming habitual.

In the general body of knowledge in modern times the facts are apprehended in terms of causal sequence.　This is especially true of that knowledge of brute facts which is shaped by the exigencies of the modern mechanical industry.　To men thoroughly imbued with this matter-of-fact habit of mind the laws and theorems of economics, and of the other sciences that treat of the normal course of things, have a character of " unreality " and futility that bars out any serious interest in their discussion.　The laws and theorems are " unreal " to them because they are not to be apprehended in the terms which these men make use of in handling the facts with which they are perforce habitually occupied.　The same matter-of-fact spiritual attitude and mode of procedure have now made their way well up into the higher levels of scientific knowledge, even in the sciences which deal in a more elementary way with the same human material that makes the subject-matter of economics, and the economists themselves are beginning to feel the unreality of their theorems about " normal " cases. Provided the practical exigencies of modern industrial life continue of the same character as they now are, and so continue to enforce the impersonal method of knowledge, it is only a question of time when that (substantially animistic) habit of mind which proceeds on the notion of a definitive normality shall be displaced in the field of economic inquiry by that (substantially materialistic) habit of mind which seeks a comprehension of facts in terms of a cumulative sequence.

The later method of apprehending and assimilating facts and handling them for the purposes of knowledge may be better or worse, more or less worthy or adequate, than the earlier; it may be of greater or less ceremonial or æsthetic effect; we may be moved to regret the incursion of underbred habits of thought into the scholar's domain. But all that is beside the present point. Under the stress of modern technological exigencies, men's everyday habits of thought are falling into the lines that in the sciences constitute the evolutionary method; and knowledge which proceeds on a higher, more archaic plane is becoming alien and meaningless to them. The social and political sciences must follow the drift, for they are already caught in it.

THE PRECONCEPTIONS OF ECONOMIC SCIENCE [1]

I

In an earlier paper [2] the view has been expressed that the economics handed down by the great writers of a past generation is substantially a taxonomic science. A view of much the same purport, so far as concerns the point here immediately in question, is presented in an admirably lucid and cogent way by Professor Clark in a recent number of this journal. [3] There is no wish hereby to burden Professor Clark with a putative sponsorship of any ungraceful or questionable generalisations reached in working outward from this main position, but expression may not be denied the comfort which his unintended authentication of the main position affords. It is true, Professor Clark does not speak of taxonomy, but employs the term " statics," which is perhaps better suited to his immediate purpose. Nevertheless, in spite of the high authority given the term " statics," in this connection, through its use by Professor Clark and by other writers eminent in the science, it is fairly to be questioned whether the term can legitimately be used to characterize the received economic theories. The word is borrowed from the jargon of physics, where it is used to designate the theory of

[1] Reprinted by permission from *The Quarterly Journal of Economics*, vol. xiii, Jan., 1899.

[2] " Why is Economics not an Evolutionary Science? " *Quarterly Journal of Economics*, July, 1898.

[3] " The Future of Economic Theory," *ibid.*, October, 1898.

bodies at rest or of forces in equilibrium. But there is much in the received economic theories to which the analogy of bodies at rest or of forces in equilibrium will not apply. It is perhaps not too much to say that those articles of economic theory that do not lend themselves to this analogy make up the major portion of the received doctrines. So, for instance, it seems scarcely to the point to speak of the statics of production, exchange, consumption, circulation. There are, no doubt, appreciable elements in the theory of these several processes that may fairly be characterized as statical features of the theory; but the doctrines handed down are after all, in the main, theories of the process discussed under each head, and the theory of a process does not belong in statics. The epithet " statical " would, for instance, have to be wrenched somewhat ungently to make it apply to Quesnay's classic *Tableau Économique* or to the great body of Physiocratic speculations that take their rise from it. The like is true for Books II. and III. of Adam Smith's *Wealth of Nations,* as also for considerable portions of Ricardo's work, or, to come down to the present generation, for much of Marshall's *Principles,* and for such a modern discussion as Smart's *Studies in Economics,* as well as for the fruitful activity of the Austrians and of the later representatives of the Historical School.

But to return from this terminological digression. While economic science in the remoter past of its history has been mainly of a taxonomic character, later writers of all schools show something of a divergence from the taxonomic line and an inclination to make the science a genetic account of the economic life process, sometimes even without an ulterior view to the taxonomic value of the results obtained. This divergence from the ancient canons of theoretical formulation is to be taken as an

episode of the movement that is going forward in latter-
day science generally; and the progressive change which
thus affects the ideals and the objective point of the modern sciences seems in its turn to be an expression of that
matter-of-fact habit of mind which the prosy but exacting
exigencies of life in a modern industrial community
breed in men exposed to their unmitigated impact.

In speaking of this matter-of-fact character of the modern sciences it has been broadly characterized as " evolutionary "; and the evolutionary method and the evolutionary ideals have been placed in antithesis to the taxonomic
methods and ideals of pre-evolutionary days.   But the
characteristic attitude, aims, and ideals which are so desig-
nated here are by no means peculiar to the group of sciences that are professedly occupied with a process of development, taking that term in its most widely accepted
meaning.   The latter-day inorganic sciences are in this
respect like the organic.   They occupy themselves with
" dynamic " relations and sequences.   The question which
they ask is always, What takes place next, and why?
Given a situation wrought out by the forces under inquiry,
what follows as the consequence of the situation so
wrought out? or what follows upon the accession of a
further element of force?   Even in so non-evolutionary a
science as inorganic chemistry the inquiry consistently
runs on a process, an active sequence, and the value of the
resulting situation as a point of departure for the next
step in an interminable cumulative sequence.   The last
step in the chemist's experimental inquiry into any substance is, What comes of the substance determined?
What will it do?   What will it lead to, when it is made
the point of departure in further chemical action?   There
is no ultimate term, and no definitive solution except in
terms of further action.   The theory worked out is al-

ways a theory of a genetic succession of phenomena, and the relations determined and elaborated into a body of doctrine are always genetic relations. In modern chemistry no cognisance is taken of the honorific bearing of reactions or molecular formulæ. The modern chemist, as contrasted with his ancient congener, knows nothing of the worth, elegance, or cogency of the relations that may subsist between the particles of matter with which he busies himself, for any other than the genetic purpose. The spiritual element and the elements of worth and propensity no longer count. Alchemic symbolism and the hierarchical glamour and virtue that once hedged about the nobler and more potent elements and reagents are almost altogether a departed glory of the science. Even the modest imputation of propensity involved in the construction of a scheme of coercive normality, for the putative guidance of reactions, finds little countenance with the later adepts of chemical science. The science has outlived that phase of its development at which the taxonomic feature was the dominant one.

In the modern sciences, of which chemistry is one, there has been a gradual shifting of the point of view from which the phenomena which the science treats of are apprehended and passed upon; and to the historian of chemical science this shifting of the point of view must be a factor of great weight in the development of chemical knowledge. Something of a like nature is true for economic science; and it is the aim here to present, in outline, some of the successive phases that have passed over the spiritual attitude of the adepts of the science, and to point out the manner in which the transition from one point of view to the next has been made.

As has been suggested in the paper already referred to,

the characteristic spiritual attitude or point of view of a given generation or group of economists is shown not so much in their detail work as in their higher syntheses — the terms of their definitive formulations — the grounds of their final valuation of the facts handled for purpose of theory. This line of recondite inquiry into the spiritual past and antecedents of the science has not often been pursued seriously or with singleness of purpose, perhaps because it is, after all, of but slight consequence to the practical efficiency of the present-day science. Still, not a little substantial work has been done towards this end by such writers as Hasbach, Oncken, Bonar, Cannan, and Marshall. And much that is to the purpose is also due to writers outside of economics, for the aims of economic speculation have never been insulated from the work going forward in other lines of inquiry. As would necessarily be the case, the point of view of economists has always been in large part the point of view of the enlightened common sense of their time. The spiritual attitude of a given generation of economists is therefore in good part a special outgrowth of the ideals and preconceptions current in the world about them.

So, for instance, it is quite the conventional thing to say that the speculations of the Physiocrats were dominated and shaped by the preconception of Natural Rights. Account has been taken of the effect of natural-rights preconceptions upon the Physiocratic schemes of policy and economic reform, as well as upon the details of their doctrines.[4] But little has been said of the significance of these preconceptions for the lower courses of the Physiocrats' theoretical structure. And yet that habit of mind

[4] See, for instance, Hasbach, *Allgemeine philosophische Grundlagen der von François Quesnay und Adam Smith begründeten politischen Oekonomie.*

to which the natural-rights view is wholesome and adequate is answerable both for the point of departure and for the objective point of the Physiocratic theories, both for the range of facts to which they turned and for the terms in which they were content to formulate their knowledge of the facts which they handled. The failure of their critics to place themselves at the Physiocratic point of view has led to much destructive criticism of their work; whereas, when seen through Physiocratic eyes, such doctrines as those of the net product and of the barrenness of the artisan class appear to be substantially true.

The speculations of the Physiocrats are commonly accounted the first articulate and comprehensive presentation of economic theory that is in line with later theoretical work. The Physiocratic point of view may, therefore, well be taken as the point of departure in an attempt to trace that shifting of aims and norms of procedure that comes into view in the work of later economists when compared with earlier writers.

Physiocratic economics is a theory of the working-out of the Law of Nature (*loi naturelle*) in its economic bearing, and this Law of Nature is a very simple matter.

Les lois naturelles sont ou physiques ou morales.

On entend ici, par loi physique, *le cours réglé de tout évènement physique de l'ordre naturel, évidemment le plus avantageux au genre humain.*

On entend ici, par loi morale, *la règle de toute action humaine de l'ordre morale, conforme à l'ordre physique évidemment le plus avantageux au genre humain.*

Ces lois forment ensemble ce qu'on appelle la *loi naturelle.* Tous les hommes et toutes les puissances humaines doivent être soumis à ces lois souveraines, instituées par l'Être-Suprême: elles sont immuables et irréfragables, et les meilleures lois possibles.[5]

[5] Quesnay, *Droit Naturel,* ch. v. (Ed. Daire, *Physiocrates,* pp. 52–53).

The settled course of material facts tending beneficently to the highest welfare of the human race,— this is the final term in the Physiocratic speculations. This is the touchstone of substantiality. Conformity to these " immutable and unerring " laws of nature is the test of economic truth. The laws are immutable and unerring, but that does not mean that they rule the course of events with a blind fatality that admits of no exception and no divergence from the direct line. Human nature may, through infirmity or perversity, willfully break over the beneficent trend of the laws of nature; but to the Physiocrat's sense of the matter the laws are none the less immutable and irrefragable on that account. They are not empirical generalisations on the course of phenomena, like the law of falling bodies or of the angle of reflection; although many of the details of their action are to be determined only by observation and experience, helped out, of course, by interpretation of the facts of observation under the light of reason. So, for instance, Turgot, in his *Réflections,* empirically works out a doctrine of the reasonable course of development through which wealth is accumulated and reaches the existing state of unequal distribution; so also his doctrines of interest and of money. The immutable natural laws are rather of the nature of canons of conduct governing nature than generalisations of mechanical sequence, although in a general way the phenomena of mechanical sequence are details of the conduct of nature working according to these canons of conduct. The great law of the order of nature is of the character of a propensity working to an end, to the accomplishment of a purpose. The processes of nature working under the quasi-spiritual stress of this immanent propensity may be characterised as nature's habits of life. Not that nature is conscious of its travail, and knows and desires the worthy end of its endeavors; but for all that there is a quasi-spiritual nexus between antecedent and consequent in the scheme of operation in which nature is engaged. Nature is not uneasy about interruptions of its course or occasional deflections from the direct line through an untoward conjunction of mechanical causes, nor does the validity of the great overruling law suffer through such an episode. The introduction of a mere mechanically effective causal factor cannot thwart the course of Nature from reaching the goal to which she animistically tends. Nothing can thwart this teleological propensity of nature except counter-activity or divergent activity of a similarly teleological kind. Men can break over the law, and have short-sightedly and willfully done so; for men are also agents who guide their actions by an end to be achieved. Human conduct is activity of the same kind — on the same plane of spiritual reality or competency — as the course of Nature, and it may therefore traverse the latter. The remedy for this short-sighted traffic of misguided human nature is enlightenment,—" instruction publique et privée des lois de l'ordre naturel." [6]

The nature in terms of which all knowledge of phenomena — for the present purpose economic phenomena — is to be finally synthesised is, therefore, substantially of a quasi-spiritual or animistic character. The laws of nature are in the last resort teleological: they are of the nature of a propensity. The substantial fact in all the sequences of nature is the end to which the sequence naturally tends, not the brute fact of mechanical compulsion or causally effective forces. Economic theory is accordingly the theory (1) of how the efficient causes of the *ordre naturel* work in an orderly unfolding sequence, guided

[6] Quesnay, *Droit Naturel,* ch. v (Ed. Daire, *Physiocrates,* p. 53).

by the underlying natural laws — the propensity immanent in nature to establish the highest well-being of mankind, and (2) of the conditions imposed upon human conduct by these natural laws in order to reach the ordained
goal of supreme human welfare. The conditions so imposed on human conduct are as definitive as the laws and
the order by force of which they are imposed; and the
theoretical conclusions reached, when these laws and this
order are known, are therefore expressions of absolute
economic truth. Such conclusions are an expression of
reality, but not necessarily of fact.

Now, the objective end of this propensity that determines the course of nature is human well-being. But economic speculation has to do with the workings of nature
only so far as regards the *ordre physique.* And the laws
of nature in the *ordre physique,* working through mechanical sequence, can only work out the physical well-being
of man, not necessarily the spiritual. This propensity to
the physical well-being of man is therefore the law of
nature to which economic science must bring its generalisations, and this law of physical beneficence is the substantial ground of economic truth. Wanting this, all our
speculations are vain; but having its authentication they
are definitive. The great, typical function, to which all
the other functioning of nature is incidental if not subsidiary, is accordingly that of the alimentation, nutrition
of mankind. In so far, and only in so far as the physical
processes contribute to human sustenance and fullness of
life, can they, therefore, further the great work of nature.
Whatever processes contribute to human sustenance by
adding to the material available for human assimilation
and nutrition, by increasing the substance disposable for
human comfort, therefore count towards the substantial
end. All other processes, however serviceable in other

than this physiological respect, lack the substance of economic reality. Accordingly, human industry is productive, economically speaking, if it heightens the effectiveness of the natural processes out of which the material of human sustenance emerges; otherwise not. The test of productivity, of economic reality in material facts, is the increase of nutritive material. Whatever employment of time or effort does not afford an increase of such material is unproductive, however profitable it may be to the person employed, and however useful or indispensable it may be to the community. The type of such productive industry is the husbandman's employment, which yields a substantial (nutritive) gain. The artisan's work may be useful to the community and profitable to himself, but its economic effect does not extend beyond an alteration of the form in which the material afforded by nature already lies at hand. It is formally productive only, not really productive. It bears no part in the creative or generative work of nature; and therefore it lacks the character of economic substantiality. It does not enhance nature's output of vital force. The artisan's labors, therefore, yield no net product, whereas the husbandman's labors do.

Whatever constitutes a material increment of this output of vital force is wealth, and nothing else is. The theory of value contained in this position has not to do with value according to men's appraisement of the valuable article. Given items of wealth may have assigned to them certain relative values at which they exchange, and these conventional values may differ more or less widely from the natural or intrinsic value of the goods in question; but all that is beside the substantial point. The point in question is not the degree of predilection shown by certain individuals or bodies of men for certain goods. That is a matter of caprice and convention, and it does not directly touch the substantial ground of the economic life. The question of value is a question of the extent to which the given item of wealth forwards the end of nature's unfolding process. It is valuable, intrinsically and really, in so far as it avails the great work which nature has in hand.

Nature, then, is the final term in the Physiocratic speculations. Nature works by impulse and in an unfolding process, under the stress of a propensity to the accomplishment of a given end. This propensity, taken as the final cause that is operative in any situation, furnishes the basis on which to coördinate all our knowledge of those efficient causes through which Nature works to her ends. For the purpose of economic theory proper, this is the ultimate ground of reality to which our quest of economic truth must penetrate. But back of Nature and her works there is, in the Physiocratic scheme of the universe, the Creator, by whose all-wise and benevolent power the order of nature has been established in all the strength and beauty of its inviolate and immutable perfection. But the Physiocratic conception of the Creator is essentially a deistic one: he stands apart from the course of nature which he has established, and keeps his hands off. In the last resort, of course, "Dieu seul est producteur. Les hommes travaillent, receuillent, économisent, conservent; mais *économiser* n'est pas *produire.*"[7] But this last resort does not bring the Creator into economic theory as a fact to be counted with in formulating economic laws. He serves a homiletical purpose in the Physiocratic speculations rather than fills an office essential to the theory. He comes within the purview of the theory by way of authentication rather than as a subject of inquiry

[7] Dupont de Nemours, *Correspondance avec J.-B. Say* (Ed. Daire, *Physiocrates,* première partie, p. 399).

or a term in the formulation of economic knowledge. The Physiocratic God can scarcely be said to be an economic fact, but it is otherwise with that Nature whose ways and means constitute the subject-matter of the Physiocratic inquiry.

When this natural system of the Physiocratic speculation is looked at from the side of the psychology of the investigators, or from that of the logical premises employed, it is immediately recognised as essentially animistic. It runs consistently on animistic ground; but it is animism of a high grade,— highly integrated and enlightened, but, after all, retaining very much of that primitive force and naïveté which characterise the animistic explanations of phenomena in vogue among the untroubled barbarians. It is not the disjected animism of the vulgar, who see a willful propensity — often a willful perversity — in given objects or situations to work towards a given outcome, good or bad. It is not the gambler's haphazard sense of fortuitous necessity or the housewife's belief in lucky days, numbers or phases of the moon. The Physiocrat's animism rests on a broader outlook, and does not proceed by such an immediately impulsive imputation of propensity. The teleological element — the element of propensity — is conceived in a large way, unified and harmonised, as a comprehensive order of nature as a whole. But it vindicates its standing as a true animism by never becoming fatalistic and never being confused or confounded with the sequence of cause and effect. It has reached the last stage of integration and definition, beyond which the way lies downward from the high, quasi-spiritual ground of animism to the tamer levels of normality and causal uniformities.

There is already discernible a tone of dispassionate and colorless " tendency " about the Physiocratic animism, such as to suggest a wavering towards the side of normality. This is especially visible in such writers as the half-protestant Turgot. In his discussion of the development of farming, for instance, Turgot speaks almost entirely of human motives and the material conditions under which the growth takes place. There is little metaphysics in it, and that little does not express the law of nature in an adequate form. But, after all has been said, it remains true that the Physiocrat's sense of substantiality is not satisfied until he reaches the animistic ground; and it remains true also that the arguments of their opponents made little impression on the Physiocrats so long as they were directed to other than this animistic ground of their doctrine. This is true in great measure even of Turgot, as witness his controversy with Hume. Whatever criticism is directed against them on other grounds is met with impatience, as being inconsequential, if not disingenuous.[8]

To an historian of economic theory the source and the line of derivation whereby this precise form of the order-of-nature preconception reached the Physiocrats are of first-rate importance; but it is scarcely a question to be taken up here,— in part because it is too large a question to be handled here, in part because it has met with adequate treatment at more competent hands,[9] and in part because it is somewhat beside the immediate point under discussion. This point is the logical, or perhaps better the psychological, value of the Physiocrats' preconception, as a factor in shaping their point of view and the terms of their definitive formulation of economic knowledge. For this purpose it may be sufficient to point out that the pre-

[8] See, for instance, the concluding chapters of La Rivière's *Ordre Naturel des Sociétés Politiques.*

[9] E. g., Hasbach, *loc. cit.;* Bonar, *Philosophy and Political Economy,* Book II; Ritchie, *Natural Rights.*

conception in question belongs to the generation in which the Physiocrats lived, and that it is the guiding norm of all serious thought that found ready assimilation into the common-sense views of that time. It is the characteristic and controlling feature of what may be called the common-sense metaphysics of the eighteenth century, especially so far as concerns the enlightened French community.

It is to be noted as a point bearing more immediately on the question in hand that this imputation of final causes to the course of phenomena expresses a spiritual attitude which has prevailed, one might almost say, always and everywhere, but which reached its finest, most effective development, and found its most finished expression, in the eighteenth-century metaphysics. It is nothing recondite; for it meets us at every turn, as a matter of course, in the vulgar thinking of to-day,— in the pulpit and in the market place,— although it is not so ingenuous, nor does it so unquestionedly hold the primacy in the thinking of any class to-day as it once did. It meets us likewise, with but little change of features, at all past stages of culture, late or early. Indeed, it is the most generic feature of human thinking, so far as regards a theoretical or speculative formulation of knowledge. Accordingly, it seems scarcely necessary to trace the lineage of this characteristic preconception of the era of enlightenment, through specific channels, back to the ancient philosophers or jurists of the empire. Some of the specific forms of its expression — as, for instance, the doctrine of Natural Rights — are no doubt traceable through mediæval channels to the teachings of the ancients; but there is no need of going over the brook for water, and tracing back to specific teachings the main features of that habit of mind or spiritual attitude of which the doctrines of Natural

Rights and the Order of Nature are specific elaborations only. This dominant habit of mind came to the generation of the Physiocrats on the broad ground of group inheritance, not by lineal devolution from any one of the great thinkers of past ages who had thrown its deliverances into a similarly competent form for the use of his own generation.

In leaving the Physiocratic discipline and the immediate sphere of Physiocratic influence for British ground, we are met by the figure of Hume. Here, also, it will be impracticable to go into details as to the remoter line of derivation of the specific point of view that we come upon on making the transition, for reasons similar to those already given as excuse for passing over the similar question with regard to the Physiocratic point of view. Hume is, of course, not primarily an economist; but that placid unbeliever is none the less a large item in any inventory of eighteenth-century economic thought. Hume was not gifted with a facile acceptance of the group inheritance that made the habit of mind of his generation. Indeed, he was gifted with an alert, though somewhat histrionic, skepticism touching everything that was well received. It is his office to prove all things, though not necessarily to hold fast that which is good.

Aside from the strain of affectation discernible in Hume's skepticism, he may be taken as an accentuated expression of that characteristic bent which distinguishes British thinking in his time from the thinking of the Continent, and more particularly of the French. There is in Hume, and in the British community, an insistence on the prosy, not to say the seamy, side of human affairs. He is not content with formulating his knowledge of things in terms of what ought to be or in terms of the objective

point of the course of things. He is not even content
with adding to the teleological account of phenomena a
chain of empirical, narrative generalisations as to the
usual course of things. He insists, in season and out of
season, on an exhibition of the efficient causes engaged in
any sequence of phenomena; and he is skeptical — irrev-
erently skeptical — as to the need or the use of any formulation of knowledge that outruns the reach of his own
matter-of-fact, step-by-step argument from cause to
effect.

In short, he is too modern to be wholly intelligible to
those of his contemporaries who are most neatly abreast
of their time. He out-Britishes the British; and, in his
footsore quest for a perfectly tame explanation of things,
he finds little comfort, and indeed scant courtesy, at the
hands of his own generation. He is not in sufficiently
naïve accord with the range of preconceptions then in
vogue.

But, while Hume may be an accentuated expression of
a national characteristic, he is not therefore an untrue
expression of this phase of British eighteenth-century
thinking. The peculiarity of point of view and of method
for which he stands has sometimes been called the critical
attitude, sometimes the inductive method, sometimes the
materialistic or mechanical, and again, though less aptly,
the historical method. Its characteristic is an insistence
on matter of fact.

This matter-of-fact animus that meets any historian of
economic doctrine on his introduction to British economics is a large, but not the largest, feature of the British
scheme of early economic thought. It strikes the attention because it stands in contrast with the relative absence of this feature in the contemporary speculations of
the Continent. The most potent, most formative habit of

thought concerned in the early development of economic teaching on British ground is best seen in the broader generalisations of Adam Smith, and this more potent factor in Smith is a bent that is substantially identical with that which gives consistency to the speculations of the Physiocrats. In Adam Smith the two are happily combined, not to say blended; but the animistic habit still holds the primacy, with the matter-of-fact as a subsidiary though powerful factor. He is said to have combined deduction with induction. The relatively great prominence given the latter marks the line of divergence of British from French economics, not the line of coincidence; and on this account it may not be out of place to look more narrowly into the circumstances to which the emergence of this relatively greater penchant for a matter-of-fact explanation of things in the British community is due.

To explain the characteristic animus for which Hume stands, on grounds that might appeal to Hume, we should have to inquire into the peculiar circumstances — ultimately material circumstances — that have gone to shape the habitual view of things within the British community, and that so have acted to differentiate the British preconceptions from the French, or from the general range of preconceptions prevalent on the Continent. These peculiar formative circumstances are no doubt to some extent racial peculiarities; but the racial complexion of the British community is not widely different from the French, and especially not widely different from certain other Continental communities which are for the present purpose roughly classed with the French. Race difference can therefore not wholly, nor indeed for the greater part, account for the cultural difference of which this difference in preconceptions is an outcome. Through its cumulative effect on institutions the race difference must be held to have had a considerable effect on the habit of mind of the community; but, if the race difference is in this way taken as the remoter ground of an institutional peculiarity, which in its turn has shaped prevalent habits of thought, then the attention may be directed to the proximate causes, the concrete circumstances, through which this race difference has acted, in conjunction with other ulterior circumstances, to work out the psychological phenomena observed. Race differences, it may be remarked, do not so nearly coincide with national lines of demarcation as differences in the point of view from which things are habitually apprehended or differences in the standards according to which facts are rated.

If the element of race difference be not allowed definitive weight in discussing national peculiarities that underlie the deliverances of common sense, neither can these national peculiarities be confidently traced to a national difference in the transmitted learning that enters into the common-sense view of things. So far as concerns the concrete facts embodied in the learning of the various nations within the European culture, these nations make up but a single community. What divergence is visible does not touch the character of the positive information with which the learning of the various nations is occupied. Divergence is visible in the higher syntheses, the methods of handling the material of knowledge, the basis of valuation of the facts taken up, rather than in the material of knowledge. But this divergence must be set down to a cultural difference, a difference of point of view, not to a difference in inherited information. When a given body of information passes the national frontiers it acquires a new complexion, a new national, cultural physiognomy. It is this cultural physiognomy of learning that is here

under inquiry, and a comparison of early French economics (the Physiocrats) with early British economics (Adam
Smith) is here entered upon merely with a view to making
out what significance this cultural physiognomy of the
science has for the past progress of economic speculation.

The broad features of economic speculation, as it stood
at the period under consideration, may be briefly summed
up, disregarding the element of policy, or expediency,
which is common to both groups of economists, and at-
tending to their theoretical work alone. With the Physiocrats, as with Adam Smith, there are two main points of
view from which economic phenomena are treated: (*a*)
the matter-of-fact point of view or preconception, which
yields a discussion of causal sequences and correlations;
and (*b*) what, for want of a more expressive word, is here
called the animistic point of view or preconception, which
yields a discussion of teleological sequences and correlations,— a discussion of the function of this and that
" organ," of the legitimacy of this or the other range of
facts. The former preconception is allowed a larger
scope in the British than in the French economics: there
is more of " induction " in the British. The latter preconception is present in both, and is the definitive element
in both; but the animistic element is more colorless in the
British, it is less constantly in evidence, and less able to
stand alone without the support of arguments from cause
to effect. Still, the animistic element is the controlling
factor in the higher syntheses of both; and for both alike
it affords the definitive ground on which the argument
finally comes to rest. In neither group of thinkers is the
sense of substantiality appeased until this quasi-spiritual
ground, given by the natural propensity of the course of
events, is reached. But the propensity in events, the natural or normal course of things, as appealed to by the

British speculators, suggests less of an imputation of will-power, or personal force, to the propensity in question. It may be added, as has already been said in another place, that the tacit imputation of will-power or spiritual consistency to the natural or normal course of events has progressively weakened in the later course of economic speculation, so that in this respect, the British economists of the eighteenth century may be said to represent a later phase of economic inquiry than the Physiocrats.

Unfortunately, but unavoidably, if this question as to the cultural shifting of the point of view in economic science is taken up from the side of the causes to which the shifting is traceable, it will take the discussion back to ground on which an economist must at best feel himself to be but a raw layman, with all a layman's limitations and ineptitude, and with the certainty of doing badly what might be done well by more competent hands. But, with a reliance on charity where charity is most needed, it is necessary to recite summarily what seems to be the psychological bearing of certain cultural facts.

A cursory acquaintance with any of the more archaic phases of human culture enforces the recognition of this fact,— that the habit of construing the phenomena of the inanimate world in animistic terms prevails pretty much universally on these lower levels. Inanimate phenomena are apprehended to work out a propensity to an end; the movements of the elements are construed in terms of quasi-personal force. So much is well authenticated by the observations on which anthropologists and ethnologists draw for their materials. This animistic habit, it may be said, seems to be more effectual and far-reaching among those primitive communities that lead a predatory life.

But along with this feature of archaic methods of

thought or of knowledge, the picturesqueness of which has drawn the attention of all observers, there goes a second feature, no less important for the purpose in hand, though less obtrusive. The latter is of less interest to the men who have to do with the theory of cultural development, because it is a matter of course. This second feature of archaic thought is the habit of also apprehending facts in non-animistic, or impersonal, terms. The imputation of propensity in no case extends to all the mechanical facts in the case. There is always a substratum of matter of fact, which is the outcome of an habitual imputation of causal sequence, or, perhaps better, an imputation of mechanical continuity, if a new term be permitted. The agent, thing, fact, event, or phenomenon, to which propensity, will-power, or purpose, is imputed, is always apprehended to act in an environment which is accepted as spiritually inert. There are always opaque facts as well as self-directing agents. Any agent acts through means which lend themselves to his use on other grounds than that of spiritual compulsion, although spiritual compulsion may be a large feature in any given case.

The same features of human thinking, the same two complementary methods of correlating facts and handling them for the purposes of knowledge, are similarly in constant evidence in the daily life of men in our own community. The question is, in great part, which of the two bears the greater part in shaping human knowledge at any given time and within any given range of knowledge or of facts.

Other features of the growth of knowledge, which are remoter from the point under inquiry, may be of no less consequence to a comprehensive theory of the development of culture and of thought; but it is of course out of the question here to go farther afield. The present inquiry will have enough to do with these two. No other features are correlative with these, and these merit discussion on account of their intimate bearing on the point of view of economics. The point of interest with respect to these two correlative and complementary habits of thought is the question of how they have fared under the changing exigencies of human culture; in what manner they come, under given cultural circumstances, to share the field of knowledge between them; what is the relative part of each in the composite point of view in which the two habits of thought express themselves at any given cultural stage.

The animistic preconception enforces the apprehension of phenomena in terms generically identical with the terms of personality or individuality. As a certain modern group of psychologists would say, it imputes to objects and sequences an element of habit and attention similar in kind, though not necessarily in degree, to the like spiritual attitude present in the activities of a personal agent. The matter-of-fact preconception, on the other hand, enforces a handling of facts without imputation of personal force or attention, but with an imputation of mechanical continuity, substantially the preconception which has reached a formulation at the hands of scientists under the name of conservation of energy or persistence of quantity. Some appreciable resort to the latter method of knowledge is unavoidable at any cultural stage, for it is indispensable to all industrial efficiency. All technological processes and all mechanical contrivances rest, psychologically speaking, on this ground. This habit of thought is a selectively necessary consequence of industrial life, and, indeed, of all human experience in making use of the material means of life. It should therefore follow that, in a general way, the higher the culture, the greater the share of the mechanical preconception in shaping human thought and knowledge, since, in a general way, the stage of culture attained depends on the efficiency of industry. The rule, while it does not hold with anything like extreme generality, must be admitted to hold to a good extent; and to that extent it should hold also that, by a selective adaptation of men's habits of thought to the exigencies of those cultural phases that have actually supervened, the mechanical method of knowledge should have gained in scope and range. Something of the sort is borne out by observation.

A further consideration enforces the like view. As the community increases in size, the range of observation of the individuals in the community also increases; and continually wider and more far-reaching sequences of a mechanical kind have to be taken account of. Men have to adapt their own motives to industrial processes that are not safely to be construed in terms of propensity, predilection, or passion. Life in an advanced industrial community does not tolerate a neglect of mechanical fact; for the mechanical sequences through which men, at an appreciable degree of culture, work out their livelihood, are no respecters of persons or of will-power. Still, on all but the higher industrial stages, the coercive discipline of industrial life, and of the scheme of life that inculcates regard for the mechanical facts of industry, is greatly mitigated by the largely haphazard character of industry, and by the great extent to which man continues to be the prime mover in industry. So long as industrial efficiency is chiefly a matter of the handicraftsman's skill, dexterity, and diligence, the attention of men in looking to the industrial process is met by the figure of the workman, as the chief and characteristic factor; and thereby it comes to run on the personal element in industry.

But, with or without mitigation, the scheme of life which men perforce adopt under exigencies of an advanced industrial situation shapes their habits of thought on the side of their behavior, and thereby shapes their habits of thought to some extent for all purposes. Each individual is but a single complex of habits of thought, and the same psychical mechanism that expresses itself in one direction as conduct expresses itself in another direction as knowledge. The habits of thought formed in the one connection, in response to stimuli that call for a response in terms of conduct, must, therefore, have their effect when the same individual comes to respond to stimuli that call for a response in terms of knowledge. The scheme of thought or of knowledge is in good part a reverberation of the scheme of life. So that, after all has been said, it remains true that with the growth of industrial organization and efficiency there must, by selection and by adaptation, supervene a greater resort to the mechanical or dispassionate method of apprehending facts.

But the industrial side of life is not the whole of it, nor does the scheme of life in vogue in any community or at any cultural stage comprise industrial conduct alone. The social, civic, military, and religious interests come in for their share of attention, and between them they commonly take up by far the larger share of it. Especially is this true so far as concerns those classes among whom we commonly look for a cultivation of knowledge for knowledge's sake. The discipline which these several interests exert does not commonly coincide with the training given by industry. So the religious interest, with its canons of truth and of right living, runs exclusively on personal relations and the adaptation of conduct to the predilections of a superior personal agent. The weight of its discipline, therefore, falls wholly on the animistic side. It acts to heighten our appreciation of the spiritual bearing of phenomena and to discountenance a matter-of-fact apprehension of things. The skeptic of the type of Hume has never been in good repute with those who stand closest to the accepted religious truths. The bearing of this side of our culture upon the development of economics is shown by what the mediæval scholars had to say on economic topics.

The disciplinary effects of other phases of life, outside of the industrial and the religious, is not so simple a matter; but the discussion here approaches nearer to the point of immediate inquiry,— namely, the cultural situation in the eighteenth century, and its relation to economic speculation,— and this ground of interest in the question may help to relieve the topic of the tedium that of right belongs to it.

In the remoter past of which we have records, and even in the more recent past, Occidental man, as well as man elsewhere, has eminently been a respecter of persons. Wherever the warlike activity has been a large feature of the community's life, much of human conduct in society has proceeded on a regard for personal force. The scheme of life has been a scheme of personal aggression and subservience, partly in the naïve form, partly conventionalised in a system of status. The discipline of social life for the present purpose, in so far as its canons of conduct rest on this element of personal force in the unconventionalised form, plainly tends to the formation of a habit of apprehending and coördinating facts from the animistic point of view. So far as we have to do with life under a system of status, the like remains true, but with a difference. The régime of status inculcates an unremitting and very nice discrimination and observance of distinctions of personal superiority and inferiority. To the criterion of personal force, or will-power, taken in its immediate bearing on conduct, is added the criterion of personal excellence-in-general, regardless of the first-hand potency of the given person as an agent. This criterion of conduct requires a constant and painstaking imputation of personal value, regardless of fact. The discrimination enjoined by the canons of status proceeds on an invidious comparison of persons in respect of worth, value, potency, virtue, which must, for the present purpose, be taken as putative. The greater or less personal value assigned a given individual or a given class under the canons of status is not assigned on the ground of visible effciency, but on the ground of a dogmatic allegation accepted on the strength of an uncontradicted categorical affirmation simply. The canons of status hold their ground by force of preëmption. Where distinctions of status are based on a putative worth transmitted by descent from honorable antecedents, the sequence of transmission to which appeal is taken as the arbiter of honor is of a putative and animistic character rather than a visible mechanical continuity. The habit of accepting as final what is prescriptively right in the affairs of life has as its reflex in the affairs of knowledge the formula, *Quid ab omnibus, quid ubique creditur credendum est.*

Even this meager account of the scheme of life that characterises a régime of status should serve to indicate what is its disciplinary effect in shaping habits of thought, and therefore in shaping the habitual criteria of knowledge and of reality. A culture whose institutions are a framework of invidious comparisons implies, or rather involves and comprises, a scheme of knowledge whose definitive standards of truth and substantiality are of an animistic character; and, the more undividedly the canons of status and ceremonial honor govern the conduct of the community, the greater the facility with which the sequence of cause and effect is made to yield before the higher claims of a spiritual sequence or guidance in the course of events. Men consistently trained to an unremitting discrimination of honor, worth, and personal force in their daily conduct, and to whom these criteria afford the definitive ground of sufficiency in coördinating facts for the purposes of life, will not be satisfied to fall short of the like definitive ground of sufficiency when they come to coördinate facts for the purposes of knowledge simply. The habits formed in unfolding his activity in one direction, under the impulse of a given interest, assert themselves when the individual comes to unfold his activity in any other direction, under the impulse of any other interest. If his last resort and highest criterion of truth in conduct is afforded by the element of personal force and invidious comparison, his sense of substantiality or truth in the quest of knowledge will be satisfied only when a like definitive ground of animistic force and invidious comparison is reached. But when such ground is reached he rests content and pushes the inquiry no farther. In his practical life he has acquired the habit of resting his case on an authentic deliverance as to what is absolutely right. This absolutely right and good final term in conduct has the character of finality only when conduct is construed in a ceremonial sense; that is to say, only when life is conceived as a scheme of conformity to a purpose outside and beyond the process of living. Under the régime of status this ceremonial finality is found in the concept of worth or honor. In the religious domain it is the concept of virtue, sanctity, or tabu. Merit lies in what one is, not in what one does. The habit of appeal to ceremonial finality, formed in the school of status, goes with the individual in his quest of knowledge, as a dependence upon a similarly authentic norm of absolute truth,— a similar seeking of a final term outside and beyond the range of knowledge.

The discipline of social and civic life under a régime of status, then, reënforces the discipline of the religious life; and the outcome of the resulting habituation is that the canons of knowledge are cast in the animistic mold and converge to a ground of absolute truth, and this absolute truth is of a ceremonial nature. Its subject-matter is a reality regardless of fact.

The outcome, for science, of the religious and social life of the civilisation of status, in Occidental culture, was a structure of quasi-spiritual appreciations and explanations, of which astrology, alchemy, and mediæval theology and metaphysics are competent, though somewhat one-sided, exponents. Throughout the range of this early learning the ground of correlation of phenomena is in part the supposed relative potency of the facts correlated; but it is also in part a scheme of status, in which facts are scheduled according to a hierarchical gradation of worth or merit, having only a ceremonial relation to the observed phenomena. Some elements (some metals, for instance) are noble, others base; some planets, on grounds of ceremonial efficacy, have a sinister influence, others a beneficent one; and it is a matter of serious consequence whether they are in the ascendant, and so on.

The body of learning through which the discipline of animism and invidious comparison transmitted its effects to the science of economics was what is known as natural theology, natural rights, moral philosophy, and natural law. These several disciplines or bodies of knowledge had wandered far from the naïve animistic standpoint at the time when economic science emerged, and much the same is true as regards the time of the emergence of other modern sciences. But the discipline which makes for an animistic formulation of knowledge continued to hold the primacy in modern culture, although its dominion was never altogether undivided or unmitigated. Occidental culture has long been largely an industrial culture; and, as already pointed out, the discipline of industry, and of life in an industrial community, does not favor the animistic preconception. This is especially true as regards industry which makes large use of mechanical contrivances. The difference in these respects between Occidental industry and science, on the one hand, and the industry and science of other cultural regions, on the other hand, is worth noting in this connection. The result has been that the sciences, as that word is understood in later usage, have come forward gradually, and in a certain rough parallelism with the development of industrial processes and industrial organisation. It is possible to hold that both modern industry (of the mechanical sort) and modern science center about the region of the North Sea. It is still more palpably true that within this general area the sciences, in the recent past, show a family likeness to the civil and social institutions of the communities in which they have been cultivated, this being true to the greatest extent of the higher or speculative sciences; that is, in that range of knowledge in which the animistic preconception can chiefly and most effectively find application. There is, for instance, in the eighteenth century a perceptible parallelism between the divergent character of British and Continental culture and institutions, on the one hand, and the dissimilar aims of British and Continental speculation, on the other hand.

Something has already been said of the difference in preconceptions between the French and the British economists of the eighteenth century. It remains to point out

the correlative cultural difference between the two communities, to which it is conceived that the difference in
scientific animus is in great measure due. It is, of course,
only the general features, the general attitude of the speculators, that can be credited to the difference in culture.
Differences of detail in the specific doctrines held could
be explained only on a much more detailed analysis than
can be entered on here, and after taking account of facts
which cannot here be even allowed for in detail.

Aside from the greater resort to mechanical contrivances and the larger scale of organisation in British industry, the further cultural peculiarities of the British community run in the same general direction. British religious life and beliefs had less of the element of fealty —
personal or discretionary mastery and subservience — and
more of a tone of fatalism. The civil institutions of the
British had not the same rich personal content as those
of the French. The British subject owned allegiance to
an impersonal law rather than to the person of a superior. Relatively, it may be said that the sense of status, as
a coercive factor, was in abeyance in the British community. Even in the warlike enterprise of the British community a similar characteristic is traceable. Warfare is,
of course, a matter of personal assertion. Warlike communities and classes are necessarily given to construing
facts in terms of personal force and personal ends. They
are always superstitious. They are great sticklers for
rank and precedent, and zealously cultivate those distinctions and ceremonial observances in which a system of
status expresses itself. But, while warlike enterprise has
by no means been absent from the British scheme of life,
the geographical and strategic isolation of the British community has given a characteristic turn to their military
relations. In recent times British warlike operations have

been conducted abroad. The military class has consequently in great measure been segregated out from the body of the community, and the ideals and prejudices of the class have not been transfused through the general body with the same facility and effect that they might otherwise have had. The British community at home has seen the campaign in great part from the standpoint of the " sinews of war."

The outcome of all these national peculiarities of circumstance and culture has been that a different scheme of life has been current in the British community from what has prevailed on the Continent. There has resulted the formation of a different body of habits of thought and a different animus in their handling of facts. The preconception of causal sequence has been allowed larger scope in the correlation of facts for purposes of knowledge ; and, where the animistic preconception has been resorted to, as it always has in the profounder reaches of learning, it has commonly been an animism of a tamer kind.

Taking Adam Smith as an exponent of this British attitude in theoretical knowledge, it is to be noted that, while he formulates his knowledge in terms of a propensity (natural laws) working teleologically to an end, the end or objective point which controls the formulation has not the same rich content of vital human interest or advantage as is met with in the Physiocratic speculations. There is perceptibly less of an imperious tone in Adam Smith's natural laws than in those of the contemporary French economists. It is true, he sums up the institutions with which he deals in terms of the ends which they should subserve, rather than in terms of the exigencies and habits of life out of which they have arisen; but he does not with the same tone of finality appeal to the end subserved as a final cause through whose coercive guidance the complex of phenomena is kept to its appointed task. Under his hands the restraining, compelling agency retires farther into the background, and appeal is taken to it neither so directly nor on so slight provocation.

But Adam Smith is too large a figure to be disposed of in a couple of concluding paragraphs. At the same time his work and the bent which he gave to economic speculation are so intimately bound up with the aims and bias that characterise economics in its next stage of development that he is best dealt with as the point of departure for the Classical School rather than merely as a British counterpart of Physiocracy. Adam Smith will accordingly be considered in immediate connection with the bias of the classical school and the incursion of utilitarianism into economics.

THE PRECONCEPTIONS OF ECONOMIC SCIENCE [1]

II

ADAM SMITH's animistic bent asserts itself more plainly and more effectually in the general trend and aim of his discussion than in the details of theory. "Adam Smith's *Wealth of Nations* is, in fact, so far as it has one single purpose, a vindication of the unconscious law present in the separate actions of men when these actions are directed by a certain strong personal motive." [2]   Both in the *Theory of the Moral Sentiments* and in the *Wealth of Nations* there are many passages that testify to his abiding conviction that there is a wholesome trend in the natural course of things, and the characteristically optimistic tone in which he speaks for natural liberty is but an expression of this conviction.   An extreme resort to this animistic ground occurs in his plea for freedom of investment. [3]

[1] Reprinted by permission from *The Quarterly Journal of Economics*, Vol. XIII, July. 1899.

[2] Bonar, *Philosophy and Political Economy*, pp. 177, 178.

[3] "Every individual is continually exerting himself to find out the most advantageous employment for whatever capital he can command.   It is his own advantage, and not that of the society, which he has in view.   But the study of his own advantage naturally, or rather necessarily, leads him to prefer that employment which is most advantageous to the society. . . . By directing that industry in such a manner as its produce may be of the greatest value, he intends only his own gain; and he is in this, as in many other cases, led by an invisible hand to promote an end which was no part of his intention.   Nor is it always the worse for society that it was no part of it.   By pursuing his own

In the proposition that men are " led by an invisible hand," Smith does not fall back on a meddling Providence who is to set human affairs straight when they are in danger of going askew. He conceives the Creator to be very continent in the matter of interference with the natural course of things. The Creator has established the natural order to serve the ends of human welfare; and he has very nicely adjusted the efficient causes comprised in the natural order, including human aims and motives, to this work that they are to accomplish. The guidance of the invisible hand takes place not by way of interposition, but through a comprehensive scheme of contrivances established from the beginning. For the purpose of economic theory, man is conceived to be consistently self-seeking; but this economic man is a part of the mechanism of nature, and his self-seeking traffic is but a means whereby, in the natural course of things, the general welfare is worked out. The scheme as a whole is guided by the end to be reached, but the sequence of events through which the end is reached is a causal sequence which is not broken into episodically. The benevolent work of guidance was performed in first establishing an ingenious mechanism of forces and motives capable of accomplishing an ordained result, and nothing beyond the enduring constraint of an established trend remains to enforce the divine purpose in the resulting natural course of things.

The sequence of events, including human motives and human conduct, is a causal sequence; but it is also something more, or, rather, there is also another element of continuity besides that of brute cause and effect, present even in the step-by-step process whereby the natural

interest he frequently promotes that of the society more effectually than when he really intends to promote it." *Wealth of Nations,* Book IV, chap. ii.

course of things reaches its final term. The presence of such a quasi-spiritual or non-causal element is evident from two (alleged) facts. (1) The course of things may be deflected from the direct line of approach to that consummate human welfare which is its legitimate end. The natural trend of things may be overborne by an untoward conjuncture of causes. There is a distinction, often distressingly actual and persistent, between the legitimate and the observed course of things. If "natural," in Adam Smith's use, meant necessary, in the sense of causally determined, no divergence of events from the natural or legitimate course of things would be possible. If the mechanism of nature, including man, were a mechanically competent contrivance for achieving the great artificer's design, there could be no such episodes of blundering and perverse departure from the direct path as Adam Smith finds in nearly all existing arrangements. Institutional facts would then be "natural." [4] (2) When things have gone wrong, they will right themselves if interference with the natural course ceases; whereas, in the case of a causal sequence simply, the mere cessation of interference will not leave the outcome the same as if no interference had taken place. This recuperative power of nature is of an extra-mechanical character. The continuity of sequence by force of which the natural course of things prevails is, therefore, not of the nature of cause and effect, since it bridges intervals and interruptions in the causal sequence. [5] Adam Smith's use of the term "real"

[4] The discrepancy between the actual, causally determined situation and the divinely intended consummation is the metaphysical ground of all that inculcation of morality and enlightened policy that makes up so large a part of Adam Smith's work. The like, of course, holds true for all moralists and reformers who proceed on the assumption of a providential order.

[5] "In the political body, however, the wisdom of nature has

in statements of theory — as, for example, " real value,"
" real price " [6]— is evidence to this effect. " Natural "
commonly has the same meaning as " real " in this connection.[7] Both " natural " and " real " are placed in contrast with the actual; and, in Adam Smith's apprehension,
both have a substantiality different from and superior to
facts. The view involves a distinction between reality
and fact, which survives in a weakened form in the
theories of " normal " prices, wages, profits, costs, in
Adam Smith's successors.

This animistic prepossession seems to pervade the earlier of his two monumental works in a greater degree
than the later. In the *Moral Sentiments* recourse is had
to the teleological ground of the natural order more freely
and with perceptibly greater insistence. There seems to
be reason for holding that the animistic preconception
weakened or, at any rate, fell more into the background
as his later work of speculation and investigation proceeded. The change shows itself also in some details of
his economic theory, as first set forth in the *Lectures,* and
afterwards more fully developed in the *Wealth of Nations.* So, for instance, in the earlier presentation of the

fortunately made ample provision for remedying many of the bad
effects of the folly and injustice of man; in the same manner
as it has done in the natural body, for remedying those of his
sloth and intemperance." *Wealth of Nations,* Book IV, chap. ix.

[6] *E.g.,* " the real measure of the exchangeable value of all
commodities." *Wealth of Nations,* Book I, chap. v, and repeatedly in the like connection.

[7] *E.g.,* Book I, chap. vii: " When the price of any commodity is
neither more nor less than what is sufficient to pay the rent of the
land, the wages of the labor, and the profits of the stock employed
in raising, preparing, and bringing it to market, according to their
natural rates, the commodity is then sold for what may be called
its *natural* price." " The actual price at which any commodity is
commonly sold is called its market price. It may be either above
or below or exactly the same with its natural price."

matter, "the division of labor is the immediate cause of opulence"; and this division of labor, which is the chief condition of economic well-being, "flows from a direct propensity in human nature for one man to barter with another."[8] The "propensity" in question is here appealed to as a natural endowment immediately given to man with a view to the welfare of human society, and without any attempt at further explanation of how man has come by it. No causal explanation of its presence or character is offered. But the corresponding passage of the *Wealth of Nations* handles the question more cautiously.[9] Other parallel passages might be compared, with much the same effect. The guiding hand has withdrawn farther from the range of human vision.

However, these and other like filial expressions of a devout optimism need, perhaps, not be taken as integral features of Adam Smith's economic theory, or as seriously affecting the character of his work as an economist. They are the expression of his general philosophical and theological views, and are significant for the present purpose chiefly as evidences of an animistic and optimistic bent. They go to show what is Adam Smith's accepted ground of finality,— the ground to which all his speculations on human affairs converge; but they do not in any

[8] *Lectures of Adam Smith* (Ed. Cannan, 1896), p. 169.

[9] "This division of labor, from which so many advantages are derived, is not originally the effect of any human wisdom, which foresees and intends that general opulence to which it gives occasion. It is the necessary though very slow and gradual consequence of a certain propensity in human nature which has in view no such extensive utility,— the propensity to truck, barter, and exchange one thing for another. Whether this propensity be one of those original principles in human nature of which no further account can be given, or whether, as seems more probable, it be the necessary consequence of the faculties of n and speech, it belongs not to our present subject to in-
" *Wealth of Nations*, Book I, chap. ii.

great degree show the teleological bias guiding his formulation of economic theory in detail.

The effective working of the teleological bias is best seen in Smith's more detailed handling of economic phenomena — in his discussion of what may loosely be called economic institutions — and in the criteria and principles of procedure by which he is guided in incorporating these features of economic life into the general structure of his theory. A fair instance, though perhaps not the most telling one, is the discussion of the " real and nominal price," and of the " natural and market price " of commodities, already referred to above.[10] The " real " price of commodities is their value in terms of human life. At this point Smith differs from the Physiocrats, with whom the ultimate terms of value are afforded by human sustenance taken as a product of the functioning of brute nature; the cause of the difference being that the Physiocrats conceived the natural order which works towards the material well-being of man to comprise the non-human environment only, whereas Adam Smith includes man in this concept of the natural order, and, indeed, makes him the central figure in the process of production. With the Physiocrats, production is the work of nature: with Adam Smith, it is the work of man and nature, with man in the foreground. In Adam Smith, therefore, labor is the final term in valuation. This " real " value of commodities is the value imputed to them by the economist under the stress of his teleological preconception. It has little, if any, place in the course of economic events, and no bearing on human affairs, apart from the sentimental influence which such a preconception in favor of a " real value " in things may exert upon men's notions of what is the good and equitable course to pursue in their trans-

[10] *Wealth of Nations,* Book I, chaps. v.–vii.

actions. It is impossible to gauge this real value of goods; it cannot be measured or expressed in concrete terms. Still, if labor exchanges for a varying quantity of goods, " it is their value which varies, not that of the labor which purchases them." [11] The values which practically attach to goods in men's handling of them are conceived to be determined without regard to the real value which Adam Smith imputes to the goods; but, for all that, the substantial fact with respect to these market values is their presumed approximation to the real values teleologically imputed to the goods under the guidance of inviolate natural laws. The real, or natural, value of articles has no causal relation to the value at which they exchange. The discussion of how values are determined in practice runs on the motives of the buyers and sellers, and the relative advantage enjoyed by the parties to the transaction.[12] It is a discussion of a process of valuation, quite unrelated to the " real," or " natural," price of things, and quite unrelated to the grounds on which things are held to come by their real, or natural, price; and yet, when the complex process of valuation has been traced out in terms of human motives and the exigencies of the market, Adam Smith feels that he has only cleared the ground. He then turns to the serious business of accounting for value and price theoretically, and making the ascertained facts articulate with his teleological theory of economic life.[13]

[11] *Wealth of Nations,* Book I, chap. v.

[12] As, *e.g.,* the entire discussion of the determination of Wages, Profits and Rent, in Book I, chaps. viii.–xi.

[13] " There is in every society or neighborhood an ordinary or average rate both of wages and profit in every different employment of labor and stock This rate is naturally regulated, . . . partly by the general circumstances of the society. . . . There is, likewise, in every society or neighborhood an ordinary or average rate of rent, which is regulated, too. . . . These ordinary or average rates may be called the natural rates of wages, profit,

The occurrence of the words " ordinary " and " average " in this connection need not be taken too seriously. The context makes it plain that the equality which commonly subsists between the ordinary or average rates, and the natural rates, is a matter of coincidence, not of identity. Not only are there temporary deviations, but there may be a permanent divergence between the ordinary and the natural price of a commodity; as in case of a monopoly or of produce grown under peculiar circumstances of soil or climate.[14]

The natural price coincides with the price fixed by competition, because competition means the unimpeded play of those efficient forces through which the nicely adjusted mechanism of nature works out the design to accomplish which it was contrived. The natural price is reached through the free interplay of the factors of production, and it is itself an outcome of production. Nature, including the human factor, works to turn out the goods; and the natural value of the goods is their appraisement from the standpoint of this productive process of nature. Natural value is a category of production: whereas, notoriously exchange value or market price is a category of distribution. And Adam Smith's theoretical handling of market price aims to show how the factors of human predilection and human wants at work in the higgling of the

and rent, at the time and place in which they commonly prevail. When the price of any commodity is neither more nor less than what is sufficient to pay the rent of the land, the wages of the labor, and the profits of the stock employed in raising, preparing, and bringing it to market, according to their natural rates, the commodity is then sold for what may be called its natural price." *Wealth of Nations,* Book I, chap. vii.

[14] " Such commodities may continue for whole centuries together to be sold at this high price; and that part of it which resolves itself into the rent of land is, in this case, the part which is generally paid above its natural rate." Book I, chap. vii.

market bring about a result in passable consonance with the natural laws that are conceived to govern production.

The natural price is a composite result of the blending of the three " component parts of the price of commodities,"— the natural wages of laborer, the natural profits of stock, and the natural rent of land; and each of these three components is in its turn the measure of the productive effect of the factor to which it pertains. The further discussion of these shares in distribution aims to account for the facts of distribution on the ground of the productivity of the factors which are held to share the product between them. That is to say, Adam Smith's preconception of a productive natural process as the basis of his economic theory dominates his aims and procedure, when he comes to deal with phenomena that cannot be stated in terms of production. The causal sequence in the process of distribution is, by Adam Smith's own showing, unrelated to the causal sequence in the process of production; but, since the latter is the substantial fact, as viewed from the standpoint of a teleological natural order, the former must be stated in terms of the latter before Adam Smith's sense of substantiality, or " reality," is satisfied. Something of the same kind is, of course, visible in the Physiocrats and in Cantillon. It amounts to an extension of the natural-rights preconception to economic theory. Adam Smith's discussion of distribution as a function of productivity might be traced in detail through his handling of Wages, Profits, and Rent; but, since the aim here is a brief characterisation only, and not an exposition, no farther pursuit of this point seems feasible.

It may, however, be worth while to point out another line of influence along which the dominance of the teleological preconception shows itself in Adam Smith. This is the normalisation of data, in order to bring them into consonance with an orderly course of approach to the putative natural end of economic life and development. The result of this normalisation of data is, on the one hand, the use of what James Steuart calls " conjectural history " in dealing with past phases of economic life, and, on the other hand, a statement of present-day phenomena in terms of what legitimately ought to be according to the God-given end of life rather than in terms of unconstrued observation. Account is taken of the facts (supposed or observed) ostensibly in terms of causal sequence, but the imputed causal sequence is construed to run on lines of teleological legitimacy.

A familiar instance of this " conjectural history," in a highly and effectively normalized form, is the account of " that early and rude state of society which precedes both the accumulation of stock and the appropriation of land." [15] It is needless at this day to point out that this " early and rude state," in which " the whole produce of labor belongs to the laborer," is altogether a figment. The whole narrative, from the putative origin down, is not only supposititious, but it is merely a schematic presentation of what should have been the course of past development, in order to lead up to that ideal economic situation which would satisfy Adam Smith's preconception.[16] As the narrative comes nearer the region of known latter-day facts, the normalisation of the data becomes more difficult and receives more detailed attention; but the change in method is a change of degree rather than of kind. In the " early and rude state " the coincidence of the " natural " and the actual course of events is immediate and undis-

[15] *Wealth of Nations*, Book I, chap. vi; also chap. viii.

[16] For an instance of how these early phases of industrial development appear, when not seen in the light of Adam Smith's preconception, see, among others, Bücher, *Entstehung der Volkswirtschaft*.

turbed, there being no refractory data at hand; but in the later stages and in the present situation, where refractory facts abound, the coördination is difficult, and the coincidence can be shown only by a free abstraction from phenomena that are irrelevant to the teleological trend and by a laborious interpretation of the rest. The facts of modern life are intricate, and lend themselves to statement in the terms of the theory only after they have been subjected to a " higher criticism."

The chapter " Of the Origin and Use of Money " [17] is an elegantly normalised account of the origin and nature of an economic institution, and Adam Smith's further discussion of money runs on the same lines. The origin of money is stated in terms of the purpose which money should legitimately serve in such a community as Adam Smith considered right and good, not in terms of the motives and exigencies which have resulted in the use of money and in the gradual rise of the existing method of payment and accounts. Money is " the great wheel of circulation," which effects the transfer of goods in process of production and the distribution of the finished goods to the consumers. It is an organ of the economic commonwealth rather than an expedient of accounting and a conventional repository of wealth.

It is perhaps superfluous to remark that to the " plain man," who is not concerned with the " natural course of things " in a consummate *Geldwirtschaft,* the money that passes his hand is not a " great wheel of circulation." To the Samoyed, for instance, the reindeer which serves him as unit of value is wealth in the most concrete and tangible form. Much the same is true of coin, or even of bank-notes, in the apprehension of unsophisticated people among ourselves to-day. And yet it is in terms of the

[17] Book I, chap. iv.

habits and conditions of life of these " plain people " that
the development of money will have to be accounted for
if it is to be stated in terms of cause and effect.

The few scattered passages already cited may serve to
illustrate how Adam Smith's animistic or teleological bent
shapes the general structure of his theory and gives it
consistency. The principle of definitive formulation in
Adam Smith's economic knowledge is afforded by a putative purpose that does not at any point enter causally into
the economic life process which he seeks to know. This
formative or normative purpose or end is not freely conceived to enter as an efficient agent in the events discussed,
or to be in any way consciously present in the process.
It can scarcely be taken as an animistic agency engaged
in the process. It sanctions the course of things, and
gives legitimacy and substance to the sequence of events,
so far as this sequence may be made to square with the
requirements of the imputed end. It has therefore a
ceremonial or symbolical force only, and lends the discussion a ceremonial competency; although with economists
who have been in passable agreement with Adam Smith as
regards the legitimate end of economic life this ceremonial consistency, or consistency *de jure,* has for many purposes been accepted as the formulation of a causal continuity in the phenomena that have been interpreted in
its terms. Elucidations of what normally ought to happen, as a matter of ceremonial necessity, have in this way
come to pass for an account of matters of fact.

But, as has already been pointed out, there is much more
to Adam Smith's exposition of theory than a formulation
of what ought to be. Much of the advance he achieved
over his predecessors consists in a larger and more painstaking scrutiny of facts, and a more consistent tracing

out of causal continuity in the facts handled. No doubt, his superiority over the Physiocrats, that characteristic of his work by virtue of which it superseded theirs in the farther growth of economic science, lies to some extent in his recourse to a different, more modern ground of normality,— a ground more in consonance with the body of preconceptions that have had the vogue in later generations. It is a shifting of the point of view from which the facts are handled; but it comes in great part to a substitution of a new body of preconceptions for the old, or a new adaptation of the old ground of finality, rather than an elimination of all metaphysical or animistic norms of valuation. With Adam Smith, as with the Physiocrats, the fundamental question, the answer to which affords the point of departure and the norm of procedure, is a question of substantiality or economic " reality." With both, the answer to this question is given naïvely, as a deliverance of common sense. Neither is disturbed by doubts as to this deliverance of common sense or by any need of scrutinising it. To the Physiocrats this substantial ground of economic reality is the nutritive process of Nature. To Adam Smith it is Labor. His reality has the advantage of being the deliverance of the common sense of a more modern community, and one that has maintained itself in force more widely and in better consonance with the facts of latter-day industry. The Physiocrats owe their preconception of the productiveness of nature to the habits of thought of a community in whose economic life the dominant phenomenon was the owner of agricultural land. Adam Smith owes his preconception in favor of labor to a community in which the obtrusive economic feature of the immediate past was handicraft and agriculture, with commerce as a scarcely secondary phenomenon.

So far as Adam Smith's economic theories are a tracing out of the causal sequence in economic phenomena, they are worked out in terms given by these two main directions of activity,— human effort directed to the shaping of the material means of life, and human effort and discretion directed to a pecuniary gain. The former is the great, substantial productive force : the latter is not immediately, or proximately, productive.[18] Adam Smith still has too lively a sense of the nutritive purpose of the order of nature freely to extend the concept of productiveness to any activity that does not yield a material increase of the creature comforts. His instinctive appreciation of the substantial virtue of whatever effectually furthers nutrition, even leads him into the concession that " in agriculture nature labors along with man," although the general tenor of his argument is that the productive force with which the economist always has to count is human labor. This recognised substantiality of labor as productive is, as has already been remarked, accountable for his effort to reduce to terms of productive labor such a category of distribution as exchange value.

With but slight qualification, it will hold that, in the causal sequence which Adam Smith traces out in his economic theories proper (contained in the first three books of the *Wealth of Nations*), the causally efficient factor is conceived to be human nature in these two relations,— of productive efficiency and pecuniary gain through exchange. Pecuniary gain — gain in the material means of life through barter — furnishes the motive force to the economic activity of the individual; although productive efficiency is the legitimate, normal end of the community's economic life. To such an extent does this

[18] See *Wealth of Nations*, Book II, chap. v, " Of the Different Employment of Capitals."

concept of man's seeking his ends through " truck, barter, and exchange " pervade Adam Smith's treatment of economic processes that he even states production in its terms, and says that " labor was the first price, the original purchase-money, that was paid for all things." [19]   The human nature engaged in this pecuniary traffic is conceived in somewhat hedonistic terms, and the motives and movements of men are normalised to fit the requirements of a hedonistically conceived order of nature.   Men are very much alike in their native aptitudes and propensities; [20] and, so far as economic theory need take account of these aptitudes and propensities, they are aptitudes for the production of the " necessaries and conveniences of life," and propensities to secure as great a share of these creature comforts as may be.

Adam Smith's conception of normal human nature — that is to say, the human factor which enters causally in the process which economic theory discusses — comes, on the whole, to this: Men exert their force and skill in a mechanical process of production, and their pecuniary sagacity in a competitive process of distribution, with a view to individual gain in the material means of life. These material means are sought in order to the satisfaction of men's natural wants through their consumption. It is true, much else enters into men's endeavors in the struggle for wealth, as Adam Smith points out; but this consumption comprises the legitimate range of incentives,

[19] *Wealth of Nations*, Book I, chap. v.   See also the plea for free trade, Book IV, chap. ii: " But the annual revenue of every society is always precisely equal to the exchangeable value of the whole annual produce of its industry, or, rather, is precisely the same thing with that exchangeable value."

[20] " The difference of natural talents in different men is in reality much less than we are aware of."   *Wealth of Nations*, Book I, chap. ii.

and a theory which concerns itself with the natural course
of things need take but incidental account of what does
not come legitimately in the natural course. In point of
fact, there are appreciable " actual," though scarcely
" real," departures from this rule. They are spurious and
insubstantial departures, and do not properly come within
the purview of the stricter theory. And, since human nature is strikingly uniform, in Adam Smith's apprehension,
both the efforts put forth and the consumptive effect
accomplished may be put in quantitative terms and treated
algebraically, with the result that the entire range of phenomena comprised under the head of consumption need be
but incidentally considered; and the theory of production
and distribution is complete when the goods or the values
have been traced to their disappearance in the hands of
their ultimate owners. The reflex effect of consumption
upon production and distribution is, on the whole, quantitative only.

Adam Smith's preconception of a normal teleological
order of procedure in the natural course, therefore, affects
not only those features of theory where he is avowedly
concerned with building up a normal scheme of the economic process. Through his normalising the chief causal
factor engaged in the process, it affects also his arguments
from cause to effect.[21] What makes this latter feature

[21] " Mit diesen philosophischen Ueberzeugungen tritt nun Adam
Smith an die Welt der Enfahrung heran, und es ergiebt sich ihm
die Richtigkeit der Principien. Der Reiz der Smith'schen
Schriften beruht zum grossen Teile darauf, dass Smith die Principien in so innige Verbindung mit dem Thatsächlichen gebracht.
Hie und da werden dann auch die Principien, was durch diese
Verbindung veranlasst wird, an ihren Spitzen etwas abgeschliffen,
ihre allzuscharfe Ausprägung dadurch vermieden. Nichtsdesto-
weniger aber bleiben sie stets die leitenden Grundgedanken."
Richard Zeyss, *Adam Smith und der Eigennutz* (Tübingen, 1889),
p. 110.

worth particular attention is the fact that his successors carried this normalisation farther, and employed it with less frequent reference to the mitigating exceptions which Adam Smith notices by the way.

The reason for that farther and more consistent normalisation of human nature which gives us the " economic man " at the hands of Adam Smith's successors lies, in great part, in the utilitarian philosophy that entered in force and in consummate form at about the turning of the century. Some credit in the work of normalisation is due also to the farther supersession of handicraft by the " capitalistic " industry that came in at the same time and in pretty close relation with the utilitarian views.

After Adam Smith's day, economics fell into profane hands. Apart from Malthus, who, of all the greater economists, stands nearest to Adam Smith on such metaphysical heads as have an immediate bearing upon the premises of economic science, the next generation do not approach their subject from the point of view of a divinely instituted order; nor do they discuss human interests with that gently optimistic spirit of submission that belongs to the economist who goes to his work with the fear of God before his eyes. Even with Malthus the recourse to the divinely sanctioned order of nature is somewhat sparing and temperate. But it is significant for the later course of economic theory that, while Malthus may well be accounted the truest continuer of Adam Smith, it was the undevout utilitarians that became the spokesmen of the science after Adam Smith's time.

There is no wide breach between Adam Smith and the utilitarians, either in details of doctrine or in the concrete conclusions arrived at as regards questions of policy. On these heads Adam Smith might well be classed as a moderate utilitarian, particularly so far as regards his economic work. Malthus has still more of a utilitarian air,—so much so, indeed, that he is not infrequently spoken of as a utilitarian. This view, convincingly set forth by Mr. Bonar,[22] is no doubt well borne out by a detailed scrutiny of Malthus's economic doctrines. His humanitarian bias is evident throughout, and his weakness for considerations of expediency is the great blemish of his scientific work. But, for all that, in order to an appreciation of the change that came over classical economics with the rise of Benthamism, it is necessary to note that the agreement in this matter between Adam Smith and the disciples of Bentham, and less decidedly that between Malthus and the latter, is a coincidence of conclusions rather than an identity of preconceptions.[23]

With Adam Smith the ultimate ground of economic reality is the design of God, the teleological order; and his utilitarian generalisations, as well as the hedonistic character of his economic man, are but methods of the working out of this natural order, not the substantial and self-legitimating ground. Shifty as Malthus's metaphysics are, much the same is to be said for him.[24] Of the utilitarians proper the converse is true, although here, again, there is by no means utter consistency. The sub-

---

[22] See, *e.g.*, *Malthus and his Work,* especially Book III, as also the chapter on Malthus in *Philosophy and Political Economy,* Book III, Modern Philosophy: Utilitarian Economics, chap. i, " Malthus."

[23] Ricardo is here taken as a utilitarian of the Benthamite color, although he cannot be classed as a disciple of Bentham. His hedonism is but the uncritically accepted metaphysics comprised in the common sense of his time, and his substantial coincidence with Bentham goes to show how well diffused the hedonist preconception was at the time.

[24] *Cf.* Bonar, *Malthus and his Work,* pp. 323–336.

stantial economic ground is pleasure and pain: the teleological order (even the design of God, where that is admitted) is the method of its working-out.

It may be unnecessary here to go into the farther implications, psychological and ethical, which this preconception of the utilitarians involves. And even this much
may seem a taking of excessive pains with a distinction
that marks no tangible difference. But a reading of the
classical doctrines, with something of this metaphysics of
political economy in mind, will show how, and in great
part why, the later economists of the classical line diverged from Adam Smith's tenets in the early years of
the century, until it has been necessary to interpret Adam
Smith somewhat shrewdly in order to save him from
heresy.

The post-Bentham economics is substantially a theory
of value. This is altogether the dominant feature of
the body of doctrines; the rest follows from, or is
adapted to, this central discipline. The doctrine of value
is of very great importance also in Adam Smith; but
Adam Smith's economics is a theory of the production
and apportionment of the material means of life.[25] With
Adam Smith, value is discussed from the point of view of
production. With the utilitarians, production is discussed
from the point of view of value. The former makes
value an outcome of the process of production: the latter
make production the outcome of a valuation process.

The point of departure with Adam Smith is the " productive power of labor." [26] With Ricardo it is a pecuni-

[25] His work is an inquiry into "the Nature and Causes of the
Wealth of Nations."

[26] "The annual labor of every nation is the fund which
originally supplies it with all the necessaries and conveniences of
life which it annually consumes, and which consist always either
in the immediate produce of that labor or in what is purchased

ary problem concerned in the distribution of ownership;[27] but the classical writers are followers of Adam Smith, and improve upon and correct the results arrived at by him, and the difference of point of view, therefore, becomes evident in their divergence from him, and the different distribution of emphasis, rather than in a new and antagonistic departure.

The reason for this shifting of the center of gravity from production to valuation lies, proximately, in Bentham's revision of the "principles" of morals. Bentham's philosophical position is, of course, not a self-explanatory phenomenon, nor does the effect of Benthamism extend only to those who are avowed followers of Bentham; for Bentham is the exponent of a cultural change that affects the habits of thought of the entire community. The immediate point of Bentham's work, as affecting the habits of thought of the educated community, is the substitution of hedonism (utility) in place of achievement of purpose, as a ground of legitimacy and a guide in the normalisation of knowledge. Its effect is most patent in speculations on morals, where it inculcates determinism. Its close connection with determinism in ethics points the way to what may be expected of its working in economics. In both cases the result is that human action is construed in terms of the causal forces of the environment, the human agent being, at the best, taken as a mechanism of commutation, through the workings of which the sensuous effects wrought by the impinging

with that produce from other nations." *Wealth of Nations,* "Introduction and Plan," opening paragraph.

[27] "The produce of the earth — all that is derived from its surface by the united application of labor, machinery, and capital — is divided among three classes of the community. . . . To determine the laws which regulate this distribution, is the principal problem of political economy." *Political Economy,* Preface.

forces of the environment are, by an enforced process of valuation, transmuted without quantitative discrepancy into moral or economic conduct, as the case may be. In ethics and economics alike the subject-matter of the theory is this valuation process that expresses itself in conduct, resulting, in the case of economic conduct, in the pursuit of the greatest gain or least sacrifice.

Metaphysically or cosmologically considered, the human nature into the motions of which hedonistic ethics and economics inquire is an intermediate term in a causal sequence, of which the initial and the terminal members are sensuous impressions and the details of conduct. This intermediate term conveys the sensuous impulse without loss of force to its eventuation in conduct. For the purpose of the valuation process through which the impulse is so conveyed, human nature may, therefore, be accepted as uniform; and the theory of the valuation process may be formulated quantitatively, in terms of the material forces affecting the human sensory and of their equivalents in the resulting activity. In the language of economics, the theory of value may be stated in terms of the consumable goods that afford the incentive to effort and the expenditure undergone in order to procure them. Between these two there subsists a necessary equality; but the magnitudes between which the equality subsists are hedonistic magnitudes, not magnitudes of kinetic energy nor of vital force, for the terms handled are sensuous terms. It is true, since human nature is substantially uniform, passive, and unalterable in respect of men's capacity for sensuous affection, there may also be presumed to subsist a substantial equality between the psychological effect to be wrought by the consumption of goods, on the one side, and the resulting expenditure of kinetic or vital force, on the other side; but such an equality is, after all, of the nature of a coincidence, although there should be a strong presumption in favor of its prevailing on an average and in the common run of cases. Hedonism, however, does not postulate uniformity between men except in the respect of sensuous cause and effect.

The theory of value which hedonism gives is, therefore, a theory of cost in terms of discomfort. By virtue of the hedonistic equilibrium reached through the valuation process, the sacrifice or expenditure of sensuous reality involved in acquisition is the equivalent of the sensuous gain secured. An alternative statement might perhaps be made, to the effect that the measure of the value of goods is not the sacrifice or discomfort undergone, but the sensuous gain that accrues from the acquisition of the goods; but this is plainly only an alternative statement, and there are special reasons in the economic life of the time why the statement in terms of cost, rather than in terms of "utility," should commend itself to the earlier classical economists.

On comparing the utilitarian doctrine of value with earlier theories, then, the case stands somewhat as follows. The Physiocrats and Adam Smith contemplate value as a measure of the productive force that realises itself in the valuable article. With the Physiocrats this productive force is the "anabolism" of Nature (to resort to a physiological term) : with Adam Smith it is chiefly human labor directed to heightening the serviceability of the materials with which it is occupied. Production causes value in either case. The post-Bentham economics contemplates value as a measure of, or as measured by, the irksomeness of the effort involved in procuring the valuable goods. As Mr. E. C. K. Gonner has admirably pointed out,[28] Ricardo — and the like holds true of classical economics generally — makes cost the foundation of value, not its cause. This resting of value on cost takes place through a valuation. Any one who will read Adam Smith's theoretical exposition to as good purpose as Mr. Gonner has read Ricardo will scarcely fail to find that the converse is true in Adam Smith's case. But the causal relation of cost to value holds only as regards "natural" or "real" value in Adam Smith's doctrine. As regards market price, Adam Smith's theory does not differ greatly from that of Ricardo on this head. He does not overlook the valuation process by which market price is adjusted and the course of investment is guided, and his discussion of this process runs in terms that should be acceptable to any hedonist.

The shifting of the point of view that comes into economics with the acceptance of utilitarian ethics and its correlate, the associationist psychology, is in great part a shifting to the ground of causal sequence as contrasted with that of serviceability to a preconceived end. This is indicated even by the main fact already cited,— that the utilitarian economists make exchange value the central feature of their theories, rather than the conduciveness of industry to the community's material welfare. Hedonistic exchange value is the outcome of a valuation process enforced by the apprehended pleasure-giving capacities of the items valued. And in the utilitarian theories of production, arrived at from the standpoint so given by exchange value, the conduciveness to welfare is not the objective point of the argument. This objective point is rather the bearing of productive enterprise upon

[28] In the introductory essay to his edition of Ricardo's *Political Economy*. See, *e.g.*, paragraphs 9 and 24.

the individual fortunes of the agents engaged, or upon the fortunes of the several distinguishable classes of beneficiaries comprised in the industrial community; for the great immediate bearing of exchange values upon the life of the collectivity is their bearing upon the distribution of wealth. Value is a category of distribution. The result is that, as is well shown by Mr. Cannan's discussion,[29] the theories of production offered by the classical economists have been sensibly scant, and have been carried out with a constant view to the doctrines on distribution. An incidental but telling demonstration of the same facts is given by Professor Bücher;[30] and in illustration may be cited Torrens's *Essay on the Production of Wealth,* which is to a good extent occupied with discussions of value and distribution. The classical theories of production have been theories of the production of "wealth"; and " wealth," in classical usage, consists of material things having exchange value. During the vogue of the classical economics the accepted characteristic by which " wealth " has been defined has been its amenability to ownership. Neither in Adam Smith nor in the Physiocrats is this amenability to ownership made so much of, nor is it in a similar degree accepted as a definite mark of the subject-matter of the science.

As their hedonistic preconception would require, then, it is to the pecuniary side of life that the classical economists give their most serious attention, and it is the pecuniary bearing of any given phenomenon or of any institution that commonly shapes the issue of the argument. The causal sequence about which the discussion

[29] *Theories of Production and Distribution,* 1776–1848.
[30] *Entstehung der Volkswirtschaft* (second edition). *Cf.* especially chaps. ii, iii, vi, and vii.

centers is a process of pecuniary valuation. It runs on distribution, ownership, acquisition, gain, investment, exchange.[31] In this way the doctrines on production come to take a pecuniary coloring; as is seen in a less degree also in Adam Smith, and even in the Physiocrats, although these earlier economists very rarely, if ever, lose touch with the concept of generic serviceability as the characteristic feature of production. The tradition derived from Adam Smith, which made productivity and serviceability the substantial features of economic life, was not abruptly put aside by his successors, though the emphasis was differently distributed by them in following out the line of investigation to which the tradition pointed the way. In the classical economics the ideas of production and of acquisition are not commonly held apart, and very much of what passes for a theory of production is occupied with phenomena of investment and acquisition. Torrens's *Essay* is a case in point, though by no means an extreme case.

This is as it should be; for to the consistent hedonist the sole motive force concerned in the industrial process is the self-regarding motive of pecuniary gain, and industrial activity is but an intermediate term between the expenditure or discomfort undergone and the pecuniary gain sought. Whether the end and outcome is an invidious gain for the individual (in contrast with or at the cost of his neighbors), or an enhancement of the facility of human life on the whole, is altogether a by-question in

[31] " Even if we put aside all questions which involve a consideration of the effects of industrial institutions in modifying the habits and character of the classes of the community, . . . that enough still remains to constitute a separate science, the mere enumeration of the chief terms of economics — wealth, value, exchange, credit, money, capital, and commodity — will suffice to show." Shirres, *Analysis of the Ideas of Economics* (London, 1893), pp. 8 and 9.

any discussion of the range of incentives by which men are prompted to their work or the direction which their efforts take. The serviceability of the given line of activity, for the life purposes of the community or for one's neighbors, "is not of the essence of this contract." These features of serviceability come into the account chiefly as affecting the vendibility of what the given individual has to offer in seeking gain through a bargain.[32]

In hedonistic theory the substantial end of economic life is individual gain; and for this purpose production and acquisition may be taken as fairly coincident, if not identical. Moreover, society, in the utilitarian philosophy, is the algebraic sum of the individuals; and the interest of the society is the sum of the interests of the individuals. It follows by easy consequence, whether strictly true or not, that the sum of individual gains is the gain of the society, and that, in serving his own interest in the way of acquisition, the individual serves the collective interest of the community. Productivity or serviceability is, therefore, to be presumed of any occupation or enterprise that looks to a pecuniary gain; and so, by a roundabout path, we get back to the ancient conclusion of Adam Smith, that the remuneration of classes or persons engaged in industry coincides with their productive contribution to the output of services and consumable goods.

A felicitous illustration of the working of this hedonistic norm in classical economic doctrine is afforded by the theory of the wages of superintendence,— an element in distribution which is not much more than suggested in

---

[32] "If a commodity were in no way useful, . . . it would be destitute of exchangeable value; . . . (but), possessing utility, commodities derive their exchangeable value from two sources," etc. Ricardo, *Political Economy*, chap. i, sect. i.

Adam Smith, but which receives ampler and more pains-taking attention as the classical body of doctrines reaches a fuller development. The "wages of superintendence" are the gains due to pecuniary management. They are the gains that come to the director of the "business,"— not those that go to the director of the mechanical process or to the foreman of the shop. The latter are wages simply. This distinction is not altogether clear in the earlier writers, but it is clearly enough· contained in the fuller development of the theory.

The undertaker's work is the management of investment. It is altogether of a pecuniary character, and its proximate aim is "the main chance." If it leads, indirectly, to an enhancement of serviceability or a heightened aggregate output of consumable goods, that is a fortuitous circumstance incident to that heightened vendibility on which the investor's gain depends. Yet the classical doctrine says frankly that the wages of superintendence are the remuneration of superior productivity,[33] and the classical theory of production is in good part a doctrine of investment in which the identity of production and pecuniary gain is taken for granted.

The substitution of investment in the place of industry as the central and substantial fact in the process of production is due not to the acceptance of hedonism simply, but rather to the conjunction of hedonism with an economic situation of which the investment of capital and

---

[33] *Cf.*, for instance, Senior, *Political Economy* (London, 1872), particularly pp. 88, 89, and 130–135, where the wages of superintendence are, somewhat reluctantly, classed under profits; and the work of superintendence is thereupon conceived as being, immediately or remotely, an exercise of "abstinence" and a productive work. The illustration of the bill-broker is particularly apt. The like view of the wages of superintendence is an article of theory with more than one of the later descendants of the classical line.

its management for gain was the most obvious feature.
The situation which shaped the common-sense apprehension of economic facts at the time was what has since been
called a capitalistic system, in which pecuniary enterprise
and the phenomena of the market were the dominant and
tone-giving facts. But this economic situation was also
the chief ground for the vogue of hedonism in economics;
so that hedonistic economics may be taken as an interpretation of human nature in terms of the market-place.
The market and the " business world," to which the business man in his pursuit of gain was required to adapt his
motives, had by this time grown so large that the course
of business events was beyond the control of any one
person; and at the same time those far-reaching organisations of invested wealth which have latterly come to
prevail and to coerce the market were not then in the
foreground. The course of market events took its passionless way without traceable relation or deference to
any man's convenience and without traceable guidance
towards an ulterior end. Man's part in this pecuniary
world was to respond with alacrity to the situation, and
so adapt his vendible effects to the shifting demand as to
realise something in the outcome. What he gained in
his traffic was gained without loss to those with whom he
dealt, for they paid no more than the goods were worth
to them. One man's gain need not be another's loss;
and, if it is not, then it is net gain to the community.

Among the striking remoter effects of the hedonistic
preconception, and its working out in terms of pecuniary
gain, is the classical failure to discriminate between capital
as investment and capital as industrial appliances. This
is, of course, closely related to the point already spoken
of. The appliances of industry further the production of
goods, therefore capital (invested wealth) is productive;

and the rate of its average remuneration marks the degree of its productiveness.[34] The most obvious fact limiting the pecuniary gain secured by means of invested wealth is the sum invested. Therefore, capital limits the productiveness of industry; and the chief and indispensable condition to an advance in material well-being is the accumulation of invested wealth. In discussing the conditions of industrial improvement, it is usual to assume that " the state of the arts remains unchanged," which is, for all purposes but that of a doctrine of profits per cent., an exclusion of the main fact. Investments may, further, be transferred from one enterprise to another. Therefore, and in that degree, the means of production are " mobile."

Under the hands of the great utilitarian writers, therefore, political economy is developed into a science of wealth, taking that term in the pecuniary sense, as things amenable to ownership. The course of things in economic life is treated as a sequence of pecuniary events, and economic theory becomes a theory of what should happen in that consummate situation where the permutation of pecuniary magnitudes takes place without disturbance and without retardation. In this consummate situation the pecuniary motive has its perfect work, and guides all the acts of economic man in a guileless, colorless, unswerving quest of the greatest gain at the least sacrifice. Of course, this perfect competitive system, with its untainted " economic man," is a feat of the scientific imagination, and is not intended as a competent expression of fact. It is an expedient of abstract reasoning;

[34] *Cf.* Böhm-Bawerk, *Capital and Interest,* Books II and IV, as well as the Introduction and chaps. iv and v of Book I. Böhm-Bawerk's discussion bears less immediately on the present point than the similarity of the terms employed would suggest.

and its avowed competency extends only to the abstract principles, the fundamental laws of the science, which hold only so far as the abstraction holds. But, as happens in such cases, having once been accepted and assimilated as real, though perhaps not as actual, it becomes an effective constituent in the inquirer's habits of thought, and goes to shape his knowledge of facts. It comes to serve as a norm of substantiality or legitimacy; and facts in some degree fall under its constraint, as is exemplified by many allegations regarding the " tendency " of things.

To this consummation, which Senior speaks of as " the natural state of man," [35] human development tends by force of the hedonistic character of human nature; and in terms of its approximation to this natural state, therefore, the immature actual situation had best be stated. The pure theory, the " hypothetical science " of Cairnes, " traces the phenomena of the production and distribution of wealth up to their causes, in the principles of human nature and the laws and events — physical, political, and social — of the external world." [36] But since the principles of human nature that give the outcome in men's economic conduct, so far as it touches the production and distribution of wealth, are but the simple and constant sequence of hedonistic cause and effect, the element of human nature may fairly be eliminated from the problem, with great gain in simplicity and expedition. Human nature being eliminated, as being a constant intermediate term, and all institutional features of the situation being also eliminated (as being similar constants under that natural or consummate pecuniary *régime* with which the

[35] *Political Economy*, p. 87.

[36] *Character and Logical Method of Political Economy* (New York, 1875), p. 71. Cairnes may not be altogether representative of the high tide of classicism, but his characterisation of the science is none the less to the point.

pure theory is concerned), the laws of the phenomena of wealth may be formulated in terms of the remaining factors. These factors are the vendible items that men handle in these processes of production and distribution; and economic laws come, therefore, to be expressions of the algebraic relations subsisting between the various elements of wealth and investment,— capital, labor, land, supply and demand of one and the other, profits, interest, wages. Even such items as credit and population become dissociated from the personal factor, and figure in the computation as elemental factors acting and reacting though a permutation of values over the heads of the good people whose welfare they are working out.

To sum up: the classical economics, having primarily to do with the pecuniary side of life, is a theory of a process of valuation. But since the human nature at whose hands and for whose behoof the valuation takes place is simple and constant in its reaction to pecuniary stimulus, and since no other feature of human nature is legitimately present in economic phenomena than this reaction to pecuniary stimulus, the valuer concerned in the matter is to be overlooked or eliminated; and the theory of the valuation process then becomes a theory of the pecuniary interaction of the facts valued. It is a theory of valuation with the element of valuation left out,— a theory of life stated in terms of the normal paraphernalia of life.

In the preconceptions with which classical economics set out were comprised the remnants of natural rights and of the order of nature, infused with that peculiarly mechanical natural theology that made its way into popular vogue on British ground during the eighteenth century and was reduced to a neutral tone by the British penchant for the commonplace — stronger at this time than at any earlier period. The reason for this growing penchant for the commonplace, for the explanation of things in causal terms, lies partly in the growing resort to mechanical processes and mechanical prime movers in industry, partly in the (consequent) continued decline of the aristocracy and the priesthood, and partly in the growing density of population and the consequent greater specialisation and wider organisation of trade and business. The spread of the discipline of the natural sciences, largely incident to the mechanical industry, counts in the same direction; and obscurer factors in modern culture may have had their share.

The animistic preconception was not lost, but it lost tone; and it partly fell into abeyance, particularly so far as regards its avowal. It is visible chiefly in the unavowed readiness of the classical writers to accept as imminent and definitive any possible outcome which the writer's habit or temperament inclined him to accept as right and good. Hence the visible inclination of classical economists to a doctrine of the harmony of interests, and their somewhat uncircumspect readiness to state their generalisations in terms of what ought to happen according to the ideal requirements of that consummate *Geldwirtschaft* to which men " are impelled by the provisions of nature." [37] By virtue of their hedonistic preconceptions, their habituation to the ways of a pecuniary culture, and their unavowed animistic faith that nature is in the right, the classical economists knew that the consummation to which, in the nature of things, all things tend, is the frictionless and beneficent competitive system. This competitive ideal, therefore, affords the normal, and conformity to its requirements affords the test of absolute

[37] Senior, *Political Economy*, p. 87.

economic truth. The standpoint so gained selectively guides the attention of the classical writers in their observation and apprehension of facts, and they come to see evidence of conformity or approach to the normal in the most unlikely places. Their observation is, in great part, interpretative, as observation commonly is. What is peculiar to the classical economists in this respect is their particular norm of procedure in the work of interpretation. And, by virtue of having achieved a standpoint of absolute economic normality, they became a " deductive " school, so called, in spite of the patent fact that they were pretty consistently employed with an inquiry into the causal sequence of economic phenomena.

The generalisation of observed facts becomes a normalisation of them, a statement of the phenomena in terms of their coincidence with, or divergence from, that normal tendency that makes for the actualisation of the absolute economic reality. This absolute or definitive ground of economic legitimacy lies beyond the causal sequence in which the observed phenomena are conceived to be interlinked. It is related to the concrete facts neither as cause nor as effect in any such way that the causal relation may be traced in a concrete instance. It has little causally to do either with the " mental " or with the " physical " data with which the classical economist is avowedly employed. Its relation to the process under discussion is that of an extraneous — that is to say, a ceremonial — legitimation. The body of knowledge gained by its help and under its guidance is, therefore, a taxonomic science.

So, by way of a concluding illustration, it may be pointed out that money, for instance, is normalised in terms of the legitimate economic tendency. It becomes a measure of value and a medium of exchange. It has become primarily an instrument of pecuniary commutation, instead of being, as under the earlier normalisation of Adam Smith, primarily a great wheel of circulation for the diffusion of consumable goods. The terms in which the laws of money, as of the other phenomena of pecuniary life, are formulated, are terms which connote its normal function in the life history of objective values as they live and move and have their being in the consummate pecuniary situation of the " natural " state. To a similar work of normalisation we owe those creatures of the myth-maker, the quantity theory and the wages-fund.

THE PRECONCEPTIONS OF ECONOMIC
SCIENCE [1]

III

In what has already been said, it has appeared that the changes which have supervened in the preconceptions of the earlier economists constitute a somewhat orderly succession. The feature of chief interest in this development has been a gradual change in the received grounds of finality to which the successive generations of economists have brought their theoretical output, on which they have been content to rest their conclusions, and beyond which they have not been moved to push their analysis of events or their scrutiny of phenomena. There has been a fairly unbroken sequence of development in what may be called the canons of economic reality; or, to put it in other words, there has been a precession of the point of view from which facts have been handled and valued for the purpose of economic science.

The notion which has in its time prevailed so widely, that there is in the sequence of events a consistent trend which it is the office of the science to ascertain and turn to account,— this notion may be well founded or not. But that there is something of such a consistent trend in the sequence of the canons of knowledge under whose guidance the scientist works is not only a generalisation from the past course of things, but lies in the nature of the case; for the canons of knowledge are of the nature

[1] Reprinted by permission from *The Quarterly Journal of Economics.* Vol. XIV, Feb., 1900.

of habits of thought, and habit does not break with the past, nor do the hereditary aptitudes that find expression in habit vary gratuitously with the mere lapse of time. What is true in this respect, for instance, in the domain of law and institutions is true, likewise, in the domain of science. What men have learned to accept as good and definitive for the guidance of conduct and of human relations remains true and definitive and unimpeachable until the exigencies of a later, altered situation enforce a variation from the norms and canons of the past, and so give rise to a modification of the habits of thought that decide what is, for the time, right in human conduct. So in science the ancient ground of finality remains a good and valid test of scientific truth until the altered exigencies of later life enforce habits of thought that are not wholly in consonance with the received notions as to what constitutes the ultimate, self-legitimating term — the substantial reality — to which knowledge in any given case must penetrate.

This ultimate term or ground of knowledge is always of a metaphysical character. It is something in the way of a preconception, accepted uncritically, but applied in criticism and demonstration of all else with which the science is concerned. So soon as it comes to be criticised, it is in a way to be superseded by a new, more or less altered formulation; for criticism of it means that it is no longer fit to survive unaltered in the altered complex of habits of thought to which it is called upon to serve as fundamental principle. It is subject to natural selection and selective adaptation, as are other conventions. The underlying metaphysics of scientific research and purpose, therefore, changes gradually and, of course, incompletely, much as is the case with the metaphysics underlying the common law and the schedule of civil rights. As in the

legal framework the now avowedly useless and meaningless preconceptions of status and caste and precedent are
even yet at the most metamorphosed and obsolescent
rather than overpassed,— witness the facts of inheritance,
vested interests, the outlawry of debts through lapse of
time, the competence of the State to coerce individuals
into support of a given policy,— so in the science the living generation has not seen an abrupt and traceless disappearance of the metaphysics that fixed the point of
view of the early classical political economy. This is true
even for those groups of economists who have most incontinently protested against the absurdity of the classical
doctrines and methods. In Professor Marshall's words,
" There has been no real breach of continuity in the development of the science."

But, while there has been no breach, there has none the
less been change,— more far-reaching change than some
of us are glad to recognise ; for who would not be glad to
read his own modern views into the convincing words of
the great masters?

Seen through modern eyes and without effort to turn
past gains to modern account, the metaphysical or preconceptional furniture of political economy as it stood about
the middle of this century may come to look quite curious.
The two main canons of truth on which the science proceeded, and with which the inquiry is here concerned,
were : (*a*) a hedonistic-associational psychology, and (*b*)
an uncritical conviction that there is a meliorative trend
in the course of events, apart from the conscious ends of
the individual members of the community. This axiom
of a meliorative developmental trend fell into shape as a
belief in an organic or quasi-organic (physiological) [2] life

---

[2] So, *e.g.*, Roscher, Comte, the early socialists, J. S. Mill, and
later Spencer, Schaeffle, Wagner.

process on the part of the economic community or of the nation; and this belief carried with it something of a constraining sense of self-realising cycles of growth, maturity and decay in the life history of nations or communities.

Neglecting what may for the immediate purpose be negligible in this outline of fundamental tenets, it will bear the following construction. (a) On the ground of the hedonistic or associational psychology, all spiritual continuity and any consequent teleological trend is tacitly denied so far as regards individual conduct, where the later psychology, and the sciences which build on this later psychology, insist upon and find such a teleological trend at every turn. (b) Such a spiritual or quasi-spiritual continuity and teleological trend is uncritically affirmed as regards the non-human sequence or the sequence of events in the affairs of collective life, where the modern sciences diligently assert that nothing of the kind is discernible, or that, if it is discernible, its recognition is beside the point, so far as concerns the purposes of the science.

This position, here outlined with as little qualification as may be admissible, embodies the general metaphysical ground of that classical political economy that affords the point of departure for Mill and Cairnes, and also for Jevons. And what is to be said of Mill and Cairnes in this connection will apply to the later course of the science, though with a gradually lessening force.

By the middle of the century the psychological premises of the science are no longer so neat and succinct as they were in the days of Bentham and James Mill. At J. S. Mill's hands, for instance, the naïvely quantitative hedonism of Bentham is being supplanted by a sophisticated hedonism, which makes much of an assumed qualitative divergence between the different kinds of pleasures that afford the motives of conduct. This revision of hedonistic dogma, of course, means a departure from the strict hedonistic ground. Correlated with this advance more closely in the substance of the change than in the assignable dates, is a concomitant improvement — at least, set forth as an improvement — upon the received associational psychology, whereby " similarity " is brought in to supplement " contiguity " as a ground of connection between ideas. This change is well shown in the work of J. S. Mill and Bain. In spite of all the ingenuity spent in maintaining the associational legitimacy of this new article of theory, it remains a patent innovation and a departure from the ancient standpoint. As is true of the improved hedonism, so it is true of the new theory of association that it is no longer able to construe the process which it discusses as a purely mechanical process, a concatenation of items simply. Similarity of impressions implies a comparison of impressions by the mind in which the association takes place, and thereby it implies some degree of constructive work on the part of the perceiving subject. The perceiver is thereby construed to be an agent in the work of perception; therefore, he must be possessed of a point of view and an end dominating the perceptive process. To perceive the similarity, he must be guided by an interest in the outcome, and must " attend." The like applies to the introduction of qualitative distinctions into the hedonistic theory of conduct. Apperception in the one case and discretion in the other cease to be the mere registration of a simple and personally uncolored sequence of permutations enforced by the factors of the external world. There is implied a spiritual — that is to say, active —" teleological " continuity of process on the part of the perceiving or of the discretionary agent, as the case may be.

It is on the ground of their departure from the stricter hedonistic premises that Mill and, after him, Cairnes are able, for instance, to offer their improvement upon the earlier doctrine of cost of production as determining value. Since it is conceived that the motives which guide men in their choice of employments and of domicile differ from man to man and from class to class, not only in degree, but in kind, and since varying antecedents, of heredity and of habit, variously influence men in their choice of a manner of life, therefore the mere quantitative pecuniary stimulus cannot be depended on to decide the outcome without recourse. There are determinable variations in the alacrity with which different classes or communities respond to the pecuniary stimulus; and in so far as this condition prevails, the classes or communities in question are non-competing. Between such non-competing groups the norm that determines values is not the unmitigated norm of cost of production taken absolutely, but only taken relatively. The formula of cost of production is therefore modified into a formula of reciprocal demand. This revision of the cost-of-production doctrine is extended only sparingly, and the emphasis is thrown on the pecuniary circumstances on which depend the formation and maintenance of non-competing groups. Consistency with the earlier teaching is carefully maintained, so far as may be; but extra-pecuniary factors are, after all, even if reluctantly, admitted into the body of the theory. So also, since there are higher and lower motives, higher and lower pleasures,— as well as motives differing in degree,— it follows that an unguided response even to the mere quantitative pecuniary stimuli may take different directions, and so may result in activities of widely differing outcome.   Since activities set up in this way through appeal to higher and lower motives are no longer conceived to represent simply a mechanically adequate effect of the stimuli, working under the control of natural laws that tend to one beneficent consummation, therefore the outcome of activity set up even by the normal pecuniary stimuli may take a form that may or may not be serviceable to the community.   Hence *laissez-faire* ceases to be a sure remedy for the ills of society.   Human interests are still conceived normally to be at one; but the detail of individual conduct need not, therefore, necessarily serve these generic human interests.[3]   Therefore, other inducements than the unmitigated impact of pecuniary exigencies may be necessary to bring about a coincidence of class or individual endeavor with the interests of the community.   It becomes incumbent on the advocate of *laissez-faire* to " prove his minor premise."   It is no longer self-evident that: " Interests left to themselves tend to harmonious combinations, and to the progressive preponderance of the general good." [4]

The natural-rights preconception begins to fall away as soon as the hedonistic mechanics have been seriously tampered with.   Fact and right cease to coincide, because the individual in whom the rights are conceived to inhere has

[3] " Let us not confound the statement that *human* interests are at one with the statement that *class* interests are at one.   The latter I believe to be as false as the former is true. . . . But accepting the major premises of the syllogism, that the interests of human beings are fundamentally the same, how as to the minor? — how as to the assumption that people know their interests in the sense in which they are identical with the interests of others, and that they spontaneously follow them *in this sense?* "— Cairnes, Essays in Political Economy (London, 1873), p. 245.   This question cannot consistently be asked by an adherent of the stricter hedonism.

[4] Bastiat, quoted by Cairnes, *Essays,* p. 319.

come to be something more than the field of intersection of natural forces that work out in human conduct. The mechanics of natural liberty — that assumed constitution of things by force of which the free hedonistic play of the laws of nature across the open field of individual choice is sure to reach the right outcome — is the hedonistic psychology; and the passing of the doctrine of natural rights and natural liberty, whether as a premise or as a dogma, therefore coincides with the passing of that mechanics of conduct on the validity of which the theoretical acceptance of the dogma depends. It is, therefore, something more than a coincidence that the half-century which has seen the disintegration of the hedonistic faith and of the associational psychology has also seen the dissipation, in scientific speculations, of the concomitant faith in natural rights and in that benign order of nature of which the natural-rights dogma is a corollary.

It is, of course, not hereby intended to say that the later psychological views and premises imply a less close dependence of conduct on environment than do the earlier ones. Indeed, the reverse may well be held to be true. The pervading characteristic of later thinking is the constant recourse to a detailed analysis of phenomena in causal terms. The modern catchword, in the present connection, is " response to stimulus "; but the manner in which this response is conceived has changed. The fact, and ultimately the amplitude, at least in great part, of the reaction to stimulus, is conditioned by the forces in impact; but the constitution of the organism, as well as its attitude at the moment of impact, in great part decides what will serve as a stimulus, as well as what the manner and direction of the response will be.

The later psychology is biological, as contrasted with the metaphysical psychology of hedonism. It does not conceive the organism as a causal hiatus. The causal sequence in the "reflex arc" is, no doubt, continuous; but the continuity is not, as formerly, conceived in terms of spiritual substance transmitting a shock: it is conceived in terms of the life activity of the organism. Human conduct, taken as the reaction of such an organism under stimulus, may be stated in terms of tropism, involving, of course, a very close-knit causal sequence between the impact and the response, but at the same time imputing to the organism a habit of life and a self-directing and selective attention in meeting the complex of forces that make up its environment. The selective play of this tropismatic complex that constitutes the organism's habit of life under the impact of the forces of the environment counts as discretion.

So far, therefore, as it is to be placed in contrast with the hedonistic phase of the older psychological doctrines, the characteristic feature of the newer conception is the recognition of a selectively self-directing life process in the agent. While hedonism seeks the causal determinant of conduct in the (probable) outcome of action, the later conception seeks this determinant in the complex of propensities that constitutes man a functioning agent, that is to say, a personality. Instead of pleasure ultimately determining what human conduct shall be, the tropismatic propensities that eventuate in conduct ultimately determine what shall be pleasurable. For the purpose in hand, the consequence of the transition to the altered conception of human nature and its relation to the environment is that the newer view formulates conduct in terms of personality, whereas the earlier view was content to formulate it in terms of its provocation and its by-product. Therefore, for the sake of brevity, the older preconceptions of the science are here spoken of as construing human nature in inert terms, as contrasted with the newer, which construes it in terms of functioning.

It has already appeared above that the second great article of the metaphysics of classical political economy — the belief in a meliorative trend or a benign order of nature — is closely connected with the hedonistic conception of human nature; but this connection is more intimate and organic than appears from what has been said above. The two are so related as to stand or fall together, for the latter is but the obverse of the former. The doctrine of a trend in events imputes purpose to the sequence of events; that is, it invests this sequence with a discretionary, teleological character, which asserts itself in a constraint over all the steps in the sequence by which the supposed objective point is reached. But discretion touching a given end must be single, and must alone cover all the acts by which the end is to be reached. Therefore, no discretion resides in the intermediate terms through which the end is worked out. Therefore, man being such an intermediate term, discretion cannot be imputed to him without violating the supposition. Therefore, given an indefeasible meliorative trend in events, man is but a mechanical intermediary in the sequence. It is as such a mechanical intermediate term that the stricter hedonism construes human nature.[5] Accordingly, when more of teleological activity came to be imputed to man, less was thereby allowed to the complex of events. Or it may be put in the converse form: When less of a teleological continuity came to be imputed to the course of events, more was thereby imputed to man's life process. The latter form of state-

[5] It may be remarked, by the way, that the use of the differential calculus and similar mathematical expedients in the discussion of marginal utility and the like, proceeds on this psychological ground, and that the theoretical results so arrived at are valid to the full extent only if this hedonistic psychology is accepted.

ment probably suggests the direction in which the causal relation runs, more nearly than the former. The change whereby the two metaphysical premises in question have lost their earlier force and symmetry, therefore, amounts to a (partial) shifting of the seat of putative personality from inanimate phenomena to man.

It may be mentioned in passing, as a detail lying perhaps afield, yet not devoid of significance for latter-day economic speculation, that this elimination of personality, and so of teleological content, from the sequence of events, and its increasing imputation to the conduct of the human agent, is incident to a growing resort to an apprehension of phenomena in terms of process rather than in terms of outcome, as was the habit in earlier schemes of knowledge. On this account the categories employed are, in a gradually increasing degree, categories of process,—" dynamic " categories. But categories of process applied to conduct, to discretionary action, are teleological categories: whereas categories of process applied in the case of a sequence where the members of the sequence are not conceived to be charged with discretion, are, by the force of this conception itself, non-teleological, quantitative categories. The continuity comprised in the concept of process as applied to conduct is consequently a spiritual, teleological continuity: whereas the concept of process under the second head, the non-teleological sequence, comprises a continuity of a quantitative, causal kind, substantially the conservation of energy. In its turn the growing resort to categories of process in the formulation of knowledge is probably due to the epistemological discipline of modern mechanical industry, the technological exigencies of which enforce a constant recourse to the apprehension of phenomena in terms of process, differing therein from the earlier forms of industry, which neither obtruded visible mechanical process so constantly upon the apprehension nor so imperatively demanded an articulate recognition of continuity in the processes actually involved. The contrast in this respect is still more pronounced between the discipline of modern life in an industrial community and the discipline of life under the conventions of status and exploit that formerly prevailed.

To return to the benign order of nature, or the meliorative trend,— its passing, as an article of economic faith, was not due to criticism leveled against it by the later classical economists on grounds of its epistemological incongruity. It was tried on its merits, as an alleged account of facts; and the weight of evidence went against it. The belief in a self-realising trend had no sooner reached a competent and exhaustive statement — *e.g.*, at Bastiat's hands, as a dogma of the harmony of interests specifically applicable to the details of economic life — than it began to lose ground. With his usual concision and incisiveness, Cairnes completed the destruction of Bastiat's special dogma, and put it forever beyond a rehearing. But Cairnes is not a destructive critic of the classical political economy, at least not in intention: he is an interpreter and continuer — perhaps altogether the clearest and truest continuer — of the classical teaching. While he confuted Bastiat and discredited Bastiat's peculiar dogma, he did not thereby put the order of nature bodily out of the science. He qualified and improved it, very much as Mill qualified.and improved the tenets of the hedonistic psychology. As Mill and the ethical speculation of his generation threw more of personality into the hedonistic psychology, so Cairnes and the speculators on scientific method (such as Mill and Jevons) attenuated the imputation of personality or teleological content to the process of material cause and effect. The work is of course, by no means, an achievement of Cairnes alone; but he is, perhaps, the best exponent of this advance in economic theory. In Cairnes's redaction this foundation of the science became the concept of a colorless normality.

It was in Cairnes's time the fashion for speculators in other fields than the physical sciences to look to those sciences for guidance in method and for legitimation of the ideals of scientific theory which they were at work to realize. More than that, the large and fruitful achievements of the physical sciences had so far taken men's attention captive as to give an almost instinctive predilection for the methods that had approved themselves in that field. The ways of thinking which had on this ground become familiar to all scholars occupied with any scientific inquiry, had permeated their thinking on any subject whatever. This is eminently true of British thinking.

It had come to be a commonplace of the physical sciences that " natural laws " are of the nature of empirical generalisations simply, or even of the nature of arithmetical averages. Even the underlying preconception of the modern physical sciences — the law of the conservation of energy, or persistence of quantity — was claimed to be an empirical generalisation, arrived at inductively and verified by experiment. It is true the alleged proof of the law took the whole conclusion for granted at the start, and used it constantly as a tacit axiom at every step in the argument which was to establish its truth; but that fact serves rather to emphasise than to call in question the abiding faith which these empiricists had in the sole efficacy of empirical generalisation. Had they been able overtly to admit any other than an associational origin of knowledge, they would have seen the impossibility of accounting on the mechanical grounds of association for the premise on which all experience of mechanical fact rests.

That any other than a mechanical origin should be assigned to experience, or that any other than a so-conceived empirical ground was to be admitted for any general principle, was incompatible with the prejudices of men trained in the school of the associational psychology, however widely they perforce departed from this ideal in practice. Nothing of the nature of a personal element was to be admitted into these fundamental empirical generalisations; and nothing, therefore, of the nature of a discretionary or teleological movement was to be comprised in the generalisations to be accepted as " natural laws." Natural laws must in no degree be imbued with personality, must say nothing of an ulterior end; but for all that they remained " laws " of the sequences subsumed under them. So far is the reduction to colorless terms carried by Mill, for instance, that he formulates the natural laws as empirically ascertained sequences simply, even excluding or avoiding all imputation of causal continuity, as that term is commonly understood by the unsophisticated. In Mill's ideal no more of organic connection or continuity between the members of a sequence is implied in subsuming them under a law of causal relationship than is given by the ampersand. He is busied with dynamic sequences, but he persistently confines himself to static terms.

Under the guidance of the associational psychology, therefore, the extreme of discontinuity in the deliverances of inductive research is aimed at by those economists — Mill and Cairnes being taken as typical — whose names have been associated with deductive methods in modern science. With a fine sense of truth they saw that the notion of causal continuity, as a premise of scientific generalisation, is an essentially metaphysical postulate; and they avoided its treacherous ground by denying it, and construing causal sequence to mean a uniformity of co-existences and successions simply. But, since a strict uniformity is nowhere to be observed at first hand in the phenomena with which the investigator is occupied, it has to be found by a laborious interpretation of the phenomena and a diligent abstraction and allowance for disturbing circumstances, whatever may be the meaning of a disturbing circumstance where causal continuity is denied. In this work of interpretation and expurgation the investigator proceeds on a conviction of the orderliness of the natural sequence. " Natura non facit saltum ": a maxim which has no meaning within the stricter limits of the associational theory of knowledge.

Before anything can be said as to the orderliness of the sequence, a point of view must be chosen by the speculator, with respect to which the sequence in question does or does not fulfill this condition of orderliness; that is to say, with respect to which it is a sequence. The endeavor to avoid all metaphysical premises fails here as everywhere. The associationists, to whom economics owes its transition from the older classical phase to the modern or quasi-classical, chose as their guiding point of view the metaphysical postulate of congruity,— in substance, the " similarity " of the associationist theory of knowledge. This must be called their *proton pseudos,* if associationism pure and simple is to be accepted. The notion of congruity works out in laws of resemblance and equivalence, in both of which it is plain to the modern psychologist that a metaphysical ground of truth, antecedent to and controlling empirical data, is assumed. But the use of the postulate of congruence as a test of scientific truth has the merit of avoiding all open dealing with an imputed substantiality of the data handled, such as would be involved in the overt use of the concept of causation. The data are congruous among themselves, as items of knowledge; and they may therefore be handled in a logical synthesis and concatenation on the basis of this congruence alone, without committing the scientist to an imputation of a kinetic or motor relation between them. The metaphysics of process is thereby avoided, in appearance. The sequences are uniform or consistent with one another, taken as articles of theoretical synthesis simply; and so they become elements of a system or discipline of knowledge in which the test of theoretical truth is the congruence of the system with its premises.

In all this there is a high-wrought appearance of matter-of-fact, and all metaphysical subreption of a non-empirical or non-mechanical standard of reality or substantiality is avoided in appearance. The generalisations which make up such a system of knowledge are, in this way, stated in terms of the system itself; and when a competent formulation of the alleged uniformities has been so made in terms of their congruity or equivalence with the prime postulates of the system, the work of theoretical inquiry is done.

The concrete premises from which proceeds the systematic knowledge of this generation of economists are certain very concise assumptions concerning human nature, and certain slightly less concise generalisations of physical fact,[6] presumed to be mechanically empirical generalisations. These postulates afford the standard of normality. Whatever situation or course of events can be shown to express these postulates without mitigation is normal; and wherever a departure from this normal course of things occurs, it is due to disturbing causes,—that is to say, to causes not comprised in the main prem-

[6] See, *e.g.*, Cairnes, *Character and Logical Method* (New York), p. 71.

ises of the science,— and such departures are to be taken
account of by way of qualification. Such departures and
such qualification are constantly present in the facts to be
handled by the science; but, being not congruous with the
underlying postulates, they have no place in the body of
the science. The laws of the science, that which makes
up the economist's theoretical knowledge, are laws of the
normal case. The normal case does not occur in concrete
fact. These laws are, therefore, in Cairnes's terminology,
" hypothetical " truths; and the science is a " hypothetical " science. They apply to concrete facts only as the
facts are interpreted and abstracted from, in the light of
the underlying postulates. The science is, therefore, a
theory of the normal case, a discussion of the concrete
facts of life in respect of their degree of approximation
to the normal case. That is to say, it is a taxonomic
science.

Of course, in the work actually done by these economists this standpoint of rigorous normality is not consistently maintained; nor is the unsophisticated imputation of causality to the facts under discussion consistently
avoided. The associationist postulate, that causal sequence means empirical uniformity simply, is in great
measure forgotten when the subject-matter of the science
is handled in detail. Especially is it true that in Mill the
dry light of normality is greatly relieved by a strong common sense. But the great truths or laws of the science
remain hypothetical laws; and the test of scientific reality
is congruence with the hypothetical laws, not coincidence
with matter-of-fact events.

The earlier, more archaic metaphysics of the science,
which saw in the orderly correlation and sequence of
events a constraining guidance of an extra-causal, teleological kind, in this way becomes a metaphysics of nor-

mality which asserts no extra-causal constraint over events, but contents itself with establishing correlations, equivalencies, homologies, and theories concerning the conditions of an economic equilibrium. The movement, the process of economic life, is not overlooked, and it may even be said that it is not neglected; but the pure theory, in its final deliverances, deals not with the dynamics, but with the statics of the case. The concrete subject-matter of the science is, of course, the process of economic life,— that is unavoidably the case,— and in so far the discussion must be accepted as work bearing on the dynamics of the phenomena discussed; but even then it remains true that the aim of this work in dynamics is a determination and taxis of the outcome of the process under discussion rather than a theory of the process as such. The process is rated in terms of the equilibrium to which it tends or should tend, not conversely. The outcome of the process, taken in its relation of equivalence within the system, is the point at which the inquiry comes to rest. It is not primarily the point of departure for an inquiry into what may follow. The science treats of a balanced system rather than of a proliferation. In this lies its characteristic difference from the later evolutionary sciences. It is this characteristic bent of the science that leads its spokesman, Cairnes, to turn so kindly to chemistry rather than to the organic sciences, when he seeks an analogy to economics among the physical sciences.[7] What Cairnes has in mind in his appeal to chemistry is, of course, the received, extremely taxonomic (systematic) chemistry of his own time, not the tentatively genetic theories of a slightly later day.

It may seem that in the characterisation just offered of

[7] *Character and Logical Method*, p. 62.

the standpoint of normality in economics there is too
strong an implication of colorlessness and impartiality.
The objection holds as regards much of the work of the
modern economists of the classical line. It will hold true
even as to much of Cairnes's work; but it cannot be admitted as regards Cairnes's ideal of scientific aim and
methods. The economists whose theories Cairnes received and developed, assuredly did not pursue the discussion of the normal case with an utterly dispassionate
animus. They had still enough of the older teleological
metaphysics left to give color to the accusation brought
against them that they were advocates of *laissez-faire.*
The preconception of the utilitarians,— in substance the
natural-rights preconception,— that unrestrained human
conduct will result in the greatest human happiness, re-
tains so much of its force in Cairnes's time as is implied
in the then current assumption that what is normal is also
right. The economists, and Cairnes among them, not
only are concerned to find out what is normal and to determine what consummation answers to the normal, but they
also are at pains to approve that consummation. It is
this somewhat uncritical and often unavowed identification of the normal with the right that gives colorable
ground for the widespread vulgar prejudice, to which
Cairnes draws attention,[8] that political economy " sanctions " one social arrangement and " condemns " another.
And it is against this uncritical identification of two essentially unrelated principles or categories that Cairnes's
essay on " Political Economy and Laissez-faire," and in
good part also that on Bastiat, are directed. But, while
this is one of the many points at which Cairnes has substantially advanced the ideals of the science, his own concluding argument shows him to have been but half-way

[8] *Essays in Political Economy,* pp. 260–264.

emancipated from the prejudice, even while most effectively combating it.[9] It is needless to point out that the like prejudice is still present in good vigor in many later economists who have had the full benefit of Cairnes's teachings on this head.[10] Considerable as Cairnes's achievement in this matter undoubtedly was, it effected a mitigation rather than an elimination of the untenable metaphysics against which he contended.

The advance in the general point of view from animistic teleology to taxonomy is shown in a curiously succinct manner in a parenthetical clause of Cairnes's in the chapter on Normal Value.[11] With his acceptance of the later point of view involved in the use of the new term, Cairnes becomes the interpreter of the received theoretical results. The received positions are not subjected to a destructive criticism. The aim is to complete them where they fall short and to cut off what may be needless or what may run beyond the safe ground of scientific generalisation. In his work of redaction, Cairnes does not avow — probably he is not sensible of — any substantial shifting of the point of view or any change in the accepted ground of theoretic reality. But his advance to an unteleological taxonomy none the less changes the scope and aim of his theoretical discussion. The discussion of Normal Value may be taken in illustration.

Cairnes is not content to find (with Adam Smith) that value will " naturally " coincide with or be measured by

[9] See especially *Essays,* pp. 263, 264.

[10] It may be interesting to point out that the like identification of the categories of normality and right gives the dominant note of Mr. Spencer's ethical and social philosophy, and that later economists of the classical line are prone to be Spencerians.

[11] " Normal value (called by Adam Smith and Ricardo 'natural value,' and by Mill 'necessary value,' but best expressed, it seems to me, by the term which I have used)." *Leading Principles* (New York), p. 45.

cost of production, or even (with Mill) that cost of production must, in the long run, "necessarily" determine value. "This . . . is to take a much too limited view of the range of this phenomenon." [12] He is concerned to determine not only this general tendency of values to a normal, but all those characteristic circumstances as well which condition this tendency and which determine the normal to which values tend. His inquiry pursues the phenomena of value in a normal economic system rather than the manner and rate of approach of value relations to a teleologically or hedonistically defensible consummation. It therefore becomes an exhaustive but very discriminating analysis of the circumstances that bear upon market values, with a view to determine what circumstances are normally present; that is to say, what circumstances conditioning value are commonly effective and at the same time in consonance with the premises of economic theory. These effective conditions, in so far as they are not counted anomalous and, therefore, to be set aside in the theoretical discussion, are the circumstances under which a hedonistic valuation process in any modern industrial community is held perforce to take place,— the circumstances which are held to enforce a recognition and rating of the pleasure-bearing capacity of facts. They are not, as under the earlier cost-of-production doctrines, the circumstances which determine the magnitude of the forces spent in the production of the valuable article. Therefore, the normal (natural) value is no longer (as with Adam Smith, and even to some extent with his classical successors) the primary or initial fact in value theory, the substantial fact of which the market value is an approximate expression and by which the latter is controlled. The argument does not, as formerly, set out

[12] *Leading Principles,* p. 45.

from that expenditure of personal force which was once
conceived to constitute the substantial value of goods, and
then construe market value to be an approximate and
uncertain expression of this substantial fact. The direction in which the argument runs is rather the reverse of
this. The point of departure is taken from the range of
market values and the process of bargaining by which
these values are determined. This latter is taken to be
a process of discrimination between various kinds and
degrees of discomfort, and the average or consistent outcome of such a process of bargaining constitutes normal
value. It is only by virtue of a presumed equivalence
between the discomfort undergone and the concomitant
expenditure, whether of labor or of wealth, that the normal value so determined is conceived to be an expression
of the productive force that goes into the creation of the
valuable goods. Cost being only in uncertain equivalence
with sacrifice or discomfort, as between different persons,
the factor of cost falls into the background; and the
process of bargaining, which is in the foreground, being
a process of valuation, a balancing of individual demand
and supply, it follows that a law of reciprocal demand
comes in to supplant the law of cost. In all this the
proximate causes at work in the determination of values
are plainly taken account of more adequately than in
earlier cost-of-production doctrines; but they are taken
account of with a view to explaining the mutual adjustment and interrelation of elements in a system rather than
to explain either a developmental sequence or the working
out of a foreordained end.

This revision of the cost-of-production doctrine,
whereby it takes the form of a law of reciprocal demand,
is in good part effected by a consistent reduction of cost
to terms of sacrifice,— a reduction more consistently car-

ried through by Cairnes than it had been by earlier hedonists, and extended by Cairnes's successors with even more
far-reaching results. By this step the doctrine of cost is
not only brought into closer accord with the neo-hedon-
istic premises, in that it in a greater degree throws the
stress upon the factor of personal discrimination, but it
also gives the doctrine a more general bearing upon
economic conduct and increases its serviceability as a comprehensive principle for the classification of economic
phenomena. In the further elaboration of the hedonistic
theory of value at the hands of Jevons and the Austrians
the same principle of sacrifice comes to serve as the chief
ground of procedure.

Of the foundations of later theory, in so far as the postulates of later economists differ characteristically from
those of Mill and Cairnes, little can be said in this place.
Nothing but the very general features of the later development can be taken up; and even these general features of
the existing theoretic situation can not be handled with
the same confidence as the corresponding features of a
past phase of speculation. With respect to writers of the
present or the more recent past the work of natural selection, as between variants of scientific aim and animus and
between more or less divergent points of view, has not yet
taken effect; and it would be over-hazardous to attempt
an anticipation of the results of the selection that lies in
great part yet in the future. As regards the directions of
theoretical work suggested by the names of Professor
Marshall, Mr. Cannan, Professor Clark, Mr. Pierson,
Professor Loria, Professor Schmoller, the Austrian
group,— no off-hand decision is admissible as between
these candidates for the honor, or, better, for the work,
of continuing the main current of economic speculation

and inquiry. No attempt will here be made even to pass a verdict on the relative claims of the recognised two or three main "schools" of theory, beyond the somewhat obvious finding that, for the purpose in hand, the so-called Austrian school is scarcely distinguishable from the neo-classical, unless it be in the different distribution of emphasis. The divergence between the modernised classical views, on the one hand, and the historical and Marxist schools, on the other hand, is wider,— so much so, indeed, as to bar out a consideration of the postulates of the latter under the same head of inquiry with the former. The inquiry, therefore, confines itself to the one line standing most obviously in unbroken continuity with that body of classical economics whose life history has been traced in outline above. And, even for this phase of modernised classical economics, it seems necessary to limit discussion, for the present, to a single strain, selected as standing peculiarly close to the classical source, at the same time that it shows unmistakable adaptation to the later habits of thought and methods of knowledge.

For this later development in the classical line of political economy, Mr. Keynes's book may fairly be taken as the maturest exposition of the aims and ideals of the science; while Professor Marshall excellently exemplifies the best work that is being done under the guidance of the classical antecedents. As, after a lapse of a dozen or fifteen years from Cairnes's days of full conviction, Mr. Keynes interprets the aims of modern economic science, it has less of the "hypothetical" character assigned it by Cairnes; that is to say, it confines its inquiry less closely to the ascertainment of the normal case and the interpretative subsumption of facts under the normal. It takes fuller account of the genesis and developmental continuity of all features of modern economic life, gives more and closer attention to institutions and their history. This is, no doubt, due, in part at least, to impulse received from German economists; and in so far it also reflects the peculiarly vague and bewildered attitude of protest that characterises the earlier expositions of the historical school. To the same essentially extraneous source is traceable the theoretic blur embodied in Mr. Keynes's attitude of tolerance towards the conception of economics as a "normative" science having to do with "economic ideals," or an "applied economics" having to do with "economic precepts." [13]   An inchoate departure from the consistent taxonomic ideals shows itself in the tentative resort to historical and genetic formulations, as well as in Mr. Keynes's pervading inclination to define the scope of the science, not by exclusion of what are conceived to be non-economic phenomena, but by disclosing a point of view from which all phenomena are seen to be economic facts.   The science comes to be characterised not by the delimitation of a range of facts, as in Cairnes,[14] but as an inquiry into the bearing which all facts have upon men's economic activity.   It is no longer that certain phenomena belong within the science, but rather that the science is concerned with any and all phenomena as seen from the point of view of the economic interest.   Mr. Keynes does not go fully to the length which this last proposition indicates.   He finds [15] that political economy "treats of the phenomena arising out of the economic activities of mankind in society"; but, while the discussion by which he

[13] *Scope and Method of Political Economy* (London, 1891), chaps. i and ii.

[14] *Character and Logical Method; e.g.,* Lecture II, especially pp. 53, 54, and 71.

[15] *Scope and Method of Political Economy,* chap. iii, particularly p. 97.

leads up to this definition might be construed to say that all the activities of mankind in society have an economic bearing, and should therefore come within the view of the science, Mr. Keynes does not carry out his elucidation of the matter to that broad conclusion. Neither can it be said that modern political economy has, in practice, taken on the scope and character which this extreme position would assign it.

The passage from which the above citation is taken is highly significant also in another and related bearing, and it is at the same time highly characteristic of the most effective modernised classical economics. The subject-matter of the science has come to be the " economic activities " of mankind, and the phenomena in which these activities manifest themselves. So Professor Marshall's work, for instance, is, in aim, even if not always in achievement, a theoretical handling of human activity in its economic bearing,— an inquiry into the multiform phases and ramifications of that process of valuation of the material means of life by virtue of which man is an economic agent. And still it remains an inquiry directed to the determination of the conditions of an equilibrium of activities and a quiescent normal situation. It is not in any eminent degree an inquiry into cultural or institutional development as affected by economic exigencies or by the economic interest of the men whose activities are analysed and portrayed. Any sympathetic reader of Professor Marshall's great work — and that must mean every reader — comes away with a sense of swift and smooth movement and interaction of parts; but it is the movement of a consummately conceived and self-balanced mechanism, not that of a cumulatively unfolding process or an institutional adaptation to cumulatively unfolding exigencies. The taxonomic bearing is, after all, the dominant feature. It is significant of the same point that even in his discussion of such vitally dynamic features of the economic process as the differential effectiveness of different laborers or of different industrial plants, as well as of the differential advantages of consumers, Professor Marshall resorts to an adaptation of so essentially taxonomic a category as the received concept of rent. Rent is a pecuniary category, a category of income, which is essentially a final term, not a category of the motor term, work or interest.[16] It is not a factor or a feature of the process of industrial life, but a phenomenon of the pecuniary situation which emerges from this process under given conventional circumstances. However far-reaching and various the employment of the rent concept in economic theory has been, it has through all permutations remained, what it was to begin with, a rubric in the classification of incomes. It is a pecuniary, not an industrial category. In so far as resort is had to the rent concept in the formulation of a theory of the industrial process,— as in Professor Marshall's work,— it comes to a statement of the process in terms of its residue. Let it not seem presumptuous to say that, great and permanent as is the value of Professor Marshall's exposition of quasi-rents and the like, the endeavor which it involves to present in terms of a concluded system what is of the nature of a fluent process has made the exposition unduly bulky, unwieldy, and inconsequent.

There is a curious reminiscence of the perfect taxonomic day in Mr. Keynes's characterisation of political economy as a " positive science," " the sole province of which is to establish economic uniformities ";[17] and, in

---

[16] "Interest" is, of course, here used in the sense which it has in modern psychological discussion.
[17] *Scope and Method of Political Economy,* p. 46.

this resort to the associationist expedient of defining a natural law as a " uniformity," Mr. Keynes is also borne out by Professor Marshall.[18] But this and other survivals of the taxonomic terminology, or even of the taxonomic canons of procedure, do not hinder the economists of the modern school from doing effective work of a character that must be rated as genetic rather than taxonomic. Professor Marshall's work in economics is not unlike that of Asa Gray in botany, who, while working in great part within the lines of " systematic botany " and adhering to its terminology, and on the whole also to its point of view, very materially furthered the advance of the science outside the scope of taxonomy.

Professor Marshall shows an aspiration to treat economic life as a development; and, at least superficially, much of his work bears the appearance of being a discussion of this kind. In this endeavor his work is typical of what is aimed at by many of the later economists. The aim shows itself with a persistent recurrence in his *Principles*. His chosen maxim is, " Natura non facit saltum," — a maxim that might well serve to designate the prevailing attitude of modern economists towards questions of economic development as well as towards questions of classification or of economic policy. His insistence on the continuity of development and of the economic structure of communities is a characteristic of the best work along the later line of classical political economy. All this gives an air of evolutionism to the work. Indeed, the work of the neo-classical economics might be compared, probably without offending any of its adepts, with that of the early generation of Darwinians, though such a comparison might somewhat shrewdly have to avoid any but super-

[18] *Principles of Economics,* Vol. I, Book I, chap. vi, sect. 6, especially p. 105 (3d edition).

ficial features. Economists of the present day are commonly evolutionists, in a general way. They commonly accept, as other men do, the general results of the evolutionary speculation in those directions in which the evolutionary method has made its way. But the habit of handling by evolutionist methods the facts with which their own science is concerned has made its way among the economists to but a very uncertain degree.

The prime postulate of evolutionary science, the preconception constantly underlying the inquiry, is the notion of a cumulative causal sequence; and writers on economics are in the habit of recognising that the phenomena with which they are occupied are subject to such a law of development. Expressions of assent to this proposition abound. But the economists have not worked out or hit upon a method by which the inquiry in economics may consistently be conducted under the guidance of this postulate. Taking Professor Marshall as exponent, it appears that, while the formulations of economic theory are not conceived to be arrived at by way of an inquiry into the developmental variation of economic institutions and the like, the theorems arrived at are held, and no doubt legitimately, to apply to the past,[19] and with due reserve also to the future, phases of the development. But these theorems apply to the various phases of the development not as accounting for the developmental sequence, but as limiting the range of variation. They say little, if anything, as to the order of succession, as to the derivation and the outcome of any given phase, or as to the causal relation of one phase of any given economic convention or scheme of relations to any other. They indicate the conditions of survival to which any innovation is subject, sup-

[19] See, *e.g.*, Professor Marshall's "Reply" to Professor Cunningham in the *Economic Journal* for 1892, pp. 508–113.

posing the innovation to have taken place, not the conditions of variational growth. The economic laws, the "statements of uniformity," are therefore, when construed in an evolutionary bearing, theorems concerning the superior or the inferior limit of persistent innovations, as the case may be.[20] It is only in this negative, selective bearing that the current economic laws are held to be laws of developmental continuity; and it should be added that they have hitherto found but relatively scant application at the hands of the economists, even for this purpose.

Again, as applied to economic activities under a given situation, as laws governing activities in equilibrium, the economic laws are, in the main, laws of the limits within which economic action of a given purpose runs. They are theorems as to the limits which the economic (commonly the pecuniary) interest imposes upon the range of activities to which the other life interests of men incite, rather than theorems as to the manner and degree in which the economic interest creatively shapes the general scheme of life. In great part they formulate the normal inhibitory effect of economic exigencies rather than the cumulative modification and diversification of human activities through the economic interest, by initiating and guiding habits of life and of thought. This, of course, does not go to say that economists are at all slow to credit the economic exigencies with a large share in the growth of culture; but, while claims of this kind are large and recurrent, it remains true that the laws which make up the framework of economic doctrine are, when construed as generalisations of causal relation, laws of conservation and selection, not of genesis and proliferation. The truth of this, which is but a commonplace generalisa-

[20] This is well illustrated by what Professor Marshall says of the Ricardian law of rent in his "Reply," cited above.

tion, might be shown in detail with respect to such fundamental theorems as the laws of rent, of profits, of wages, of the increasing or diminishing returns of industry, of population, of competitive prices, of cost of production.

In consonance with this quasi-evolutionary tone of the neo-classical political economy, or as an expression of it, comes the further clarified sense that nowadays attaches to the terms " normal " and economic " laws." The laws have gained in colorlessness, until it can no longer be said that the concept of normality implies approval of the phenomena to which it is applied.[21] They are in an increasing degree laws of conduct, though they still continue to formulate conduct in hedonistic terms; that is to say, conduct is construed in terms of its sensuous effect, not in terms of its teleological content. The light of the science is a drier light than it was, but it continues to be shed upon the accessories of human action rather than upon the process itself. The categories employed for the purpose of knowing this economic conduct with which the scientists occupy themselves are not the categories under which the men at whose hands the action takes place themselves apprehend their own action at the instant of acting. Therefore, economic conduct still continues to be somewhat mysterious to the economists; and they are forced to content themselves with adumbrations whenever the discussion touches this central, substantial fact.

All this, of course, is intended to convey no dispraise of the work done, nor in any way to disparage the theories which the passing generation of economists have elaborated, or the really great and admirable body of knowledge which they have brought under the hand of the science;

[21] See, *e.g.*, Marshall, *Principles,* Book I, chap. vi, sect. 6, pp. 105–108. The like dispassionateness is visible in most other modern writers on theory; as, *e.g.,* Clark, Cannan, and the Austrians.

but only to indicate the direction in which the inquiry in its later phases — not always with full consciousness — is shifting as regards its categories and its point of view. The discipline of life in a modern community, particularly the industrial life, strongly reënforced by the modern sciences, has divested our knowledge of non-human phenomena of that fullness of self-directing life that was once imputed to them, and has reduced this knowledge to terms of opaque causal sequence. It has thereby narrowed the range of discretionary, teleological action to the human agent alone; and so it is compelling our knowledge of human conduct, in so far as it is distinguished from the non-human, to fall into teleological terms. Foot-pounds, calories, geometrically progressive procreation, and doses of capital, have not been supplanted by the equally uncouth denominations of habits, propensities, aptitudes, and conventions, nor does there seem to be any probability that they will be; but the discussion which continues to run in terms of the former class of concepts is in an increasing degree seeking support in concepts of the latter class.

PROFESSOR CLARK'S ECONOMICS [1]

FOR some time past economists have been looking with lively anticipation for such a comprehensive statement of Mr. Clark's doctrines as is now offered. The leading purpose of the present volume [2] is "to offer a brief and provisional statement of the more general laws of progress"; although it also comprises a more abridged re-statement of the laws of "Economic Statics" already set forth in fuller form in his *Distribution of Wealth*. Though brief, this treatise is to be taken as systematically complete, as including in due correlation all the "essentials" of Mr. Clark's theoretical system. As such, its publication is an event of unusual interest and consequence.

Mr. Clark's position among this generation of economists is a notable and commanding one. No serious student of economic theory will, or can afford to, forego a pretty full acquaintance with his development of doctrines. Nor will any such student avoid being greatly influenced by the position which Mr. Clark takes on any point of theory on which he may speak, and many look confidently to him for guidance where it is most needed. Very few of those interested in modern theory are under no obligations to him. He has, at the same time, in a singular degree the gift of engaging the affections as well as the attention of students in his field. Yet the critic is

[1] Reprinted by permission from *The Quarterly Journal of Economics,* Vol. XXII, Feb., 1908.
[2] *The Essentials of Economic Theory, as Applied to Modern Problems of Industry and Public Policy.* By John Bates Clark. New York: The Macmillan Company. 1907.

required to speak impersonally of Mr. Clark's work as a phase of current economic theory.

In more than one respect Mr. Clark's position among economists recalls the great figures in the science a hundred years ago. There is the same rigid grasp of the principles, the " essentials," out of which the broad theorems of the system follow in due sequence and correlation; and like the leaders of the classical era, while Mr. Clark is always a theoretician, never to be diverted into an inconsistent makeshift, he is moved by an alert and sympathetic interest in current practical problems. While his aim is a theoretical one, it is always with a view to the theory of current affairs; and his speculations are animated with a large sympathy and an aggressive interest in the amelioration of the lot of man.

His relation to the ancient adepts of the science, however, is something more substantial than a resemblance only. He is, by spiritual consanguinity, a representative of that classical school of thought that dominated the science through the better part of the nineteenth century. This is peculiarly true of Mr. Clark, as contrasted with many of those contemporaries who have fought for the marginal-utility doctrines. Unlike these spokesmen of the Austrian wing, he has had the insight and courage to see the continuity between the classical position and his own, even where he advocates drastic changes in the classical body of doctrines. And although his system of theory embodies substantially all that the consensus of theorists approves in the Austrian contributions to the science, yet he has arrived at his position on these heads not under the guidance of the Austrian school, but, avowedly, by an unbroken development out of the position given by the older generation of economists.[3] Again, in

---

[3] *Cf., e.g., The Distribution of Wealth*, p. 376, note.

the matter of the psychological postulates of the science, he accepts a hedonism as simple, unaffected, and uncritical as that of Jevons or of James Mill. In this respect his work is as true to the canons of the classical school as the best work of the theoreticians of the Austrian observance. There is the like unhesitating appeal to the calculus of pleasure and pain as the indefeasible ground of action and solvent of perplexities, and there is the like readiness to reduce all phenomena to terms of a " normal," or " natural," scheme of life constructed on the basis of this hedonistic calculus. Even in the ready recourse to " conjectural history," to use Steuart's phrase, Mr. Clark's work is at one with both the early classical and the late (Jevons-Austrian) marginal-utility school. It has the virtues of both, coupled with the graver shortcomings of both. But, as his view exceeds theirs in breadth and generosity, so his system of theory is a more competent expression of current economic science than what is offered by the spokesmen of the Jevons-Austrian wing. It is as such, as a competent and consistent system of current economic theory, that it is here intended to discuss Mr. Clark's work, not as a body of doctrines peculiar to Mr. Clark or divergent from the main current.

Since hedonism came to rule economic science, the science has been in the main a theory of distribution,— distribution of ownership and of income. This is true both of the classical school and of those theorists who have taken an attitude of ostensible antagonism to the classical school. The exceptions to the rule are late and comparatively few, and they are not found among the economists who accept the hedonistic postulate as their point of departure. And, consistently with the spirit of hedonism, this theory of distribution has centered about a doctrine of exchange value (or price) and has worked out its scheme of (normal) distribution in terms of (normal) price. The normal economic community, upon which theoretical interest has converged, is a business community, which centers about the market, and whose scheme of life is a scheme of profit and loss. Even when some considerable attention is ostensibly devoted to theories of consumption and production, in these systems of doctrine the theories are constructed in terms of ownership, price, and acquisition, and so reduce themselves in substance to doctrines of distributive acquisition.[4] In this respect Mr. Clark's work is true to the received canons. The " Essentials of Economic Theory " are the essentials of the hedonistic theory of distribution, with sundry reflections on related topics. The scope of Mr. Clark's economics, indeed, is even more closely limited by concepts of distribution than many others, since he persistently analyses production in terms of value, and value is a concept of distribution.

As Mr. Clark justly observes (p. 4), " The primitive and general facts concerning industry . . . need to be known before the social facts can profitably be studied." In these early pages of the treatise, as in other works of its class, there is repeated reference to that more primitive and simple scheme of economic life out of which the modern complex scheme has developed, and it is repeatedly indicated that in order to an understanding of the play of forces in the more advanced stages of economic development and complication, it is necessary to apprehend these forces in their unsophisticated form as they work out in the simple scheme prevalent on the plane of

---

[4] See, *e.g.*, J. S. Mill, *Political Economy*, Book I; Marshall, *Principles of Economics*, Vol. I, Books II–V.

primitive life. Indeed, to a reader not well acquainted with Mr. Clark's scope and method of economic theorising, these early pages would suggest that he is preparing for something in the way of a genetic study,—a study of economic institutions approached from the side of their origins. It looks as if the intended line of approach to the modern situation might be such as an evolutionist would choose, who would set out with showing what forces are at work in the primitive economic community, and then trace the cumulative growth and complication of these factors as they presently take form in the institutions of a later phase of the development. Such, however, is not Mr. Clark's intention. The effect of his recourse to " primitive life " is simply to throw into the foreground, in a highly unreal perspective, those features which lend themselves to interpretation in terms of the normalised competitive system. The best excuse that can be offered for these excursions into " primitive life " is that they have substantially nothing to do with the main argument of the book, being of the nature of harmless and graceful misinformation.

In the primitive economic situation — that is to say, in savagery and the lower barbarism — there is, of course, no " solitary hunter," living either in a cave or otherwise, and there is no man who " makes by his own labor all the goods that he uses," etc. It is, in effect, a highly meretricious misrepresentation to speak in this connection of " the economy of a man who works only for himself," and say that " the inherent productive power of labor and capital is of vital concern to him," because such a presentation of the matter overlooks the main facts in the case in order to put the emphasis on a feature which is of negligible consequence. There is no reasonable doubt but that, at least since mankind reached the human plane, the economic unit has been not a " solitary hunter," but a community of some kind; in which, by the way, women seem in the early stages to have been the most consequential factor instead of the man who works for himself. The " capital " possessed by such a community — as, *e.g.,* a band of California " Digger " Indians — was a negligible quantity, more valuable to a collector of curios than to any one else, and the loss of which to the " Digger " squaws would mean very little. What was of " vital concern " to them, indeed, what the life of the group depended on absolutely, was the accumulated wisdom of the squaws, the technology of their economic situation.[5] The loss of the basket, digging-stick, and mortar, simply as physical objects, would have signified little, but the conceivable loss of the squaw's knowledge of the soil and seasons, of food and fiber plants, and of mechanical expedients, would have meant the present dispersal and starvation of the community.

This may seem like taking Mr. Clark to task for an inconsequential gap in his general information on Digger Indians, Eskimos, and palæolithic society at large. But the point raised is not of negligible consequence for economic theory, particularly not for any theory of " economic dynamics " that turns in great part about questions of capital and its uses at different stages of economic development. In the primitive culture the quantity and the value of mechanical appliances is relatively slight; and whether the group is actually possessed of more or less of such appliances at a given time is not a question of first-rate importance. The loss of these objects — tangible assets — would entail a transient inconvenience. But the accumulated, habitual knowledge of the ways and

[5] *Cf., e.g.,* such an account as Barrows, *Ethno-botany of the Coahuilla Indians.*

means involved in the production and use of these appliances is the outcome of long experience and experimentation; and, given this body of commonplace technological information, the acquisition and employment of the suitable apparatus is easily arranged. The great body of commonplace knowledge made use of in industry is the product and heritage of the group. In its essentials it is known by common notoriety, and the " capital goods " needed for putting this commonplace technological knowledge to use are a slight matter,— practically within the reach of every one. Under these circumstances the ownership of " capital-goods " has no great significance, and, as a practical fact, interest and wages are unknown, and the " earning power of capital " is not seen to be " governed by a specific power of productivity which resides in capital-goods." But the situation changes, presently, by what is called an advance " in the industrial arts." The " capital " required to put the commonplace knowledge to effect grows larger, and so its acquisition becomes an increasingly difficult matter. Through " difficulty of attainment " in adequate quantities, the apparatus and its ownership become a matter of consequence; increasingly so, until presently the equipment required for an effective pursuit of industry comes to be greater than the common man can hope to acquire in a lifetime. The commonplace knowledge of ways and means, the accumulated experience of mankind, is still transmitted in and by the body of the community at large; but, for practical purposes, the advanced " state of the industrial arts " has enabled the owners of goods to corner the wisdom of the ancients and the accumulated experience of the race. Hence " capital," as it stands at that phase of the institution's growth contemplated by Mr. Clark.

The " natural " system of free competition, or, as it was

once called, "the obvious and simple system of natural liberty," is accordingly a phase of the development of the institution of capital; and its claim to immutable dominion is evidently as good as the like claim of any other phase of cultural growth. The equity, or "natural justice," claimed for it is evidently just and equitable only in so far as the conventions of ownership on which it rests continue to be a secure integral part of the institutional furniture of the community; that is to say, so long as these conventions are part and parcel of the habits of thought of the community; that is to say, so long as these things are currently held to be just and equitable. This normalised present, or "natural," state of Mr. Clark, is, as near as may be, Senior's "Natural State of Man,"—the hypothetically perfect competitive system; and economic theory consists in the definition and classification of the phenomena of economic life in terms of this hypothetical competitive system.

Taken by itself, Mr. Clark's dealing with the past development might be passed over with slight comment, except for its negative significance, since it has no theoretical connection with the present, or even with the "natural" state in which the phenomena of economic life are assumed to arrange themselves in a stable, normal scheme. But his dealings with the future, and with the present in so far as the present situation is conceived to comprise "dynamic" factors, is of substantially the same kind. With Senior's "natural state of man" as the baseline of normality in things economic, questions of present and future development are treated as questions of departure from the normal, aberrations and excesses which the theory does not aim even to account for. What is offered in place of theoretical inquiry when these "positive perversions of the natural forces themselves" are taken up (*e.g.,* in chapters xxii.–xxix.) is an exposition of the corrections that must be made to bring the situation back to the normal static state, and solicitous advice as to what measures are to be taken with a view to this beneficent end. The problem presented to Mr. Clark by the current phenomena of economic development is: how can it be stopped? or, failing that, how can it be guided and minimised? Nowhere is there a sustained inquiry into the dynamic character of the changes that have brought the present (deplorable) situation to pass, nor into the nature and trend of the forces at work in the development that is going forward in this situation. None of this is covered by Mr. Clark's use of the word " dynamic." All that it covers in the way of theory (chapters xii.–xxi.) is a speculative inquiry as to how the equilibrium reëstablished itself when one or more of the quantities involved increases or decreases. Other than quantitive changes are not noticed, except as provocations to homiletic discourse. Not even the causes and the scope of the quantitive changes that may take place in the variables are allowed to fall within the scope of the theory of economic dynamics.

So much of the volume, then, and of the system of doctrines of which the volume is an exposition, as is comprised in the later eight chapters (pp. 372–554), is an exposition of grievances and remedies, with only sporadic intrusions of theoretical matter, and does not properly constitute a part of the theory, whether static or dynamic. There is no intention here to take exception to Mr. Clark's outspoken attitude of disapproval toward certain features of the current business situation or to quarrel with the remedial measures which he thinks proper and necessary. This phase of his work is spoken of here rather to call attention to the temperate but uncompromising tone of

Mr. Clark's writings as a spokesman for the competitive system, considered as an element in the Order of Nature, and to note the fact that this is not economic theory.[6]

The theoretical section specifically scheduled as Economic Dynamics (chapters xii.–xxi.), on the other hand, is properly to be included under the caption of Statics. As already remarked above, it presents a theory of equilibrium between variables. Mr. Clark is, indeed, barred out by his premises from any but a statical development of theory. To realise the substantially statical character of his Dynamics, it is only necessary to turn to his chapter xii. (Economic Dynamics). " A highly dynamic condition, then, is one in which the economic organism changes rapidly and yet, at any time in the course of its changes, is relatively near to a certain static model " (p. 196). " The actual shape of society at any one time is not the static model of that time; but it tends to conform to it; and in a very dynamic society is more nearly like it than it would be in one in which the forces of change are less active " (p. 197). The more " dynamic " the society, the nearer it is to the static model; until in an ideally dynamic society, with a frictionless competitive system, to use Mr. Clark's figure, the static state would be attained, except

[6] What would be the scientific rating of the work of a botanist who should spend his energy in devising ways and means to neutralize the ecological variability of plants, or of a physiologist who conceived it the end of his scientific endeavors to rehabilitate the vermiform appendix or the pineal eye, or to denounce and penalize the imitative coloring of the Viceroy butterfly? What scientific interest would attach to the matter if Mr. Loeb, *e.g.,* should devote a few score pages to canvassing the moral responsibilities incurred by him in his parental relation to his parthenogenetically developed sea-urchin eggs?

Those phenomena which Mr. Clark characterizes as " positive perversions " may be distasteful and troublesome, perhaps, but " the economic necessity of doing what is legally difficult " is not of the " essentials of theory."

for an increase in size,— that is to say, the ideally perfect " dynamic " state would coincide with the " static " state. Mr. Clark's conception of a dynamic state reduces itself to a conception of an imperfectly static state, but in such a sense that the more highly and truly " dynamic " condition is thereby the nearer to a static condition.   Neither the static nor the dynamic state, in Mr. Clark's view, it should be remarked, is a state of quiescence.   Both are states of more or less intense activity, the essential difference being that in the static state the activity goes on in perfection, without lag, leak, or friction; the movement of parts being so perfect as not to disturb the equilibrium. The static state is the more " dynamic " of the two.   The " dynamic " condition is essentially a deranged static condition: whereas the static state is the absolute perfect, " natural " taxonomic norm of competitive life.   This dynamic-static state may vary in respect of the magnitude of the several factors which hold one another in equilibrium, but these are none other than quantitive variations.   The changes which Mr. Clark discusses under the head of dynamics are all of this character,— changes in absolute or relative magnitude of the several factors comprised in the equation.

But, not to quarrel with Mr. Clark's use of the terms " static " and " dynamic," it is in place to inquire into the merits of this class of economic science apart from any adventitious shortcomings.   For such an inquiry Mr. Clark's work offers peculiar advantages.   It is lucid, concise, and unequivocal, with no temporising euphemisms and no politic affectations of sentiment.   Mr. Clark's premises, and therewith the aim of his inquiry, are the standard ones of the classical English school (including the Jevons-Austrian wing).   This school of economics stands on the pre-evolutionary ground of normality and "natural law," which the great body of theoretical science occupied in the early nineteenth century. It is like the other theoretical sciences that grew out of the rationalistic and humitarian conceptions of the eighteenth century in that its theoretical aim is taxonomy — definition and classification — with the purpose of subsuming its data under a rational scheme of categories which are presumed to make up the Order of Nature. This Order of Nature, or realm of Natural Law, is not the actual run of material facts, but the facts so interpreted as to meet the needs of the taxonomist in point of taste, logical consistency, and sense of justice. The question of the truth and adequacy of the categories is a question as to the consensus of taste and predilection among the taxonomists; *i.e.,* they are an expression of trained human nature touching the matter of what ought to be. The facts so interpreted make up the "normal," or "natural," scheme of things, with which the theorist has to do. His task is to bring facts within the framework of this scheme of "natural" categories. Coupled with this scientific purpose of the taxonomic economist is the pragmatic purpose of finding and advocating the expedient course of policy. On this latter head, again, Mr. Clark is true to the animus of the school.

The classical school, including Mr. Clark and his contemporary associates in the science, is hedonistic and utilitarian,— hedonistic in its theory and utilitarian in its pragmatic ideals and endeavors. The hedonistic postulates on which this line of economic theory is built up are of a statical scope and character, and nothing but statical theory (taxonomy) comes out of their development.[7]

---

[7] It is a notable fact that even the genius of Herbert Spencer could extract nothing but taxonomy from his hedonistic postu-

These postulates, and the theorems drawn from them, take account of none but quantitive variations, and quantitive variation alone does not give rise to cumulative change, which proceeds on changes in kind.

Economics of the line represented at its best by Mr. Clark has never entered this field of cumulative change. It does not approach questions of the class which occupy the modern sciences,— that is to say, questions of genesis, growth, variation, process (in short, questions of a dynamic import),— but confines its interest to the definition and classification of a mechanically limited range of phenomena.   Like other taxonomic sciences, hedonistic economics does not, and cannot, deal with phenomena of growth except so far as growth is taken in the quantitative sense of a variation in magnitude, bulk, mass, number, frequency.   In its work of taxonomy this economics has consistently bound itself, as Mr. Clark does, by distinctions of a mechanical, statistical nature, and has drawn its categories of classification on those grounds. Concretely, it is confined, in substance, to the determination of and refinements upon the concepts of land, labor, and capital, as handed down by the great economists of the classical era, and the correlate concepts of rent, wages, interest and profits.   Solicitously, with a painfully meticulous circumspection, the normal, mechanical metes and bounds of these several concepts are worked out, the touchstone of the absolute truth aimed at being the hedonistic calculus.   The facts of use and wont are not of the

lates; *e.g., his Social Statics.* Spencer is both evolutionist and hedonist, but it is only by recourse to other factors, alien to the rational hedonistic scheme, such as habit, delusions, use and disuse, sporadic variation, environmental forces, that he is able to achieve anything in the way of genetic science, since it is only by this recourse that he is enabled to enter the field of cumulative change within which the modern post-Darwinian sciences live and move and have their being.

essence of this mechanical refinement. These several categories are mutually exclusive categories, mechanically speaking. The circumstance that the phenomena covered by them are not mechanical facts is not allowed to disturb the pursuit of mechanical distinctions among them. They nowhere overlap, and at the same time between them they cover all the facts with which this economic taxonomy is concerned. Indeed, they are in logical consistency, required to cover them. They are hedonistically "natural" categories of such taxonomic force that their elemental lines of cleavage run through the facts of any given economic situation, regardless of use and wont, even where the situation does not permit these lines of cleavage to be seen by men and recognised by use and wont; so that, *e.g.*, a gang of Aleutian Islanders slushing about in the wrack and surf with rakes and magical incantations for the capture of shell-fish are held, in point of taxonomic reality, to be engaged on a feat of hedonistic equilibration in rent, wages, and interest. And that is all there is to it. Indeed, for economic theory of this kind, that is all there is to any economic situation. The hedonistic magnitudes vary from one situation to another, but, except for variations in the arithmetical details of the hedonistic balance, all situations are, in point of economic theory, substantially alike.[8]

Taking this unfaltering taxonomy on its own recognisances, let us follow the trail somewhat more into the

---

[8] "The capital-goods have to be taken unit by unit if their value for productive purposes is to be rightly gauged. A part of a supply of potatoes is traceable to the hoes that dig them. . . . We endeavor simply to ascertain how badly the loss of one hoe would affect us or how much good the restoration of it would do us. This truth, like the foregoing ones, has a universal application in economics; for primitive men as well as civilized ones must estimate the specific productivity of the tools that they use," etc. Page 43.

arithmetical details, as it leads along the narrow ridge
of rational calculation, above the tree-tops, on the levels
of clear sunlight and moonshine. For the purpose in
hand — to bring out the character of this current economic science as a working theory of current facts, and
more particularly " as applied to modern problems of
industry and public policy " (title-page) — the sequence
to be observed in questioning the several sections into
which the theoretical structure falls is not essential. The
structure of classical theory is familiar to all students,
and Mr. Clark's redaction offers no serious departure
from the conventional lines. Such divergence from conventional lines as may occur is a matter of details, commonly of improvements in detail; and the revisions of detail do not stand in such an organic relation to one another, nor do they support and strengthen one another
in such a manner, as to suggest anything like a revolutionary trend or a breaking away from the conventional
lines.

So as regards Mr. Clark's doctrine of Capital. It does
not differ substantially from the doctrines which are
gaining currency at the hands of such writers as Mr.
Fisher or Mr. Fetter; although there are certain formal
distinctions peculiar to Mr. Clark's exposition of the
" Capital Concept." But these peculiarities are peculiarities of the method of arriving at the concept rather than
peculiarities substantial to the concept itself. The main
discussion of the nature of capital is contained in chapter
ii. (Varieties of Economic Goods). The conception of
capital here set forth is of fundamental consequence to
the system, partly because of the important place assigned
capital in this system of theory, partly because of the
importance which the conception of capital must have in
any theory that is to deal with problems of the current

(capitalistic) situation. Several classes of capital-goods are enumerated, but it appears that in Mr. Clark's apprehension — at variance with Mr. Fisher's view — persons are not to be included among the items of capital. It is also clear from the run of the argument, though not explicitly stated, that only material, tangible, mechanically definable articles of wealth go to make up capital. In current usage, in the business community, " capital " is a pecuniary concept, of course, and is not definable in mechanical terms; but Mr. Clark, true to the hedonistic taxonomy, sticks by the test of mechanical demarcation and draws the lines of his category on physical grounds; whereby it happens that any pecuniary conception of capital is out of the question. Intangible assets, or immaterial wealth, have no place in the theory; and Mr. Clark is exceptionally subtle and consistent in avoiding such modern notions. One gets the impression that such a notion as intangible assets is conceived to be too chimerical to merit attention, even by way of protest or refutation.

Here, as elsewhere in Mr. Clark's writings, much is made of the doctrine that the two facts of " capital " and " capital-goods " are conceptually distinct, though substantially identical. The two terms cover virtually the same facts as would be covered by the terms " pecuniary capital " and " industrial equipment." They are for all ordinary purposes coincident with Mr. Fisher's terms, " capital value " and " capital," although Mr. Clark might enter a technical protest against identifying his categories with those employed by Mr. Fisher.[9] " Capital is this permanent fund of productive goods, the identity of whose component elements is forever changing. Capital-

[9] *Cf.* a criticism of Mr. Fisher's conception in the *Political Science Quarterly* for February, 1908.

goods are the shifting component parts of this permanent aggregate " (p. 29). Mr. Clark admits (pp. 29–33) that capital is colloquially spoken and thought of in terms of value, but he insists that in point of substantial fact the working concept of capital is (should be) that of " a fund of productive goods," considered as an " abiding entity." The phrase itself, " a fund of productive goods," is a curiously confusing mixture of pecuniary and mechanical terms, though the pecuniary expression, " a fund," is probably to be taken in this connection as a permissible metaphor.

This conception of capital, as a physically " abiding entity " constituted by the succession of productive goods that make up the industrial equipment, breaks down in Mr. Clark's own use of it when he comes (pp. 37–38) to speak of the mobility of capital; that is to say, so soon as he makes use of it. A single illustration of this will have to suffice, though there are several points in his argument where the frailty of the conception is patent enough. " The transfer of capital from one industry to another is a dynamic phenomenon which is later to be considered. What is here important is the fact that it is in the main accomplished without entailing transfers of capital-goods. An instrument wears itself out in one industry, and instead of being succeeded by a like instrument in the same industry, it is succeeded by one of a different kind which is used in a different branch of production " (p. 38),— illustrated on the preceding page by a shifting of investment from a whaling-ship to a cotton-mill. In all this it is plain that the " transfer of capital " contemplated is a shifting of investment, and that it is, as indeed Mr. Clark indicates, not a matter of the mechanical shifting of physical bodies from one industry to the other. To speak of a transfer of " capital " which

does not involve a transfer of "capital-goods" is a contradiction of the main position, that "capital" is made
up of "capital-goods." The continuum in which the
"abiding entity" of capital resides is a continuity of
ownership, not a physical fact. The continuity, in fact,
is of an immaterial nature, a matter of legal rights, of
contract, of purchase and sale. Just why this patent
state of the case is overlooked, as it somewhat elaborately
is, is not easily seen. But it is plain that, if the concept
of capital were elaborated from observation of current
business practice, it would be found that "capital" is a
pecuniary fact, not a mechanical one; that it is an outcome of a valuation, depending immediately on the state
of mind of the valuers; and that the specific marks of
capital, by which it is distinguishable from other facts,
are of an immaterial character. This would, of course,
lead, directly, to the admission of intangible assets; and
this, in turn, would upset the law of the "natural" remuneration of labor and capital to which Mr. Clark's
argument looks forward from the start. It would also
bring in the "unnatural" phenomena of monopoly as a
normal outgrowth of business enterprise.

There is a further logical discrepancy avoided by resorting to the alleged facts of primitive industry, when
there was no capital, for the elements out of which to
construct a capital concept, instead of going to the current business situation. In a hedonistic-utilitarian
scheme of economic doctrine, such as Mr. Clark's, only
physically productive agencies can be admitted as efficient
factors in production or as legitimate claimants to a
share in distribution. Hence capital, one of the prime
factors in production and the central claimant in the current scheme of distribution, must be defined in physical
terms and delimited by mechanical distinctions. This is

necessary for reasons which appear in the succeeding chapter, on The Measure of Consumers' Wealth.

On the same page (38), and elsewhere, it is remarked that "business disasters" destroy capital in part. The destruction in question is a matter of values; that is to say, a lowering of valuation, not in any appreciable degree a destruction of material goods. Taken as a physical aggregate, capital does not appreciably decrease through business disasters, but, taken as a fact of ownership and counted in standard units of value, it decreases; there is a destruction of values and a shifting of ownership, a loss of ownership perhaps; but these are pecuniary phenomena, of an immaterial character, and so do not directly affect the material aggregate of the industrial equipment. Similarly, the discussion (pp. 301–314) of how changes of method, as, *e.g.,* labor-saving devices, "liberate capital," and at times "destroy" capital, is intelligible only on the admission that "capital" here is a matter of values owned by investors and is not employed as a synonym for industrial appliances. The appliances in question are neither liberated nor destroyed in the changes contemplated. And it will not do to say that the aggregate of "productive goods" suffers a diminution by a substitution of devices which increases its aggregate productiveness, as is implied, *e.g.,* by the passage on page 307,[10] if Mr. Clark's definition of capital

[10] "The machine itself is often a hopeless specialist. It can do one minute thing and that only, and when a new and better device appears for doing that one thing, the machine has to go, and not to some new employment, but to the junk heap. There is thus taking place a considerable waste of capital in consequence of mechanical and other progress." "Indeed, a quick throwing away of instruments which have barely begun to do their work is often the secret of the success of an enterprising manager, but it entails a destruction of capital."

is strictly adhered to. This very singular passage (pp. 306–311, under the captions, Hardships entailed on Capitalists by Progress, and the Offset for Capital destroyed by Changes of Method) implies that the aggregate of appliances of production is decreased by a change which increases the aggregate of these articles in that respect (productivity) by virtue of which they are counted in the aggregate. The argument will hold good if "productive goods" are rated by bulk, weight, number, or some such irrelevant test, instead of by their productivity or by their consequent capitalised value. On such a showing it should be proper to say that the polishing of plowshares before they are sent out from the factory diminishes the amount of capital embodied in plowshares by as much as the weight or bulk of the waste material removed from the shares in polishing them.

Several things may be said of the facts discussed in this passage. There is, presumably, a decrease, in bulk, weight, or number, of the appliances that make up the industrial equipment at the time when such a technological change as is contemplated takes place. This change, presumably, increases the productive efficiency of the equipment as a whole, and so may be said without hesitation to increase the equipment as a factor of production, while it may decrease it, considered as a mechanical magnitude. The owners of the obsolete or obsolescent appliances presumably suffer a diminution of their capital, whether they discard the obsolete appliances or not. The owners of the new appliances, or rather those who own and are able to capitalise the new technological expedients, presumably gain a corresponding advantage, which may take the form of an increase of the effective capitalisation of their outfit, as would then be shown by an increased market value of their plant. The largest theoretical outcome of the supposed changes, for an economist not bound by Mr. Clark's conception of capital, should be the generalisation that industrial capital — capital considered as a productive agent — is substantially a capitalisation of technological expedients, and that a given capital invested in industrial equipment is measured by the portion of technological expedients whose usufruct the investment appropriates. It would accordingly appear that the substantial core of all capital is immaterial wealth, and that the material objects which are formally the subject of the capitalist's ownership are, by comparison, a transient and adventitious matter. But if such a view were accepted, even with extreme reservations, Mr. Clark's scheme of the " natural " distribution of incomes between capital and labor would " go up in the air," as the colloquial phrase has it. It would be extremely difficult to determine what share of the value of the joint product of capital and labor should, under a rule of " natural " equity, go to the capitalist as an equitable return for his monopolisation of a given portion of the intangible assets of the community at large.[11] The returns actually accruing to him under competitive conditions would be a measure of the differential advantage held by him by virtue of his having become legally seized of the material contrivances by which the technological achievements of the community are put into effect.

Yet, if in this way capital were apprehended as " an historical category," as Rodbertus would say, there is at least the comfort in it all that it should leave a free

[11] The position of the laborer and his wages, in this light, would not be substantially different from that of the capitalist and his interest. Labor is no more possible, as a fact of industry, without the community's accumulated technological knowledge than is the use of " productive goods."

field for Mr. Clark's measures of repression as applied to the discretionary management of capital by the makers of trusts. And yet, again, this comforting reflection is coupled with the ugly accompaniment that by the same move the field would be left equally free of moral obstructions to the extreme proposals of the socialists. A safe and sane course for the quietist in these premises should apparently be to discard the equivocal doctrines of the passage (pp. 306–311) from which this train of questions arises, and hold fast to the received dogma, however unworkable, that " capital " is a congeries of physical objects with no ramifications or complications of an immaterial kind, and to avoid all recourse to the concept of value, or price, in discussing matters of modern business.

The center of interest and of theoretical force and validity in Mr. Clark's work is his law of " natural " distribution. Upon this law hangs very much of the rest, if not substantially the whole structure of theory. To this law of distribution the earlier portions of the theoretical development look forward, and this the succeeding portions of the treatise take as their point of departure. The law of " natural " distribution says that any productive agent " naturally " gets what it produces. Under ideally free competitive conditions — such as prevail in the " static " state, and to which the current situation approximates — each unit of each productive factor unavoidably gets the amount of wealth which it creates, — its " virtual product," as it is sometimes expressed. This law rests, for its theoretical validity, on the doctrine of " final productivity," set forth in full in the *Distribution of Wealth,* and more concisely in the *Essentials* [12] —

[12] *Cf. Distribution of Wealth,* chaps. xii, xiii, vii, viii; *Essentials,* chaps. v–x.

" one of those universal principles which govern economic life in all its stages of evolution." [13]

In combination with a given amount of capital, it is held, each succeeding unit of added labor adds a less than proportionate increment to the product. The total product created by the labor so engaged is at the same time the distributive share received by such labor as wages; and it equals the increment of product added by the " final " unit of labor, multiplied by the number of such units engaged. The law of " natural" interest is the same as this law of wages, with a change of terms. The product of each unit of labor or capital being measured by the product of the " final " unit, each gets the amount of its own product.

In all of this the argument runs in terms of value; but it is Mr. Clark's view, backed by an elaborate exposition of the grounds of his contention,[14] that the use of these terms of value is merely a matter of convenience for the argument, and that the conclusions so reached — the equality so established between productivity and remuneration — may be converted to terms of goods, or " effective utility," without abating their validity.

Without recourse to some such common denominator as value the outcome of the argument would, as Mr. Clark indicates, be something resembling the Ricardian law of differential rent instead of a law drawn in homogeneous terms of " final productivity "; and the law of " natural " distribution would then, at the best, fall short of a general formula. But the recourse to terms of value does not, as Mr. Clark recognises, dispose of the question without more ado. It smooths the way for the argument, but, unaided, it leaves it nugatory. According to Hu-

[13] *Essentials,* p. 158.
[14] *Distribution,* chap. xxiv.

dibras, " The value of a thing Is just as much as it will bring," and the later refinements on the theory of value have not set aside this dictum of the ancient authority. It answers no pertinent question of equity to say that the wages paid for labor are as much as it will bring. And Mr. Clark's chapter (xxiv.) on " The Unit for Measuring Industrial Agents and their Products " is designed to show how this tautological statement in terms of market value converts itself, under competitive conditions, into a competent formula of distributive justice. It does not conduce to intelligibility to say that the wages of labor are just and fair because they are all that is paid to labor as wages. What further value Mr. Clark's extended discussion of this matter may have will lie in his exposition of how competition converts the proposition that " the value of a thing is just as much as it will bring " into the proposition that " the market rate of wages (or interest) gives to labor (or capital) the full product of labor (or capital)."

In following up the theory at this critical point, it is necessary to resort to the fuller statement of the *Distribution of Wealth*,[15] the point being not so adequately covered in the *Essentials*. Consistently hedonistic, Mr. Clark recognises that his law of natural justice must be reduced to elementary hedonistic terms, if it is to make good its claim to stand as a fundamental principle of theory. In hedonistic theory, production of course means the production of utilities, and utility is of course utility to the consumer.[16] A product is such by virtue of and to the amount of the utility which it has for a consumer. This utility of the goods is measured, as value, by the sacrifice (disutility) which the consumer is willing to

[15] Chap. xxiv.
[16] *Essentials,* p. 40.

undergo in order to get the utility which the consumption of the goods yields him.   The unit and measure of productive labor is in the last analysis also a unit of disutility; but it is disutility to the productive laborer, not to the consumer.   The balance which establishes itself under competitive conditions is a compound balance, being a balance between the utility of the goods to the consumer and the disutility (cost) which he is willing to undergo for it, on the one hand, and, on the other hand, a balance between the disutility of the unit of labor and the utility for which the laborer is willing to undergo this disutility.   It is evident, and admitted, that there can be no balance, and no commensurability, between the laborer's disutility (pain) in producing the goods and the consumer's utility (pleasure) in consuming them, inasmuch as these two hedonistic phenomena lie each within the consciousness of a distinct person.   There is, in fact, no continuity of nervous tissue over the interval between consumer and producer, and a direct comparison, equilibrium, equality, or discrepancy in respect of pleasure and pain can, of course, not be sought except within each self-balanced individual complex of nervous tissue.[17] The wages of labor (*i.e.,* the utility of the goods received by the laborer) is not equal to the disutility undergone by him, except in the sense that he is competitively willing to accept it; nor are these wages equal to the utility got by the consumer of the goods, except in the sense

[17] Among modern economic hedonists, including Mr. Clark, there stands over from the better days of the order of nature a presumption, disavowed, but often decisive, that the sensational response to the like mechanical impact of the stimulating body is the same in different individuals. But, while this presumption stands ever in the background, and helps to many important conclusions, as in the case under discussion, few modern hedonists would question the statement in the text.

that he is competitively willing to pay them. This point is covered by the current diagrammatic arguments of marginal-utility theory as to the determination of competitive prices.

But, while the wages are not equal to or directly comparable with the disutility of the productive labor engaged, they are, in Mr. Clark's view, equal to the "productive efficiency" of that labor.[18] "Efficiency in a worker is, in reality, power to draw out labor on the part of society. It is capacity to offer that for which society will work in return." By the mediation of market price, under competitive conditions, it is held, the laborer gets, in his wages, a valid claim on the labor of other men (society) as large as they are competitively willing to allow him for the services for which he is paid his wages. The equitable balance between work and pay contemplated by the "natural" law is a balance between wages and "efficiency," as above defined; that is to say, between the wages of labor and the capacity of labor to get wages. So far, the whole matter might evidently have been left as Bastiat left it. It amounts to saying that the laborer gets what he is willing to accept and the consumers give what they are willing to pay. And this is true, of course, whether competition prevails or not.

What makes this arrangement just and right under competitive conditions, in Mr. Clark's view, lies in his further doctrine that under such conditions of unobstructed competition the prices of goods, and therefore the wages of labor, are determined, within the scope of the given market, by a quasi-consensus of all the parties in interest. There is of course no formal consensus, but what there is of the kind is implied in the fact that bargains are made, and this is taken as an appraisement by

[18] *Distribution,* p. 394.

" society " at large. The (quasi-) consensus of buyers is held to embody the righteous (quasi-) appraisement of society in the premises, and the resulting rate of wages is therefore a (quasi-) just return to the laborer.[19]   " Each man accordingly is paid an amount that equals the total product that he personally creates." [20]   If competitive conditions are in any degree disturbed, the equitable balance of prices and wages is disturbed by that much.   All this holds true for the interest of capital, with a change of terms.

The equity and binding force of this finding is evidently bound up with that common-sense presumption on which it rests; namely, that it is right and good that all men should get what they can without force or fraud and without disturbing existing property relations.   It springs from this presumption, and, whether in point of equity or of expediency, it rises no higher than its source.   It does not touch questions of equity beyond this, nor does it touch questions of the expediency or probable advent of any contemplated change in the existing conventions as to rights of ownership and initiative.   It affords a basis for those who believe in the old order — without which belief this whole structure of opinions collapses — to argue questions of wages and profits in a manner convincing to themselves, and to confirm in the faith those who already believe in the old order.   But it is not easy to see that some hundreds of pages of apparatus should be required to find one's way back to these time-worn commonplaces of Manchester.

In effect, this law of " natural " distribution says that

[19] In Mr. Clark's discussion, elsewhere, the " quasi "-character of the productive share of the laborer is indicated by saying that it is the product " imputed " or " imputable " to him.

[20] *Essentials*, p. 92. Et si sensus deficit, ad firmandum cor sincerum sola fides sufficit.

whatever men acquire without force or fraud under competitive conditions is their equitable due, no more and no less, assuming that the competitive system, with its underlying institution of ownership, is equitable and "natural." In point of economic theory the law appears on examination to be of slight consequence, but it merits further attention for the gravity of its purport. It is offered as a definitive law of equitable distribution comprised in a system of hedonistic economics which is in the main a theory of distributive acquisition only. It is worth while to compare the law with its setting, with a view to seeing how its broad declarations of economic justice shows up in contrast with the elements out of which it is constructed and among which it lies.

Among the notable chapters of the *Essentials* is one (vi.) on Value and its Relation to Different Incomes, which is not only a very substantial section of Mr. Clark's economic theory, but at the same time a type of the achievements of the latter-day hedonistic school. Certain features of this chapter alone can be taken up here. The rest may be equally worthy the student's attention, but it is the intention here not to go into the general substance of the theory of marginal utility and value, to which the chapter is devoted, but to confine attention to such elements of it as bear somewhat directly on the question of equitable distribution already spoken of. Among these latter is the doctrine of the "consumer's surplus,"— virtually the same as what is spoken of by other writers as "consumer's rent." [21] "Consumer's surplus" is the surplus of utility (pleasure) derived by the consumer of goods above the (pain) cost of the goods to him. This is held to be a very generally prevalent phenomenon. Indeed, it is held to be all but universally

[21] See pp. 102–113; also p. 172, note.

present in the field of consumption. It might, in fact, be effectively argued that even Mr. Clark's admitted exception [22] is very doubtfully to be allowed, on his own showing. Correlated with this element of utility on the consumer's side is a similar volume of disutility on the producer's side, which may be called " producer's abatement," or " producer's rent ": it is the amount of disutility by which the disutility-cost of a given article to any given producer (laborer) falls short of (or conceivably exceeds) the disutility incurred by the marginal producer. Marginal buyers or consumers and marginal sellers or producers are relatively few: the great body on both sides come in for something in the way of a " surplus " of utility or disutility.

All this bears on the law of " natural " wages and interest as follows, taking that law of just remuneration at Mr. Clark's rating of it. The law works out through the mediation of price. Price is determined, competitively, by marginal producers or sellers and marginal consumers or purchasers: the latter alone on the one side get the precise price-equivalent of the disutility incurred by them, and the latter alone on the other side pay the full price-equivalent of the utilities derived by them from the goods purchased.[23] Hence the competitive price — covering competitive wages and interest — does not reflect the consensus of all parties concerned as to the " effective utility " of the goods, on the one hand, or as to their effective (disutility) cost, on the other hand. It reflects instead, if anything of this kind, the valuations which the marginal unfortunates on each side concede under stress of competition; and it leaves on each side of the bargain relation an uncovered " surplus," which

[22] " The cheapest and poorest grades of articles."   Page 113.
[23] See p. 113.

marks the (variable) interval by which price fails to cover "effective utility." The excess utility — and the conceivable excess cost — does not appear in the market transactions that mediate between consumer and producer.[24] In the balance, therefore, which establishes itself in terms of value between the social utility of the product and the remuneration of the producer's "efficiency," the margin of utility represented by the aggregate "consumer's surplus" and like elements is not accounted for. It follows, when the argument is in this way reduced to its hedonistic elements, that no man "is paid an amount that equals the amount of the total product that he personally creates."

Supposing the marginal-utility (final-utility) theories of objective value to be true, there is no consensus, actual or constructive, as to the "effective utility" of the goods produced: there is no "social" decision in the case beyond what may be implied in the readiness of buyers to profit as much as may be by the necessities of the marginal buyer and seller. It appears that there is warrant, within these premises, for the formula: Remuneration $\gtrless$ than Product. Only by an infinitesimal chance would it hold true in any given case that, hedonistically, Remuneration $=$ Product; and, if it should ever happen to be true, there would be no finding it out.

The (hedonistic) discrepancy which so appears between remuneration and product affects both wages and interest in the same manner, but there is some (hedonistic) ground in Mr. Clark's doctrines for holding that the discrepancy does not strike both in the same degree.

[24] The disappearance, and the method of disapearance, of such elements of differential utility and disutility occupies a very important place in all marginal-utility ("final-utility") theories of market value, or "objective value."

There is indeed no warrant for holding that there is anything like an equable distribution of this discrepancy among the several industries or the several industrial concerns; but there appears to be some warrant, on Mr. Clark's argument, for thinking that the discrepancy is perhaps slighter in those branches of industry which produce the prime necessaries of life.[25] This point of doctrine throws also a faint (metaphysical) light on a, possibly generic, discrepancy between the remuneration of capitalists and that of laborers: the latter are, relatively, more addicted to consuming the necessaries of life, and it may be that they thereby gain less in the way of a consumer's surplus.

All the analysis and reasoning here set forth has an air of undue tenuity; but in extenuation of this fault it should be noted that this reasoning is made up of such matter as goes to make up the theory under review, and the fault, therefore, is not to be charged to the critic. The manner of argument required to meet this theory of the " natural law of final productivity " on its own ground is itself a sufficiently tedious proof of the futility of the whole matter in dispute. Yet it seems necessary to beg further indulgence for more of the same kind. As a needed excuse, it may be added that what immediately follows bears on Mr. Clark's application of the law of " natural distribution " to modern problems of industry and public policy, in the matter of curbing monopolies.

Accepting, again, Mr. Clark's general postulates — the postulates of current hedonistic economics — and applying the fundamental concepts, instead of their corollaries, to his scheme of final productivity, it can be shown

----

[25] " Only the simplest and cheapest things that are sold in the market at all bring just what they are worth to the buyers." Page 113.

to fail on grounds even more tenuous and hedonistically more fundamental than those already passed in review. In all final-utility (marginal-utility) theory it is of the essence of the scheme of things that successive increments of a " good " have progressively less than proportionate utility. In fact, the coefficient of decrease of utility is greater than the coefficient of increase of the stock of goods. The solitary " first loaf " is exorbitantly useful. As more loaves are successively added to the stock, the utility of each grows small by degrees and incontinently less, until, in the end, the state of the " marginal " or " final " loaf is, in respect of utility, shameful to relate. So, with a change of phrase, it fares with successive increments of a given productive factor — labor or capital — in Mr. Clark's scheme of final productivity. And so, of course, it also fares with the utility of successive increments of product created by successively adding unit after unit to the complement of a given productive factor engaged in the case. If we attend to this matter of final productivity in consistently hedonistic terms, a curious result appears.

A larger complement of the productive agent, counted by weight and tale, will, it is commonly held, create a larger output of goods, counted by weight and tale; [26] but these are not hedonistic terms and should not be allowed to cloud the argument. In the hedonistic scheme the magnitude of goods, in all the dimensions to be taken

[26] It is, *e.g.,* open to serious question whether Mr. Clark's curves of final productivity (pp. 139, 148), showing a declining output per unit in response to an increase of one of the complementary agents of production, will fit the common run of industry in case the output be counted by weight and tale. In many cases they will, no doubt; in many other cases they will not. But this is no criticism of the curves in question, since they do not, or at least should not, purport to represent the product in such terms, but in terms of utility.

account of, is measured in terms of utility, which is a different matter from weight and tale. It is by virtue of their utility that they are " goods," not by virtue of their physical dimensions, number and the like; and utility is a matter of the production of pleasure and the prevention of pain. Hedonistically speaking, the amount of the goods, the magnitude of the output, is the quantity of utility derivable from their consumption; and the utility per unit decreases faster than the number of units increases.[27] It follows that in the typical or undifferentiated case an increase of the number of units beyond a certain critical point entails a decrease of the " total effective utility " of the supply.[28] This critical point seems ordinarily to be very near the point of departure of the curve of declining utility, perhaps it frequently coincides with the latter. On the curve of declining final utility, at any point whose tangent cuts the axis of ordinates at an angle of less than 45 degrees, an increase of the number of units entails a decrease of the " total effective utility of the supply," [29] so that a gain in physical

[27] To resort to an approximation after the manner of Malthus, if the supply of goods be supposed to increase by arithmetical progression, their final utility may be said concomitantly to decrease by geometrical progression.

[28] *Cf. Essentials*, chap. iii, especially pp. 40–41.

[29] The current marginal-utility diagrams are not of much use in this connection, because the angle of the tangent with the axis of ordinates, at any point, is largely a matter of the draftsman's taste. The abscissa and the ordinate do not measure commensurable units. The units on the abscissa are units of frequency, while those on the ordinate are units of amplitude; and the greater or less segment of line allowed per unit on either axis is a matter of independently arbitrary choice. Yet the proposition in the text remains true,— as true as hedonistic propositions commonly are. The magnitude of the angle of the tangent with the axis of ordinates decides whether the total (hedonistic) productivity at a given point in the curve increases or decreases with a (mechanical) increase of the productive agent,— no student at

productivity is a loss as counted in "total effective utility." Hedonistically, therefore, the productivity in such a case diminishes, not only relatively to the (physical) magnitude of the productive agents, but absolutely. This critical point, of maximum "total effective utility," is, if the practice of shrewd business men is at all significant, commonly somewhat short of the point of maximum physical productivity, at least in modern industry and in a modern community.

The "total effective utility" may commonly be increased by decreasing the output of goods. The "total effective utility" of wages may often be increased by decreasing the amount (value) of the wages per man, particularly if such a decrease is accompanied by a rise in the price of articles to be bought with the wages. Hedonistically speaking, it is evident that the point of maximum net productivity is the point at which a perfectly shrewd business management of a perfect monopoly would limit the supply; and the point of maximum (hedonistic) remuneration (wages and interest) is the point which such a management would fix on in dealing with a wholly free, perfectly competitive supply of labor and capital.

Such a monopolistic state of things, it is true, would not answer to Mr. Clark's ideal. Each man would not be "paid an amount that equals the amount of the total product that he personally creates," but he would commonly be paid an amount that (hedonistically, in point of "effective utility") exceeds what he personally creates, because of the high final utility of what he receives.

all familiar with marginal-utility arguments will question that patent fact. But the angle of the tangent depends on the fancy of the draftsman,—no one possessed of the elemental mathematical notions will question that equally patent fact.

This is easily proven. Under the monopolistic conditions supposed, the laborers would, it is safe to assume, not be fully employed all the time; that is to say, they would be willing to work some more in order to get some more articles of consumption; that is to say, the articles of consumption which their wages offer them have so high a utility as to afford them a consumer's surplus,— the articles are worth more than they cost: [30]   Q. E. D.

The initiated may fairly doubt the soundness of the chain of argument by which these heterodox theoretical results are derived from Mr. Clark's hedonistic postulates, more particularly since the adepts of the school, including Mr. Clark, are not accustomed to draw conclusions to this effect from these premises. Yet the argument proceeds according to the rules of marginal-utility permutations. In view of this scarcely avoidable doubt, it may be permitted, even at the risk of some tedium, to show how the facts of every-day life bear out this unexpected turn of the law of natural distribution, as briefly traced above. The principle involved is well and widely accepted. The familiar practical maxim of " charging what the traffic will bear " rests on a principle of this kind, and affords one of the readiest practical illustrations of the working of the hedonistic calculus. The principle involved is that a larger aggregate return (value) may be had by raising the return per unit to such a point as to somewhat curtail the demand. In practice it is recognised, in other words, that there is a critical point at which the value obtainable per unit, multiplied by the number of units that will be taken off at that price, will give the largest net aggregate result (in value to the

[30] A similar line of argument has been followed up by Mr. Clark for capital and interest, in a different connection. See *Essentials,* pp. 340–345, 356.

seller) obtainable under the given conditions. A calculus involving the same principle is, of course, the guiding consideration in all monopolistic buying and selling; but a moment's reflection will show that it is, in fact, the ruling principle in all commercial transactions and, indeed, in all business. The maxim of " charging what the traffic will bear " is only a special formulation of the generic principle of business enterprise. Business initiative, the function of the entrepreneur (business man) is comprehended under this principle taken in its most general sense.[31] In business the buyer, it is held by the theorists, bids up to the point of greatest obtainable advantage to himself under the conditions prevailing, and the seller similarly bids down to the point of greatest obtainable net aggregate gain. For the trader (business man, entrepreneur) doing business in the open (competitive) market or for the business concern with a partial or limited monopoly, the critical point above referred to is, of course, reached at a lower point on the curve of price than would be the case under a perfect and unlimited monopoly, such as was supposed above; but the principle of charging what the traffic will bear remains intact, although the traffic will not bear the same in the one case as in the other.

Now, in the theories based on marginal (or "final") utility, value is an expression or measure of "effective utility "— or whatever equivalent term may be preferred. In operating on values, therefore, under the rule of charging what the traffic will bear, the sellers of a monopolised supply, *e.g.,* must operate through the valuations of the buyers; that is to say, they must influence the final utility of the goods or services to such effect that the

[31] *Cf. Essentials,* pp. 83–90, 118–120.

"total effective utility" of the limited supply to the consumers will be greater than would be the "total effective utility" of a larger supply, which is the point in question. The emphasis falls still more strongly on this illustration of the hedonistic calculus, if it is called to mind that in the common run of such limitations of supply by a monopolistic business management the management would be able to increase the supply at a progressively declining cost beyond the critical point by virtue of the well-known principle of increasing returns from industry. It is also to be added that, since the monopolistic business gets its enhanced return from the margin by which the "total effective utility" of the limited supply exceeds that of a supply not so limited, and since there is to be deducted from this margin the costs of monopolistic management in addition to other costs, therefore the enhancement of the "total effective utility" of the goods to the consumer in the case must be appreciably larger than the resulting net gains to the monopoly.

By a bold metaphor — a metaphor sufficiently bold to take it out of the region of legitimate figures of speech — the gains that come to enterprising business concerns by such monopolistic enhancement of the "total effective utility" of their products are spoken of as "robbery," "extortion," "plunder"; but the theoretical complexion of the case should not be overlooked by the hedonistic theorist in the heat of outraged sentiment. The monopolist is only pushing the principle of all business enterprise (free competition) to its logical conclusion; and, in point of hedonistic theory, such monopolistic gains are to be accounted the "natural" remuneration of the monopolist for his "productive" service to the community in enhancing their enjoyment per unit of consumable goods to such point as to swell their net aggregate enjoyment to a maximum.

This intricate web of hedonistic calculations might be pursued further, with the result of showing that, while the consumers of the monopolised supply of goods are gainers by virtue of the enhanced " total effective utility " of the goods, the monopolists who bring about this result do so in great part at their own cost, counting cost in terms of a reduction of " total effective utility." By injudiciously increasing their own share of goods, they lower the marginal and effective utility of their wealth to such a point as, probably, to entail a considerable (hedonistic) privation in the shrinkage of their enjoyment per unit. But it is not the custom of economists, nor does Mr. Clark depart from this custom, to dwell on the hardships of the monopolists. This much may be added, however, that this hedonistically consistent exposition of the " natural law of final productivity " shows it to be " one of those universal principles which govern economic life in all its stages of evolution," even when that evolution enters the phase of monopolistic business enterprise, — granting always the sufficiency of the hedonistic postulates from which the law is derived. Further, the considerations reviewed above go to show that, on two counts, Mr. Clark's crusade against monopoly in the later portion of his treatise is out of touch with the larger theoretical speculations of the earlier portions: (*a*) it runs counter to the hedonistic law of " natural " distribution; and (*b*) the monopolistic business against which Mr. Clark speaks is but the higher and more perfect development of that competitive business enterprise which he wishes to reinstate,— competitive business, so called, being incipiently monopolistic enterprise.

Apart from this theoretical bearing, the measures which Mr. Clark advocates for the repression of monopoly, under the head of applications "to modern problems of industry and public policy," may be good economic policy or they may not,— they are the expression of a sound common sense, an unvitiated solicitude for the welfare of mankind, and a wide information as to the facts of the situation. The merits of this policy of repression, as such, cannot be discussed here. On the other hand, the relation of this policy to the theoretical groundwork of the treatise needs also not be discussed here, inasmuch as it has substantially no relation to the theory. In this later portion of the volume Mr. Clark does not lean on doctrines of "final utility," "final productivity," or, indeed, on hedonistic economics at large. He speaks eloquently for the material and cultural interests of the community, and the references to his law of "natural distribution" might be cut bodily out of the discussion without lessening the cogency of his appeal or exposing any weakness in his position. Indeed, it is by no means certain that such an excision would not strengthen his appeal to men's sense of justice by eliminating irrelevant matter.

Certain points in this later portion of the volume, however, where the argument is at variance with specific articles of theory professed by Mr. Clark, may be taken up, mainly to elucidate the weakness of his theoretical position at the points in question. He recognises with more than the current degree of freedom that the growth and practicability of monopolies under modern conditions is chiefly due to the negotiability of securities representing capital, coupled with the joint-stock character of modern business concerns.[32] These features of the modern

[32] *Cf.* chap. xxii, especially pp. 378–392.

(capitalistic) business situation enable a sufficiently few men to control a section of the community sufficiently large to make an effective monopoly. The most effective known form of organisation for purposes of monopoly, according to Mr. Clark, is that of the holding company, and the ordinary corporation follows it closely in effectiveness in this respect. The monopolistic control is effected by means of the vendible securities covering the capital engaged. To meet the specifications of Mr. Clark's theory of capital, these vendible securities — as *e.g.,* the securities (common stock) of a holding company — should be simply the formal evidence of the ownership of certain productive goods and the like. Yet, by his own showing, the ownership of a share of productive goods proportionate to the face value, or the market value, of the securities is by no means the chief consequence of such an issue of securities.[33] One of the consequences, and for the purposes of Mr. Clark's argument the gravest consequence, of the employment of such securities, is the dissociation of ownership from the control of the industrial equipment, whereby the owners of certain securities, which stand in certain immaterial, technical relations to certain other securities, are enabled arbitrarily to control the use of the industrial equipment covered by the latter. These are facts of the modern organisation of capital, affecting the productivity of the industrial equipment and its serviceability both to its owners and to the community. They are facts, though not physically tangible objects; and they have an effect on the serviceability of industry no less decisive than the effect which any group of physically tangible objects of equal market value have. They are, moreover, facts which are bought and sold in the purchase and sale of these securities, as,

[33] *Cf.* p. 391.

e.g., the common stock of a holding company. They have a value, and therefore they have a " total effective utility."

In short, these facts are intangible assets, which are the most consequential element in modern capital, but which have no existence in the theory of capital by which Mr. Clark aims to deal with " modern problems of industry." Yet, when he comes to deal with these problems, it is, of necessity, these intangible assets that immediately engage his attention. These intangible assets are an outgrowth of the freedom of contract under the conditions imposed by the machine industry; yet Mr. Clark proposes to suppress this category of intangible assets without prejudice to freedom of contract or to the machine industry, apparently without having taken thought of the lesson which he rehearses (pp. 390–391) from the introduction of the holding company, with its " sinister perfection," to take the place of the (less efficient) " trust " when the latter was dealt with somewhat as it is now proposed to deal with the holding company. One is tempted to remark that a more naïve apprehension of the facts of modern capital would have afforded a more competent realisation of the problems of monopoly.

It appears from what has just been said of Mr. Clark's " natural " distribution and of his dealing with the problems of modern industry that the logic of hedonism is of no avail for the theory of business affairs. Yet it is held, perhaps justly, that the hedonistic interpretation may be of great avail in analysing the industrial functions of the community, in their broad, generic character, even if it should not serve so well for the intricate details of the modern business situation. It may be at least a serviceable hypothesis for the outlines of economic theory, for the first approximations to the " economic laws " sought by taxonomists. To be serviceable for this purpose, the hypothesis need perhaps not be true to fact, at least not in the final details of the community's life or without material qualification; [34] but it must at least have that ghost of actuality that is implied in consistency with its own corollaries and ramifications.

As has been suggested in an earlier paragraph, it is characteristic of hedonistic economics that the large and central element in its theoretical structure is the doctrine of distribution. Consumption being taken for granted as a quantitive matter simply,— essentially a matter of an insatiable appetite,— economics becomes a theory of acquisition; production is, theoretically, a process of acquisition, and distribution a process of distributive acquisition. The theory of production is drawn in terms of the gains to be acquired by production; and under competitive conditions this means necessarily the acquisition of a distributive share of what is available. The rest of what the facts of productive industry include, as, *e.g.,* the facts of workmanship or the " state of the industrial arts," gets but a scant and perfunctory attention. Those matters are not of the theoretical essence of the scheme. Mr. Clark's general theory of production does not differ substantially from that commonly professed by the marginal-utility school. It is a theory of competitive acquisition. An inquiry into the principles of his doctrine, therefore, as they appear, *e.g.,* in the early chapters of the *Essentials,* is, in effect, an inquiry into the competence of the main theorems of modern hedonistic economics.

" All men seek to get as much net service from material wealth as they can." " Some of the benefit received is neutralised by the sacrifice incurred; but there is a net surplus of gains not thus canceled by sacrifices, and the

[34] *Cf. Essentials,* p. 39.

generic motive which may properly be called economic
is the desire to make this surplus large." [35]   It is of the
essence of the scheme that the acquisitive activities of
mankind afford a net balance of pleasure.   It is out of
this net balance, presumably, that " the consumer's sur-
pluses " arise, or it is in this that they merge.   This optimistic conviction is a matter of presumption, of course;
but it is universally held to be true by hedonistic economists, particularly by those who cultivate the doctrines of
marginal utility.   It is not questioned and not proven.   It
seems to be a surviving remnant of the eighteenth-century
faith in a benevolent Order of Nature; that is to say, it
is a rationalistic metaphysical postulate.   It may be true
or not, as matter of fact; but it is a postulate of the school,
and its optimistic bias runs like a red thread through all
the web of argument that envelops the " normal " competitive system.   A surplus of gain is normal to the theoretical scheme.

The next great theorem of this theory of acquisition
is at cross-purposes with this one.   Men get useful goods
only at the cost of producing them, and production is
irksome, painful, as has been recounted above.   They go
on producing utilities until, at the margin, the last increment of utility in the product is balanced by the concomitant increment of disutility in the way of irksome
productive effort,— labor or abstinence.   At the margin,
pleasure-gain is balanced by pain-cost.   But the " effective utility " of the total product is measured by that of
the final unit; the effective utility of the whole is given
by the number of units of product multiplied by the
effective utility of the final unit; while the effective disutility (pain-cost) of the whole is similarly measured by
the pain-cost of the final unit.   The " total effective

[35] *Essentials,* p. 39.

utility " of the producer's product equals the "total effective disutility " of his pains of acquisition. Hence there is no net surplus of utility in the outcome.

The corrective objection is ready to hand,[36] that, while the balance of utility and disutility holds at the margin, it does not hold for the earlier units of the product, these earlier units having a larger utility and a lower cost, and so leaving a large net surplus of utility, which gradually declines as the margin is approached. But this attempted correction evades the hedonistic test. It shifts the ground from the calculus to the objects which provoke the calculation. Utility is a psychological matter, a matter of pleasurable appreciation, just as disutility, conversely, is a matter of painful appreciation. The individual who is held to count the costs and the gain in this hedonistic calculus is, by supposition, a highly reasonable person. He counts the cost to him as an individual against the gain to him as an individual. He looks before and after, and sizes the whole thing up in a reasonable course of conduct. The "absolute utility " would exceed the " effective utility " only on the supposition that the " producer " is an unreflecting sensory apparatus, such as the beasts of the field are supposed to be, devoid of that gift of appraisement and calculation which is the hypothetical hedonist's only human trait. There might on such a supposition — if the producer were an intelligent sensitive organism simply — emerge an excess of total pleasure over total pain, but there could then be no talk of utility or of disutility, since these terms imply intelligent reflection, and they are employed because they do so. The hedonistic producer looks to his own cost and gain, as an intelligent pleasure-seeker whose consciousness compasses the contrasted elements as wholes. He does not contrast

[36] *Cf. Essentials,* chap. iii, especially pp. 51–56.

the balance of pain and pleasure in the morning with the balance of pain and pleasure in the afternoon, and say that there is so much to the good because he was not so tired in the morning.   Indeed, by hypothesis, the pleasure to be derived from the consumption of the product is a future, or expected, pleasure, and can be said to be present, at the point of time at which a given unit of pain-cost is incurred, only in anticipation; and it cannot be said that the anticipated pleasure attaching to a unit of product which emerges from the effort of the producer during the relatively painless first hour's work exceeds the anticipated pleasure attaching to a similar unit emerging from the second hour's work.   Mr. Clark has, in effect, explained this matter in substantially the same way in another connection (*e.g.,* p. 42), where he shows that the magnitude on which the question of utility and cost hinges is the " total effective utility," and that the " total absolute utility " is a matter not of what hedonistically is, in respect of utility as an outcome of production, but of what might have been under different circumstances.

An equally unprofitable result may be reached from the same point of departure along a different line of argument.   Granting that increments of product should be measured, in respect of utility, by comparison with the disutility of the concomitant increment of cost, then the diagrammatic arguments commonly employed are inadequate, in that the diagrams are necessarily drawn in two dimensions only,— length and breadth: whereas they should be drawn in three dimensions, so as to take account of the intensity of application as well as of its duration.[37]

[37] This difficulty is recognized by the current marginal-utility arguments, and an allowance for intensity is made or presumed. But the allowance admitted is invariably insufficient.   It might be said to be insufficient by hypothesis, since it is by hypothesis too small to offset the factor which it is admitted to modify.

Apparently, the exigencies of graphic representation, fortified by the presumption that there always emerges a
surplus of utility, have led marginal-utility theorists, in
effect, to overlook this matter of intensity of application.

When this element is brought in with the same freedom
as the other two dimensions engaged, the argument will,
in hedonistic consistency, run somewhat as follows,— the
run of the facts being what it may.   The producer, setting
out on this irksome business, and beginning with the
production of the exorbitantly useful initial unit of product, will, by hedonistic necessity, apply himself to the
task with a correspondingly extravagant intensity, the
irksomeness (disutility) of which necessarily rises to such
a pitch as to leave no excess of utility in this initial unit
of product above the concomitant disutility of the initial
unit of productive effort.[38]   As the utility of subsequent
units of product progressively declines, so will the producer's intensity of irksome application concomitantly
decline, maintaining a nice balance between utility and
disutility throughout.   There is, therefore, no excess of
" absolute utility " above " effective utility " at any point
on the curve, and no excess of " total absolute utility "
above " total effective utility " of the product as a whole,
nor above the " total absolute disutility " or the " total
effective disutility " of the pain-cost.

A transient evasion of this outcome may perhaps be
sought by saying that the producer will act wisely, as a

[38] The limit to which the intensity rises is a margin of the same
kind as that which limits the duration. This supposition, that
the intensity of application necessarily rises to such a pitch that
its disutility overtakes and offsets the utility of the product, may
be objected to as a bit of puerile absurdity; but it is a long
time since puerility or absurdity has been a bar to any supposition in arguments on marginal utility.

good hedonist should, and save his energies during the earlier moments of the productive period in order to get the best aggregate result from his day's labor, instead of spending himself in ill-advised excesses at the outset. Such seems to be the fact of the matter, so far as the facts wear a hedonistic complexion; but this correction simply throws the argument back on the previous position and concedes the force of what was there claimed. It amounts to saying that, instead of appreciating each successive unit of product in isolated contrast with its concomitant unit of irksome productive effort, the producer, being human, wisely looks forward to his total product and rates it by contrast with his total pain-cost. Whereupon, as before, no net surplus of utility emerges, under the rule which says that irksome production of utilities goes on until utility and disutility balance.

But this revision of "final productivity" has further consequences for the optimistic doctrines of hedonism. Evidently, by a somewhat similar line of argument the "consumer's surplus" will be made to disappear, even as this that may be called the "producer's surplus" has disappeared. Production being acquisition, and the consumer's cost being cost of acquisition, the argument above should apply to the consumer's case without abatement. On considering this matter in terms of the hedonistically responsive individual concerned, with a view to determining whether there is, in his calculus of utilities and costs, any margin of uncovered utilities left over after he has incurred all the disutilities that are worth while to him,— instead of proceeding on a comparison between the pleasure-giving capacity of a given article and the market price of the article, all such alleged differential advantages within the scope of a single sensory are seen to be nothing better than an illusory diffractive effect due to a faulty instrument.

But the trouble does not end here. The equality: pain-cost = pleasure-gain, is not a competent formula. It should be: pain-cost incurred = pleasure-gain anticipated. And between these two formulas lies the old adage, " there's many a slip 'twixt the cup and the lip." In an appreciable proportion of ventures, endeavors, and enterprises, men's expectations of pleasure-gain are in some degree disappointed,— through miscalculation, through disserviceable secondary effects of their productive efforts, by " the act of God," by " fire, flood, and pestilence." In the nature of things these discrepancies fall out on the side of loss more frequently than on that of gain. After all allowance has been made for what may be called serviceable errors, there remains a margin of disserviceable error, so that pain-cost > eventual pleasure-gain = anticipated pleasure-gain — n. Hence, in general, pain-cost > pleasure-gain. Hence it appears that, in the nature of things, men's pains of production are underpaid by that much; although it may, of course, be held that the nature of things at this point is not " natural " or " normal."

To this it may be objected that the risk is discounted. Insurance is a practical discounting of risk ; but insurance is resorted to only to cover risk that is appreciated by the person exposed to it, and it is such risks as are not appreciated by those who incur them that are chiefly in question here. And it may be added that insurance has hitherto not availed to equalise and distribute the chances of success and failure. Business gains — enterpreneur's gains, the rewards of initiative and enterprise — come out of this uncovered margin of adventure, and the losses of initiative and enterprise are to be set down to the same account. In some measure this element of initiative and enterprise enters into all economic endeavor. And it is not unusual for economists to remark that the volume of unsuccessful or only partly successful enterprise is very large. There are some lines of enterprise that are, as one might say, extra hazardous, in which the average falls out habitually on the wrong side of the account. Typical of this class is the production of the precious metals, particularly as conducted under that régime of free competition for which Mr. Clark speaks. It has been the opinion, quite advisedly, of such economists of the classic age of competition as J. S. Mill and Cairnes, *e.g.,* that the world's supply of the precious metals has been got at an average or total cost exceeding their value by several fold. The producers, under free competition at least, are over-sanguine of results.

But, in strict consistency, the hedonistic theory of human conduct does not allow men to be guided in their calculation of cost and gain, when they have to do with the precious metals, by different norms from those which rule their conduct in the general quest of gain. The visible difference in this respect between the production of the precious metals and production generally should be due to the larger proportions and greater notoriety of the risks in this field rather than to a difference in the manner of response to the stimulus of expected gain. The canons of hedonistic calculus permit none but a quantitative difference in the response. What happens in the production of the precious metals is typical of what happens in a measure and more obscurely throughout the field of productive effort.

Instead of a surplus of utility of product above the disutility of acquisition, therefore, there emerges an average or aggregate net hedonistic deficit. On a consistent marginal-utility theory, all production is a losing game. The fact that Nature keeps the bank, it appears, does not take the hedonistic game of production out of the general category known of old to that class of sanguine hedonistic calculators whose day-dreams are filled with safe and sane schemes for breaking the bank. "Hope springs eternal in the human breast." Men are congenitally over-sanguine, it appears; and the production of utilities is, mathematically speaking, a function of the pig-headed optimism of mankind. It turns out that the laws of (human) nature malevolently grind out vexation for men instead of benevolently furthering the greatest happiness of the greatest number. The sooner the whole traffic ceases, the better,— the smaller will be the net balance of pain. The great hedonistic Law of Nature turns out to be simply the curse of Adam, backed by the even more sinister curse of Eve.

The remark was made in an earlier paragraph that Mr. Clark's theories have substantially no relation to his practical proposals. This broad declaration requires an equally broad qualification. While the positions reached in his theoretical development count for nothing in making or fortifying the positions taken on "problems of modern industry and public policy," the two phases of the discussion — the theoretical and the pragmatic — are the outgrowth of the same range of preconceptions and run back to the same metaphysical ground. The present canvass of items in the doctrinal system has already far overpassed reasonable limits, and it is out of the question here to pursue the exfoliation of ideas through Mr. Clark's discussion of public questions, even in the fragmentary fashion in which scattered items of the theoretical portion of his treatise have been passed in review. But a broad and rudely drawn characterisation may yet be permissible. This latter portion of the volume has the general complexion of a Bill of Rights. This is said, of course, with no intention of imputing a fault. It implies that the scope and method of the discussion is governed by the preconception that there is one right and beautiful definitive scheme of economic life, "to which the whole creation tends." Whenever and in so far as current phenomena depart or diverge from this definitive "natural" scheme or from the straight and narrow path that leads to its consummation, there is a grievance to be remedied by putting the wheels back into the rut. The future, such as it ought to be,— the only normally possible, natural future scheme of life, — is known by the light of this preconception; and men have an indefeasible right to the installation and maintenance of those specific economic relations, expedients, institutions, which this "natural" scheme comprises, and to no others. The consummation is presumed to dominate the course of things which is presumed to lead up to the consummation. The measures of redress whereby the economic Order of Nature is to renew its youth are simple, direct, and short-sighted, as becomes the proposals of pre-Darwinian hedonism, which is not troubled about the exuberant uncertainties of cumulative change. No doubt presents itself but that the community's code of right and equity in economic matters will remain unchanged under changing conditions of economic life.

THE LIMITATIONS OF MARGINAL UTILITY [1]

The limitations of the marginal-utility economics are sharp and characteristic. It is from first to last a doctrine of value, and in point of form and method it is a theory of valuation. The whole system, therefore, lies within the theoretical field of distribution, and it has but a secondary bearing on any other economic phenomena than those of distribution — the term being taken in its accepted sense of pecuniary distribution, or distribution in point of ownership. Now and again an attempt is made to extend the use of the principle of marginal utility beyond this range, so as to apply it to questions of production, but hitherto without sensible effect, and necessarily so. The most ingenious and the most promising of such attempts have been those of Mr. Clark, whose work marks the extreme range of endeavor and the extreme degree of success in so seeking to turn a postulate of distribution to account for a theory of production. But the outcome has been a doctrine of the production of values, and value, in Mr. Clark's as in other utility systems, is a matter of valuation; which throws the whole excursion back into the field of distribution. Similarly, as regards attempts to make use of this principle in an analysis of the phenomena of consumption, the best results arrived at are some formulation of the pecuniary distribution of consumption goods.

Within this limited range marginal-utility theory is of a

[1] Reprinted by permission from the *Journal of Political Economy,* Vol. XVII, No. 9 November 1909.

wholly statical character. It offers no theory of a movement of any kind, being occupied with the adjustment of
values to a given situation. Of this, again, no more convincing illustration need be had than is afforded by the
work of Mr. Clark, which is not excelled in point of ear-
nestness, perseverance, or insight. For all their use of
the term "dynamic," neither Mr. Clark nor any of his
associates in this line of research have yet contributed
anything at all appreciable to a theory of genesis, growth,
sequence, change, process, or the like, in economic life.
They have had something to say as to the bearing which
given economic changes, accepted as premises, may have
on valuation, and so on distribution; but as to the causes
of change or the unfolding sequence of the phenomena of
economic life they have had nothing to say hitherto; nor
can they, since their theory is not drawn in causal terms
but in terms of teleology.

In all this the marginal-utility school is substantially at
one with the classical economics of the nineteenth century, the difference between the two being that the former
is confined within narrower limits and sticks more consistently to its teleological premises. Both are teleological, and neither can consistently admit arguments from
cause to effect in the formulation of their main articles of
theory. Neither can deal theoretically with phenomena
of change, but at the most only with rational adjustment to change which may be supposed to have supervened.

To the modern scientist the phenomena of growth and
change are the most obtrusive and most consequential
facts observable in economic life. For an understanding
of modern economic life the technological advance of the
past two centuries — *e. g.,* the growth of the industrial
arts — is of the first importance; but marginal-utility the-

ory does not bear on this matter, nor does this matter bear on marginal-utility theory. As a means of theoretically accounting for this technological movement in the past or in the present, or even as a means of formally, technically stating it as an element in the current economic situation, that doctrine and all its works are altogether idle. The like is true for the sequence of change that is going forward in the pecuniary relations of modern life; the hedonistic postulate and its propositions of differential utility neither have served nor can serve an inquiry into these phenomena of growth, although the whole body of marginal-utility economics lies within the range of these pecuniary phenomena. It has nothing to say to the growth of business usages and expedients or to the concomitant changes in the principles of conduct which govern the pecuniary relations of men, which condition and are conditioned by these altered relations of business life or which bring them to pass.

It is characteristic of the school that wherever an element of the cultural fabric, an institution or any institutional phenomenon, is involved in the facts with which the theory is occupied, such institutional facts are taken for granted, denied, or explained away. If it is a question of price, there is offered an explanation of how exchanges may take place with such effect as to leave money and price out of the account. If it is a question of credit, the effect of credit extension on business traffic is left on one side and there is an explanation of how the borrower and lender coöperate to smooth out their respective income streams of consumable goods or sensations of consumption. The failure of the school in this respect is consistent and comprehensive. And yet these economists are lacking neither in intelligence nor in information. They are, indeed, to be credited, commonly, with a wide range of information and an exact control of materials, as well as with a very alert interest in what is going on; and apart from their theoretical pronouncements the members of the school habitually profess the sanest and most intelligent views of current practical questions, even when these questions touch matters of institutional growth and decay.

The infirmity of this theoretical scheme lies in its postulates, which confine the inquiry to generalisations of the teleological or " deductive " order.   These postulates, together with the point of view and logical method that follow from them, the marginal-utility school shares with other economists of the classical line — for this school is but a branch or derivative of the English classical economists of the nineteenth century.   The substantial difference between this school and the generality of classical economists lies mainly in the fact that in the marginal-utility economics the common postulates are more consistently adhered to at the same time that they are more neatly defined and their limitations are more adequately realized.   Both the classical school in general and its specialized variant, the marginal-utility school, in particular, take as their common point of departure the traditional psychology of the early nineteenth-century hedonists, which is accepted as a matter of course or of common notoriety and is held quite uncritically.   The central and well-defined tenet so held is that of the hedonistic calculus.   Under the guidance of this tenet and of the other psychological conceptions associated and consonant with it, human conduct is conceived of and interpreted as a rational response to the exigencies of the situation in which mankind is placed; as regards economic conduct it is such a rational and unprejudiced response to the stimulus of anticipated pleasure and pain — being, typically

and in the main, a response to the promptings of anticipated pleasure, for the hedonists of the nineteenth century and of the marginal-utility school are in the main
of an optimistic temper.[2] Mankind is, on the whole and
normally, (conceived to be) clearsighted and farsighted
in its appreciation of future sensuous gains and losses, although there may be some (inconsiderable) difference
between men in this respect. Men's activities differ,
therefore, (inconsiderably) in respect of the alertness of
the response and the nicety of adjustment of irksome
pain-cost to apprehended future sensuous gain; but, on the
whole, no other ground or line or guidance of conduct
than this rationalistic calculus falls properly within the
cognizance of the economic hedonists. Such a theory can
take account of conduct only in so far as it is rational
conduct, guided by deliberate and exhaustively intelligent
choice — wise adaptation to the demands of the main
chance.

The external circumstances which condition conduct
are variable, of course, and so they will have a varying
effect upon conduct; but their variation is, in effect, construed to be of such a character only as to vary the degree
of strain to which the human agent is subject by contact
with these external circumstances. The cultural elements involved in the theoretical scheme, elements that

[2] The conduct of mankind differs from that of the brutes in being determined by anticipated sensations of pleasure and pain,
instead of actual sensations. Hereby, in so far, human conduct
is taken out of the sequence of cause and effect and falls instead under the rule of sufficient reason. By virtue of this rational faculty in man the connection between stimulus and response is teleological instead of causal.

The reason for assigning the first and decisive place to pleasure, rather than to pain, in the determination of human conduct,
appears to be the (tacit) acceptance of that optimistic doctrine
of a beneficent order of nature which the nineteenth century inherited from the eighteenth.

are of the nature of institutions, human relations governed by use and wont in whatever kind and connection, are not subject to inquiry but are taken for granted as pre-existing in a finished, typical form and as making up a normal and definitive economic situation, under which and in terms of which human intercourse is necessarily carried on. This cultural situation comprises a few large and simple articles of institutional furniture, together with their logical implications or corollaries; but it includes nothing of the consequences or effects caused by these institutional elements. The cultural elements so tacitly postulated as immutable conditions precedent to economic life are ownership and free contract, together with such other features of the scheme of natural rights as are implied in the exercise of these. These cultural products are, for the purpose of the theory, conceived to be given a priori in unmitigated force. They are part of the nature of things; so that there is no need of accounting for them or inquiring into them, as to how they have come to be such as they are, or how and why they have changed and are changing, or what effect all this may have on the relations of men who live by or under this cultural situation.

Evidently the acceptance of these immutable premises, tacitly, because uncritically and as a matter of course, by hedonistic economics gives the science a distinctive character and places it in contrast with other sciences whose premises are of a different order. As has already been indicated, the premises in question, so far as they are peculiar to the hedonistic economics, are (*a*) a certain institutional situation, the substantial feature of which is the natural right of ownership, and (*b*) the hedonistic calculus. The distinctive character given to this system of theory by these postulates and by the point of view resulting from their acceptance may be summed up broadly and concisely in saying that the theory is confined to the ground of sufficient reason instead of proceeding on the ground of efficient cause. The contrary is true of modern science, generally (except mathematics), particularly of such sciences as have to do with the phenomena of life and growth. The difference may seem trivial. It is serious only in its consequences. The two methods of inference — from sufficient reason and from efficient cause — are out of touch with one another and there is no transition from one to the other: no method of converting the procedure or the results of the one into those of the other. The immediate consequence is that the resulting economic theory is of a teleological character —" deductive " or " a priori " as it is often called — instead of being drawn in terms of cause and effect. The relation sought by this theory among the facts with which it is occupied is the control exercised by future (apprehended) events over present conduct. Current phenomena are dealt with as conditioned by their future consequences; and in strict marginal-utility theory they can be dealt with only in respect of their control of the present by consideration of the future. Such a (logical) relation of control or guidance between the future and the present of course involves an exercise of intelligence, a taking thought, and hence an intelligent agent through whose discriminating forethought the apprehended future may affect the current course of events; unless, indeed, one were to admit something in the way of a providential order of nature or some occult line of stress of the nature of sympathetic magic. Barring magical and providential elements, the relation of sufficient reason runs by way of the interested discrimination, the forethought, of an agent who takes thought of the future and guides his present activity by regard for this future. The relation of sufficient reason runs only from the (apprehended) future into the present, and it is solely of an intellectual, subjective, personal, teleological character and force; while the relation of cause and effect runs only in the contrary direction, and it is solely of an objective, impersonal, materialistic character and force. The modern scheme of knowledge, on the whole, rests, for its definitive ground, on the relation of cause and effect; the relation of sufficient reason being admitted only provisionally and as a proximate factor in the analysis, always with the unambiguous reservation that the analysis must ultimately come to rest in terms of cause and effect. The merits of this scientific animus, of course, do not concern the present argument.

Now, it happens that the relation of sufficient reason enters very substantially into human conduct. It is this element of discriminating forethought that distinguishes human conduct from brute behavior. And since the economist's subject of inquiry is this human conduct, that relation necessarily comes in for a large share of his attention in any theoretical formulation of economic phenomena, whether hedonistic or otherwise. But while modern science at large has made the causal relation the sole ultimate ground of theoretical formulation; and while the other sciences that deal with human life admit the relation of sufficient reason as a proximate, supplementary, or intermediate ground, subsidiary, and subservient to the argument from cause to effect; economics has had the misfortune — as seen from the scientific point of view — to let the former supplant the latter. It is, of course, true that human conduct is distinguished from other natural phenomena by the human faculty for taking thought, and any science that has to do with human conduct must face the patent fact that the details of such conduct consequently fall into the teleological form; but it is the peculiarity of the hedonistic economics that by force of its postulates its attention is confined to this teleological bearing of conduct alone. It deals with this conduct only in so far as it may be construed in rationalistic, teleological terms of calculation and choice. But it is at the same time no less true that human conduct, economic or otherwise, is subject to the sequence of cause and effect, by force of such elements as habituation and conventional requirements. But facts of this order, which are to modern science of graver interest than the teleological details of conduct, necessarily fall outside the attention of the hedonistic economist, because they cannot be construed in terms of sufficient reason, such as his postulates demand, or be fitted into a scheme of teleological doctrines.

There is, therefore, no call to impugn these premises of the marginal-utility economics within their field. They commend themselves to all serious and uncritical persons at the first glance. They are principles of action which underlie the current, business-like scheme of economic life, and as such, as practical grounds of conduct, they are not to be called in question without questioning the existing law and order. As a matter of course, men order their lives by these principles and, practically, entertain no question of their stability and finality. That is what is meant by calling them institutions; they are settled habits of thought common to the generality of men. But it would be mere absentmindedness in any student of civilization therefore to admit that these or any other human institutions have this stability which is currently imputed to them or that they are in this way intrinsic to the nature of things. The acceptance by the economists of these or other institutional elements as given and immutable limits their inquiry in a particular and decisive way. It shuts off the inquiry at the point where the modern scientific interest sets in. The institutions in question are no doubt good for their purpose as institutions, but they are not good as premises for a scientific inquiry into the nature, origin, growth, and effects of these institutions and of the mutations which they undergo and which they bring to pass in the community's scheme of life.

To any modern scientist interested in economic phenomena, the chain of cause and effect in which any given phase of human culture is involved, as well as the cumulative changes wrought in the fabric of human conduct itself by the habitual activity of mankind, are matters of more engrossing and more abiding interest than the method of inference by which an individual is presumed invariably to balance pleasure and pain under given conditions that are presumed to be normal and invariable. The former are questions of the life-history of the race or the community, questions of cultural growth and of the fortunes of generations; while the latter is a question of individual casuistry in the face of a given situation that may arise in the course of this cultural growth. The former bear on the continuity and mutations of that scheme of conduct whereby mankind deals with its material means of life; the latter, if it is conceived in hedonistic terms, concerns a disconnected episode in the sensuous experience of an individual member of such a community.

In so far as modern science inquires into the phenomena of life, whether inanimate, brute, or human, it is occupied about questions of genesis and cumulative change, and it converges upon a theoretical formulation in the shape of a life-history drawn in causal terms. In so far as it is a science in the current sense of the term, any science, such as economics, which has to do with human conduct, becomes a genetic inquiry into the human scheme of life; and where, as in economics, the subject of inquiry is the conduct of man in his dealings with the material means of life, the science is necessarily an inquiry into the life-history of material civilization, on a more or less extended or restricted plan. Not that the economist's inquiry isolates material civilization from all other phases and bearings of human culture, and so studies the motions of an abstractly conceived " economic man." On the contrary, no theoretical inquiry into this material civilization that shall be at all adequate to any scientific purpose can be carried out without taking this material civilization in its causal, that is to say, its genetic, relations to other phases and bearings of the cultural complex; without studying it as it is wrought upon by other lines of cultural growth and as working its effects in these other lines. But in so far as the inquiry is economic science, specifically, the attention will converge upon the scheme of material life and will take in other phases of civilization only in their correlation with the scheme of material civilization.

Like all human culture this material civilization is a scheme of institutions — institutional fabric and institutional growth. But institutions are an outgrowth of habit. The growth of culture is a cumulative sequence of habituation, and the ways and means of it are the habitual response of human nature to exigencies that vary incontinently, cumulatively, but with something of a consistent sequence in the cumulative variations that so go forward,— incontinently, because each new move creates a new situation which induces a further new variation in the habitual manner of response; cumulatively, because each new situation is a variation of what has gone before it and embodies as causal factors all that has been effected by what went before; consistently, because the underlying traits of human nature (propensities, aptitudes, and what not) by force of which the response takes place, and on the ground of which the habituation takes effect, remain substantially unchanged.

Evidently an economic inquiry which occupies itself exclusively with the movements of this consistent, elemental human nature under given, stable institutional conditions — such as is the case with the current hedonistic economics — can reach statical results alone; since it makes abstraction from those elements that make for anything but a statical result. On the other hand an adequate theory of economic conduct, even for statical purposes, cannot be drawn in terms of the individual simply — as is the case in the marginal-utility economics — because it cannot be drawn in terms of the underlying traits of human nature simply; since the response that goes to make up human conduct takes place under institutional norms and only under stimuli that have an institutional bearing; for the situation that provokes and inhibits action in any given case is itself in great part of institutional, cultural derivation. Then, too, the phenomena of human life occur only as phenomena of the life of a group or community: only under stimuli due to contact with the group and only under the (habitual) control exercised by canons of conduct imposed by the group's scheme of life. Not only is the individual's conduct hedged about and directed by his habitual relations to his fellows in the group, but these relations, being of an institutional character, vary as the institutional scheme varies. The wants and desires, the end and aim, the ways and means, the amplitude and drift of the individual's conduct are functions of an institutional variable that is of a highly complex and wholly unstable character.

The growth and mutations of the institutional fabric are an outcome of the conduct of the individual members of the group, since it is out of the experience of the individuals, through the habituation of individuals, that institutions arise; and it is in this same experience that these institutions act to direct and define the aims and end of conduct. It is, of course, on individuals that the system of institutions imposes those conventional standards, ideals, and canons of conduct that make up the community's scheme of life. Scientific inquiry in this field, therefore, must deal with individual conduct and must formulate its theoretical results in terms of individual conduct. But such an inquiry can serve the purposes of a genetic theory only if and in so far as this individual conduct is attended to in those respects in which it counts toward habituation, and so toward change (or stability) of the institutional fabric, on the one hand, and in those respects in which it is prompted and guided by the received institutional conceptions and ideals on the other hand. The postulates of marginal utility, and the hedonistic preconceptions generally, fail at this point in that they confine the attention to such bearings of economic conduct as are conceived not to be conditioned by habitual standards and ideals and to have no effect in the way of habituation. They disregard or abstract from the causal sequence of propensity and habituation in economic life and exclude from theoretical inquiry all such interest in the facts of cultural growth, in order to attend to those features of the case that are conceived to be idle in this respect. All such facts of institutional force and growth are put on one side as not being germane to pure theory; they are to be taken account of, if at all, by afterthought, by a more or less vague and general allowance for inconsequential disturbances due to occasional human infirmity. Certain institutional phenomena, it is true, are comprised among the premises of the hedonists, as has been noted above; but they are included as postulates a priori. So the institution of ownership is taken into the inquiry not as a factor of growth or an element subject to change, but as one of the primordial and immutable facts of the order of nature, underlying the hedonistic calculus. Property, ownership, is presumed as the basis of hedonistic discrimination and it is conceived to be given in its finished (nineteenth-century) scope and force. There is no thought either of a conceivable growth of this definitive nineteenth-century institution out of a cruder past or of any conceivable cumulative change in the scope and force of ownership in the present or future. Nor is it conceived that the presence of this institutional element in men's economic relations in any degree affects or disguises the hedonistic calculus, or that its pecuniary conceptions and standards in any degree standardize, color, mitigate, or divert the hedonistic calculator from the direct and unhampered quest of the net sensuous gain. While the institution of property is included in this way among the postulates of the theory, and is even presumed to be ever-present in the economic situation, it is allowed to have no force in shaping economic conduct, which is conceived to run its course to its hedonistic outcome as if no such institutional factor intervened between the impulse and its realization. The institution of property, together with all the range of pecuniary conceptions that belong under it and that cluster about it, are presumed to give rise to no habitual or conventional canons of conduct or standards of valuation, no

proximate ends, ideals, or aspirations. All pecuniary notions arising from ownership are treated simply as expedients of computation which mediate between the pain-
cost and the pleasure-gain of hedonistic choice, without
lag, leak, or friction; they are conceived simply as the
immutably correct, God-given notation of the hedonistic
calculus.

The modern economic situation is a business situation,
in that economic activity of all kinds is commonly controlled by business considerations. The exigencies of
modern life are commonly pecuniary exigencies. That is
to say they are exigencies of the ownership of property.
Productive efficiency and distributive gain are both rated
in terms of price. Business considerations are considerations of price, and pecuniary exigencies of whatever kind
in the modern communities are exigencies of price. The
current economic situation is a price system. Economic
institutions in the modern civilized scheme of life are
(prevailingly) institutions of the price system. The accountancy to which all phenomena of modern economic
life are amenable is an accountancy in terms of price; and
by the current convention there is no other recognized
scheme of accountancy, no other rating, either in law or
in fact, to which the facts of modern life are held amenable. Indeed, so great and pervading a force has this
habit (institution) of pecuniary accountancy become
that it extends, often as a matter of course, to many facts
which properly have no pecuniary bearing and no pecuniary magnitude, as, *e. g.,* works of art, science, scholarship, and religion. More or less freely and fully, the
price system dominates the current commonsense in its
appreciation and rating of these non-pecuniary ramifications of modern culture; and this in spite of the fact that,

on reflection, all men of normal intelligence will freely admit that these matters lie outside the scope of pecuniary valuation.

Current popular taste and the popular sense of merit and demerit are notoriously affected in some degree by pecuniary considerations. It is a matter of common notoriety, not to be denied or explained away, that pecuniary ("commercial") tests and standards are habitually made use of outside of commercial interests proper. Precious stones, it is admitted, even by hedonistic economists, are more esteemed than they would be if they were more plentiful and cheaper. A wealthy person meets with more consideration and enjoys a larger measure of good repute than would fall to the share of the same person with the same habit of mind and body and the same record of good and evil deeds if he were poorer. It may well be that this current "commercialisation" of taste and appreciation has been overstated by superficial and hasty critics of contemporary life, but it will not be denied that there is a modicum of truth in the allegation. Whatever substance it has, much or little, is due to carrying over into other fields of interest the habitual conceptions induced by dealing with and thinking of pecuniary matters. These "commercial" conceptions of merit and demerit are derived from business experience. The pecuniary tests and standards so applied outside of business transactions and relations are not reducible to sensuous terms of pleasure and pain. Indeed, it may, *e. g.,* be true, as is commonly believed, that the contemplation of a wealthy neighbor's pecuniary superiority yields painful rather than pleasurable sensations as an immediate result; but it is equally true that such a wealthy neighbor is, on the whole, more highly regarded and more considerately treated than another neighbor who differs from the former only in being less enviable in respect of wealth.

It is the institution of property that gives rise to these habitual grounds of discrimination, and in modern times, when wealth is counted in terms of money, it is in terms of money value that these tests and standards of pecuniary excellence are applied. This much will be admitted. Pecuniary institutions induce pecuniary habits of thought which affect men's discrimination outside of pecuniary matters; but the hedonistic interpretation alleges that such pecuniary habits of thought do not affect men's discrimination in pecuniary matters. Although the institutional scheme of the price system visibly dominates the modern community's thinking in matters that lie outside the economic interest, the hedonistic economists insist, in effect, that this institutional scheme must be accounted of no effect within that range of activity to which it owes its genesis, growth, and persistence. The phenomena of business, which are peculiarly and uniformly phenomena of price, are in the scheme of the hedonistic theory reduced to non-pecuniary hedonistic terms and the theoretical formulation is carried out as if pecuniary conceptions had no force within the traffic in which such conceptions originate. It is admitted that preoccupation with commercial interests has " commercialised " the rest of modern life, but the " commercialisation " of commerce is not admitted. Business transactions and computations in pecuniary terms, such as loans, discounts, and capitalisation, are without hesitation or abatement converted into terms of hedonistic utility, and conversely.

It may be needless to take exception to such conversion from pecuniary into sensuous terms, for the theoretical purpose for which it is habitually made; although, if need were, it might not be excessively difficult to show that the whole hedonistic basis of such a conversion is a psychological misconception. But it is to the remoter theoretical consequences of such a conversion that exception is to be taken. In making the conversion abstraction is made from whatever elements do not lend themselves to its terms; which amounts to abstracting from precisely those elements of business that have an institutional force and that therefore would lend themselves to scientific inquiry of the modern kind — those (institutional) elements whose analysis might contribute to an understanding of modern business and of the life of the modern business community as contrasted with the assumed primordial hedonistic calculus.

The point may perhaps be made clearer. Money and the habitual resort to its use are conceived to be simply the ways and means by which consumable goods are acquired, and therefore simply a convenient method by which to procure the pleasurable sensations of consumption; these latter being in hedonistic theory the sole and overt end of all economic endeavor. Money values have therefore no other significance than that of purchasing power over consumable goods, and money is simply an expedient of computation. Investment, credit extensions, loans of all kinds and degrees, with payment of interest and the rest, are likewise taken simply as intermediate steps between the pleasurable sensations of consumption and the efforts induced by the anticipation of these sensations, other bearings of the case being disregarded. The balance being kept in terms of the hedonistic consumption, no disturbance arises in this pecuniary traffic so long as the extreme terms of this extended hedonistic equation — pain-cost and pleasure-gain — are not altered, what lies between these extreme terms being merely algebraic notation employed for convenience of accountancy. But such is not the run of the facts in modern business.

Variations of capitalization, *e. g.,* occur without its being practicable to refer them to visibly equivalent variations either in the state of the industrial arts or in the sensations of consumption. Credit extensions tend to inflation of credit, rising prices, overstocking of markets, etc., likewise without a visible or securely traceable correlation in the state of the industrial arts or in the pleasures of consumption; that is to say, without a visible basis in those material elements to which the hedonistic theory reduces all economic phenomena. Hence the run of the facts, in so far, must be thrown out of the theoretical formulation. The hedonistically presumed final purchase of consumable goods is habitually not contemplated in the pursuit of business enterprise. Business men habitually aspire to accumulate wealth in excess of the limits of practicable consumption, and the wealth so accumulated is not intended to be converted by a final transaction of purchase into consumable goods or sensations of consumption. Such commonplace facts as these, together with the endless web of business detail of a like pecuniary character, do not in hedonistic theory raise a question as to how these conventional aims, ideals, aspirations, and standards have come into force or how they affect the scheme of life in business or outside of it; they do not raise those questions because such questions cannot be answered in the terms which the hedonistic economists are content to use, or, indeed, which their premises permit them to use. The question which arises is how to explain the facts away: how theoretically to neutralize them so that they will not have to appear in the theory, which can then be drawn in direct and unambiguous terms of rational hedonistic calculation. They are explained away as being aberrations due to oversight or lapse of memory on the part of business men, or to some failure of logic or

insight.   Or they are construed and interpreted into the
rationalistic terms of the hedonistic calculus by resort to
an ambiguous use of the hedonistic concepts.   So that the
whole " money economy," with all the machinery of credit
and the rest, disappears in a tissue of metaphors to reappear theoreticaly expurgated, sterilized, and simplified
into a " refined system of barter," culminating in a net aggregate maximum of pleasurable sensations of consumption.

But since it is in just this unhedonistic, unrationalistic
pecuniary traffic that the tissue of business life consists;
since it is this peculiar conventionalism of aims and standards that differentiates the life of the modern business
community from any conceivable earlier or cruder phase
of economic life; since it is in this tissue of pecuniary intercourse and pecuniary concepts, ideals, expedients, and
aspirations that the conjunctures of business life arise and
run their course of felicity and devastation; since it is
here that those institutional changes take place which distinguish one phase or era of the business community's
life from any other; since the growth and change of these
habitual, conventional elements make the growth and
character of any business era or business community; any
theory of business which sets these elements aside or ex-
plains them away misses the main facts which it has gone
out to seek.   Life and its conjunctures and institutions
being of this complexion, however much that state of the
case may be deprecated, a theoretical account of the
phenomena of this life must be drawn in these terms in
which the phenomena occur.   It is not simply that the
hedonistic interpretation of modern economic phenomena
is inadequate or misleading; if the phenomena are subjected to the hedonistic interpretation in the theoretical
analysis they disappear from the theory; and if they

would bear the interpretation in fact they would disappear in fact. If, in fact, all the conventional relations and principles of pecuniary intercourse were subject to such a perpetual rationalized, calculating revision, so that each article of usage, appreciation, or procedure must approve itself *de novo* on hedonistic grounds of sensuous expediency to all concerned at every move, it is not conceivable that the institutional fabric would last over night.

GUSTAV SCHMOLLER'S ECONOMICS [1]

PROFESSOR SCHMOLLER'S *Grundriss* [2] is an event of the first importance in economic literature. It appears from later advices that the second and concluding volume of the work is hardly to be looked for at as early a date as the author's expressions in his preface had led us to anticipate. What lies before Professor Schmoller's readers, therefore, in this first volume of the *Outlines* is but one-half of the compendious statement which he here purposes making of his theoretical position and of his views and exemplification of the scope and method of economic science. It may accordingly seem adventurous to attempt a characterisation of his economic system on the basis of this avowedly incomplete statement. And yet such an endeavor is not altogether gratuitous, nor need it in any great measure proceed on hypothetical grounds. The introduction comprised in the present volume sketches the author's aim in an outline sufficiently full to afford a convincing view of the " system " of science for which he speaks; and the two books by which the introduction is followed show Professor Schmoller's method of inquiry consistently carried out, as well as the reach and nature of the theoretical conclusions which he considers to lie within the competency of economic science. And with regard to an economist who is so much of an innovator,— not to say so much of an iconoclast,— and whose work

---

[1] Reprinted by permission from *The Quarterly Journal of Economics,* Vol. XVI, Nov., 1901.

[2] *Grundriss der allgemeinen Volkswirtschaftslehre.* Erster Teil. Leipzig, 1900.

touches the foundations of the science so intimately and profoundly, the interest of his critics and associates must, at least for the present, center chiefly about these questions as to the scope and nature assigned to the theory by his discussion, as to the range and character of the material of which he makes use, and as to the methods of inquiry which his sagacity and experience commend. So, therefore, while the *Outlines* is yet incomplete, considered as a compendium of details of doctrine, the work in its unfinished state need not thereby be an inadequate expression of Professor Schmoller's relation to economic science.

Herewith for the first time economic readers are put in possession of a fully advised deliverance on economic science at large as seen and cultivated by that modernised historical school of which Professor Schmoller is the authoritative exponent. Valuable and characteristic as his earlier discussions on the scope and method of the science are, they are but preliminary studies and tentative formulations as compared with this maturer work, which not only avows itself a definitive formulation, but has about it an air of finality perceptible at every turn. But this comes near saying that it embodies the sole comprehensive working-out of the scientific aims of the historical school. Discussions partially covering the field, monographs and sketches there are in great number, showing the manner of economic theory that was to be looked for as an outcome of the "historical diversion." Some of these, especially some of the later ones, are extremely valuable in the results they offer, as well as significant of the trend which the science is taking under the hands of the German students.[3] But a comprehen-

[3] *E.g.*, K. Bücher's *Entstehung der Volkswirtschaft*, and *Arbeit und Rythmus;* R. Hildebrand's *Recht und Sitte;* Knapp's *Grund-*

sive work, aiming to formulate a body of economic theory on the basis afforded by the " historical method," has not hitherto been seriously attempted.

To the broad statement just made exception might perhaps be taken in favor of Schaeffle's half-forgotten work of the seventies, together possibly with several other less notable and less consistent endeavors of a similar kind, dating back to the early decades of the school. Probably none of the younger generation of economists would be tempted to cite Roscher's work as invalidating such a statement as the one made above. Although time has been allowed for the acceptance and authentication of these endeavors of the earlier historical economists in the direction of a system of economic theory,— that is to say, of an economic science,— they have failed of authentication at the hands of the students of the science ; and there seems no reason to regard this failure as less than definitive.

During the last two decades the historical school has branched into two main directions of growth, somewhat divergent, so that broad general statements regarding the historical economists can be less confidently made to-day than perhaps at any earlier time. Now, as regards the more conservative branch, the historical economists of the stricter observance,— these modern continuers of what may be called the elder line of the historical school can scarcely be said to cultivate a science at all, their aim being not theoretical work. Assuredly, the work of this elder line, of which Professor Wagner is the unquestioned head, is by no means idle. It is work of a sufficiently important and valuable order, perhaps it is indispensable to the task which the science has in hand, but, broadly

herrschaft und Rittergut; Ehrenberg's *Zeitalter der Fugger;* R. Mucke's various works.

speaking, it need not be counted with in so far as it touches directly upon economic theory. This elder line of German economics, in its numerous modern representatives, shows both insight and impartiality; but as regards economic theory their work bears the character of eclecticism rather than that of a constructive advance. Frequent and peremptory as their utterances commonly are on points of doctrine, it is only very rarely that these utterances embody theoretical views arrived at or verified by the economists who make them or by such methods of inquiry as are characteristic of these economists. Where these expressions of doctrine are not of the nature of maxims of expediency, they are, as is well known, commonly borrowed somewhat uncritically from classical sources. Of constructive scientific work — that is to say, of theory — this elder line of German economics is innocent; nor does there seem to be any prospect of an eventual output of theory on the part of that branch of the historical school, unless they should unexpectedly take advice, and make the scope, and therefore the method, of their inquiry something more than historical in the sense in which that term is currently accepted. The historical economics of the conservative kind seems to be a barren field in the theoretical respect.

So that whatever characteristic articles of general theory the historical school may enrich the science with are to be looked for at the hands of those men who, like Professor Schmoller, have departed from the strict observance of the historical method. A peculiar interest, therefore, attaches to his work as the best accepted and most authoritative spokesman of that branch of historical economics which professes to cultivate theoretical inquiry. It serves to show in what manner and degree this more scientific wing of the historical school have outgrown the original " historical " standpoint and range of conceptions, and how they have passed from a distrust of all economic theory to an eager quest of theoretical formulations that shall cover all phenomena of economic life to better purpose than the body of doctrine received from the classical writers and more in consonance with the canons of contemporary science at large. That this should have been the outcome of the half-century of development through which the school has now passed might well seem unexpected, if not incredible, to any who saw the beginning of that divergence within the school, a generation ago, out of which this modernised, theoretical historical economics has arisen.

Professor Schmoller entered the field early, in the sixties, as a protestant against the aims and ideals then in vogue in economics. His protest ran not only against the methods and results of the classical writers, but also against the views professed by the leaders of the historical school, both as regards the scope of the science and as regards the character of the laws or generalisations sought by the science. His early work, in so far as he was at variance with his colleagues, was chiefly critical; and there is no good evidence that he then had a clear conception of the character of that constructive work to which it has been his persistent aim to turn the science. Hence he came to figure in common repute as an iconoclast and an extreme exponent of the historical school, in that he was held practically to deny the feasibility of a scientific treatment of economic matters and to aim at confining economics to narrative, statistics, and description. This iconoclastic or critical phase of his economic discussion is now past, and with it the uncertainty as to the trend and outcome of his scientific activity.

To understand the significance of the diversion created by Professor Schmoller as regards the scope and method of economics, it is necessary, very briefly, to indicate the position occupied by that early generation of historical economists from which his teaching diverged, and more particularly those points of the older canon at which he has come to differ characteristically from the views previously in vogue.

As regards the situation in which the historical school, as exemplified by its leaders, was then placed, it is, of course, something of a commonplace that by the end of its first twenty years of endeavor in the reform of economic science the school had, in point of systematic results, scarcely got beyond preliminaries. And even these preliminaries were not in all respects obviously to the purpose. A new and wider scope had been indicated for economic inquiry, as well as a new aim and method for theoretical discussion. But the new ideals of theoretical advance, as well as the ways and means indicated for their attainment, still had mainly a speculative interest. Nothing substantial had been done towards the realisation of the former or the *mise en œuvre* of the latter. The historical economists can scarcely be said at that time to have put their hand to the new engines which they professed to house in their workshop. Apart from polemics and speculation concerning ideals, the serious interest and endeavors of the school had up to that time been in the field of history rather than in that of economics, except so far as the adepts of the new school continued in a fragmentary way to inculcate and, in some slight and uncertain degree, to elaborate the dogmas of the classical writers whom they sought to discredit.

The character of historical economics at the time when Professor Schmoller entered on his work of criticism and revision is fairly shown by Roscher's writings. Whatever may be thought to-day of Roscher's rank as an economist, in contrast with Knies and Hildebrand, it will scarcely be questioned that at the close of the first quarter-century of the life history of the historical school it was Roscher's conception of the scope and method of economics that found the widest acceptance and that best expressed the animus of that body of students who professed to cultivate economics by the historical method. For the purpose in hand Roscher's views may, therefore, be taken as typical, all the more readily since for the very general purpose here intended there are no serious discrepancies between Roscher and his two illustrious contemporaries. The chief difference is that Roscher is more *naïve* and more specific. He has also left a more considerable volume of results achieved by the professed use of his method.

Roscher's professed method was what he calls the " historico-physiological " method. This he contrasts with the " philosophical " or " idealistic " method. But his air of depreciation as regards " philosophical " methods in economics must not be taken to mean that Roscher's own economic speculations were devoid of all philosophical or metaphysical basis. It only means that his philosophical postulates were different from those of the economists whom he discredits, and that they were regarded by him as self-evident.

As must necessarily be the case with a writer who had neither a special aptitude for nor special training in philosophical inquiries, Roscher's metaphysical postulates are, of course, chiefly tacit. They are the common-sense, commonplace metaphysics afloat in educated German circles in the time of Roscher's youth,— during the period when his growth and education gave him his outlook on life and knowledge and laid the basis of his intellectual habits; which means that these postulates belong to what Höffding has called the " Romantic " school of thought, and are of a Hegelian complexion. Roscher being not a professed philosophical student, it is neither easy nor safe to particularise closely as regards his fundamental metaphysical tenets; but, as near as so specific an identification of his philosophical outlook is practicable, he must be classed with the Hegelian " Right." But since the Hegelian metaphysics had in Roscher's youth an unbroken vogue in reputable German circles, especially in those ultra-reputable circles within which lay the gentlemanly life and human contact of Roscher, the postulates afforded by the Hegelian metaphysics were accepted simply as a matter of course, and were not recognised as metaphysical at all. And in this his metaphysical affiliation Roscher is fairly typical of the early historical school of economics.

The Hegelian metaphysics, in so far as bears upon the matter in hand, is a metaphysics of a self-realising life process. This life process, which is the central and substantial fact of the universe, is of a spiritual nature,— " spiritual," of course, being here not contrasted with " material." The life process is essentially active, self-determining, and unfolds by inner necessity,— by necessity of its own substantially active nature. The course of culture, in this view, is an unfolding (exfoliation) of the human spirit; and the task which economic science has in hand is to determine the laws of this cultural exfoliation in its economic aspect. But the laws of the cultural development with which the social sciences, in the Hegelian view, have to do are at one with the laws of the processes of the universe at large; and, more immediately, they are at one with the laws of the life process at large. For the universe at large is itself a self-unfolding life process, substantially of a spiritual character, of which the economic life process which occupies the interest of the economist is but a phase and an aspect. Now, the course of the processes of unfolding life in organic nature has been fairly well ascertained by the students of natural history and the like; and this, in the nature of the case, must afford a clew to the laws of cultural development, in its economic as well as in any other of its aspects or bearings,— the laws of life in the universe being all substantially spiritual and substantially at one. So we arrive at a physiological conception of culture after the analogy of the ascertained physiological processes seen in the biological domain. It is conceived to be physiological after the Hegelian manner of conceiving a physiological process, which is, however, not the same as the modern scientific conception of a physiological process.[4]

[4] A physiological conception of society, or of the community, had been employed before,— *e.g.*, by the Physiocrats,— and such a concept was reached also by English speculators — *e.g.*, Herbert Spencer — during Roscher's lifetime; but these physiological conceptions of society are reached by a different line of approach from that which led up to the late-Hegelian physiological or biological conception of human culture as a spiritual structure and process. The outcome is also a different one, both as regards the use made of the analogy and as regards the theoretical results reached by its aid.

It may be remarked, by the way, that neo-Hegelianism, of the "Left," likewise gave rise to a theory of a self-determining cultural exfoliation; namely, the so-called "Materialistic Conception of History" of the Marxian socialists. This Marxian conception, too, had much of a physiological air; but Marx and his coadjutors had an advantage over Roscher and his following, in that they were to a greater extent schooled in the Hegelian philosophy, instead of being uncritical receptacles of the Romantic commonplaces left by Hegelianism as a residue in popular thought. They were therefore more fully conscious of the bearing of their postulates and less *naïve* in their assumptions of self-sufficiency.

Since this quasi-physiological process of cultural development is conceived to be an unfolding of the self-realising human spirit, whose life history it is, it is of the nature of the case that the cultural process should run through a certain sequence of phases — a certain life history prescribed by the nature of the active, unfolding spiritual substance. The sequence is determined on the whole, as regards the general features of the development, by the nature of life on the human plane. The history of cultural growth and decline necessarily repeats itself, since it is substantially the same human spirit that seeks to realise itself in every comprehensive sequence of cultural development, and since this human spirit is the only factor in the case that has substantial force. In its generic features the history of past cultural cycles is, therefore, the history of the future. Hence the importance, not to say the sole efficacy for economic science, of an historical scrutiny of culture. A well-authenticated sequence of cultural phenomena in the history of the past is conceived to have much the same binding force for the sequence of cultural phenomena in the future as a " natural law," as the term has been understood in physics or physiology, is conceived to have as regards the course of phenomena in the life history of the human body; for the onward cultural course of the human spirit, actively unfolding by inner necessity, is an organic process, following logically from the nature of this self-realising spirit. If the process is conceived to meet with obstacles or varying conditions, it adapts itself to the circumstances in any given case, and it then goes on along the line of its own logical bent until it eventuates in the consummation given by its own nature. The environment, in this view, if it is not to be conceived simply as a function of the spiritual force at work, is, at the most, of subsidiary and transient consequence only. Environmental conditions can at best give rise to minor perturbations; they do not initiate a cumulative sequence which can profoundly affect the outcome or the ulterior course of the cultural process. Hence the sole, or almost sole, importance of historical inquiry in determining the laws of cultural development, economic or other.

The working conception which this romantic-historical school had of economic life, therefore, is, in its way, a conception of development, or evolution; but it is not to be confused with Darwinism or Spencerianism. Inquiry into the cultural development under the guidance of such preconceptions as these has led to generalisations, more or less arbitrary, regarding uniformities of sequence in phenomena, while the causes which determine the course of events, and which make the uniformity or variation of the sequence, have received but scant attention. The "natural laws" found by this means are necessarily of the nature of empiricism, colored by the bias or ideals of the investigator. The outcome is a body of aphoristic wisdom, perhaps beautiful and valuable after its kind, but quite fatuous when measured by the standards and aims of modern science. As is well known, no substantial theoretical gain was made along this romantic-historical line of inquiry and speculation, for the reason, apparently, that there are no cultural laws of the kind aimed at, beyond the unprecise generalities that are sufficiently familiar beforehand to all passably intelligent adults.

It has seemed necessary to offer this much in characterisation of that "historical" aim and method which afforded a point of departure for Professor Schmoller's work of revision. When he first raised his protest against the prevailing ideals and methods, as being ill-advised and not thorough-going, he does not seem himself to have been entirely free from this Romantic, or Hegelian, bias. There is evidence to the contrary in his early writings.[5] It cannot even be said that his later theoretical work does not show something of the same animus, as, *e.g.*, when he assumes that there is a meliorative trend in the course of cultural events.[6] What has differentiated his work from that of the group of writers which has above been called the elder line of historical economics is the weakness or relative absence of this bias in his theoretical work. Particularly, he has refused to bring his researches in the field of theory definitely to rest on ground given by the Hegelian, or Romantic, school of thought. He was from the first unwilling to accept classificatory statements of uniformity or of normality as an adequate answer to questions of scientific theory. He does not commonly deny the truth or the importance of the empirical generalisations aimed at by the early historical economists. Indeed, he makes much of them and has been notoriously urgent for a full survey of historical data and a painstaking digestion of materials with a view to a comprehensive work of empirical generalisation. As is well known, in his earlier work of criticism and methodological controversy he was led to contend that for at least one generation economists must be content to spend their energies on descriptive work of this kind; and he thereby earned the reputation of aiming to reduce economics to a descriptive knowledge of details and to confine its method to the Baconian ground of generalisation by simple enumeration. But this exhaustive historical scrutiny and description of

[5] *E.g.*, in his controversy with Treitschke. See *Grundfragen der Socialpolitik und der Volkswirtschaftslehre,* particularly pp. 24, 25.

[6] *E.g., Grundriss*, pp. 225, 409, 411.

detail has always, in Professor Schmoller's view, been preliminary to an eventual theory of economic life. The survey of details and the empirical generalisations reached by its help are useful for the scientific purpose only as they serve the end of an eventual formulation of the laws of causation that work out in the process of economic life. The ulterior question, to which all else is subsidiary, is a question of the causes at work rather than a question of the historical uniformities observable in the sequence of phenomena. The scrutiny of historical details serves this end by defining the scope and character of the several factors causally at work in the growth of culture, and, what is of more immediate consequence, as they are at work in the shaping of the economic activities and the economic aims of men engaged in this unfolding cultural process as it lies before the investigator in the existing situation.

In the preliminary work, then, of defining and characterising the causes or factors of economic life, historical investigation plays a large, if not the largest, part; but it is by no means the sole line of inquiry to which recourse is had for this purpose. Nor, it may be added, is this the sole use of historical inquiry. To the like end a comparative study of the climatic, geographical, and geological features of the community's environment is drawn into the inquiry; and more particularly there is a careful study of ethnographic parallels and a scrutiny of the psychological foundations of culture and the psychological factors involved in cultural change.

Hence it appears that Professor Schmoller's work differs from that of the elder line of historical economics in respect of the scope and character of the preliminaries of economic theory no less than in the ulterior aim which he assigns the science. It is only by giving a very broad meaning to the term that this latest development of the science can be called an "historical" economics. It is Darwinian rather than Hegelian, although with the earmarks of Hegelian affiliation visible now and again; and it is "historical" only in a sense similar to that in which a Darwinian account of the evolution of economic institutions might be caller historical. For the distinguishing characteristic of Professor Schmoller's work, that wherein it differs from the earlier work of the economists of his general class, is that it aims at a Darwinistic account of the origin, growth, persistence, and variation of institutions, in so far as these institutions have to do with the economic aspect of life either as cause or as effect. In much of what he has to say, he is at one with his contemporaries and predecessors within the historical school; and he shows at many points both the excellences and weaknesses due to his "historical" antecedents. But his striking and characteristic merits lie in the direction of a post-Darwinian, causal theory of the origin and growth of species in institutions. In this line of theoretical inquiry Professor Schmoller is not alone, nor does he, perhaps, go so far or with such singleness of purpose in this direction as some others do at given points; but the seniority belongs to him, and he is also in the lead as regards the comprehensiveness of his work.

But to return to the *Grundriss,* to which recourse must be had to substantiate the characterisation here offered. The entire work as projected comprises an Introduction and four Books, of which the introduction and the first two books are contained in the volume already published. The two books yet to be published, in a second volume, promise to be of a length corresponding to the first two. The present volume should accordingly contain approximately three-fifths of the whole, counted by bulk. The scheme of the work is as follows: An Introduction (pp. 1–124) treats of (1) the Concept of Economics, (2) the Psychical, Ethical (or Conventional, *sittliche*), and Legal Foundations of Economic Life and of Culture, and (3) the Literature and Method of the Science. This is followed by Book I. (pp. 125–228) on Land, Population, and the Industrial Arts, considered as collective phenomena and factors in economic life, and Book II. (pp. 229–457), on the Constitution of Economic Society, its chief organs and the causal factors to which they are due. Books III. and IV. are to deal with the Circulation of Goods and the Distribution of Income, and to give a genetic account of the Development of Economic Society.

The course outlined differs noticeably from what has been customary in treatises on economics. The point of departure is a comprehensive general survey of the factors which enter into the growth of culture, with special reference to their economic bearing. This survey runs chiefly on psychological and ethnographic ground, historical inquiry in the stricter sense being relatively scant and obviously of secondary consequence. It is followed up with a more detailed and searching discussion of the factors engaged in the economic process in any given situation. The factors, or " collective phenomena," in question are not the time-honored Land, Labor, and Capital, but rather population, material environment, and technological conditions. Here, too, the discussion has to do with ethnographic rather than with properly historical material. The question of population concerns not the numerical force of laborers, but rather the diversity of race characteristics and the bearing of race endowment upon the growth of economic institutions. The discussion of the material environment, again, has relatively little to say of the fertility of the soil, and gives much attention to diversities of climate, geographical situation, and geological and biological conditions. And this first book closes with a survey of the growth of technological knowledge and the industrial arts.

In all this the significant innovation lies not so much in the character of the details. They are for the most part commonplace enough as details of the sciences from which they are borrowed. They are shrewdly chosen and handled in such a way as to bring out their bearing upon the ulterior questions about which the economist's interest centers; but there is, as might be expected, little attempt to go back of the returns given by specialists in the several lines of research that are laid under contribution. But the significance of it all lies rather in the fact that material of this kind should have been drawn upon for a foundation for economic theory, and that it should have seemed necessary to Professor Schmoller to make this introductory survey so comprehensive and so painstaking as it is. Its meaning is that these features of human nature and these forces of nature and circumstances of environment are the agencies out of whose interaction the economic situation has arisen by a cumulative process of change, and that it is this cumulative process of development, and its complex and unstable outcome, that are to be the economist's subject-matter. The theoretical outcome for which such a foundation is prepared is necessarily of a genetic kind. It necessarily seeks to know and explain the structure and functions of economic society in terms of how and why they have come to be what they are, not, as so many economic writers have explained them, in terms of what they are good for and what they ought to be. It means, in other words, a deliberate attempt to substitute an inquiry into the efficient causes of economic life in the place of empirical generalisations, on the one hand, and speculations as to the eternal fitness of things, on the other hand.

It follows from the nature of the case that an economics of this genetic character, working on grounds of the kind indicated, comprises nothing in the way of advice or admonition, no maxims of expediency, and no economic, political, or cultural creed. How nearly Professor Schmoller conforms to this canon of continence is another question. The above indicates the scope of such doctrines as are consistently derivable from the premises with which the work under review starts out, not the scope of its writer's speculations on economic matters.

The second book, by the help of prehistoric and ethnographic material as well as history, deals with the evolution of the methods of social organisation,— the growth of institutions in so far as this growth shapes or is shaped by the exigencies of economic life. The "organs," or social-economic institutions, whose life history is passed in review are: the family; the methods of settlement and domicile, in town and country; the political units of control and administration; differentiation of functions between industrial and other classes and groups; ownership, its growth and distribution; social classes and associations; business enterprise, industrial organisations and corporations.

As regards the singleness of purpose with which Professor Schmoller has carried out the scheme of economic theory for which he has sketched the outlines and pointed the way, it is not possible to speak with the same confidence as of his preliminary work. It goes without saying that this further work of elaboration is excellent after its kind; and this excellence, which was to be looked for at Professor Schmoller's hands, may easily divert the read-

er's attention from the shortcomings of the work in respect of kind rather than of quality. Now, while a broad
generalisation on this head may be hazardous and is to be
taken with a large margin, still, with due allowance, the
following generalisation will probably stand, so far as
regards this first volume. So long as the author is occupied with the life-history of institutions down to contemporary developments, so long his discussion proceeds by
the dry light of the scientific interest, simply, as the term
" scientific " is understood among the modern adepts of
the natural sciences; but so soon as he comes to close
quarters with the situation of to-day, and reaches the
point where a dispassionate analysis and exposition of
the causal complex at work in contemporary institutional
changes should begin, so soon the scientific light breaks
up into all the colors of the rainbow, and the author
becomes an eager and eloquent counselor, and argues the
question of what ought to be and what modern society
must do to be saved. The argument at this point loses
the character of a genetic explanation of phenomena, and
takes on the character of appeal and admonition, urged
on grounds of expediency, of morality, of good taste, of
hygiene, of political ends, and even of religion. All this,
of course, is what we are used to in the common run of
writers of the historical school; but those students whose
interest centers in the science rather than in the ways and
means of maintaining the received cultural forms of German society have long fancied they had ground to hope
for something more to the purpose when Professor
Schmoller came to put forth his great systematic work.
Brilliant and no doubt valuable in its way and for its end,
this digression into homiletics and reformatory advice
means that the argument is running into the sands just at
the stage where the science can least afford it. It is pre-

cisely at this point, where men of less years and breadth and weight would find it difficult to hold tenaciously to the course of cause and effect through the maze of jarring interests and sentiments that make up the contemporary situation,— it is precisely at this point that a genetic theory of economic life most needs the guidance of the firm, trained, dispassionate hand of the master. And at this point his guidance all but fails us.

What has just been said applies generally to Professor Schmoller's treatment of contemporary economic development, and it should be added that it applies at nearly all points with more or less of qualification. But the qualifications required are not large enough to belie the general characterisation just offered. It would be asking too large an indulgence to follow the point up in this place through all the discussions of the volume that fairly come under this criticism. The most that may be done is to point for illustration to the handling which two or three of the social-economic " organs " receive. So, for instance, Book II. opens with an account of the family and its place and function in the structure of economic society. The discussion proceeds along the beaten paths of ethnographic research, with repeated and well-directed recourse to the psychological knowledge that Professor Schmoller always has well in hand. Coming down into recent times, the discussion still proceeds to show how the large economic changes of late mediæval and early modern times acted to break down the patriarchal régime of the earlier culture ; but at the same time there comes into sight (pp. 245–249) a bias in favor of the recent as against the earlier form of the household. The author is no longer content to show the exigencies which set the earlier patriarchal household aside in favor of the modified patriarchal household of more recent times. He also offers reasons why the later, modified form is intrinsically the more desirable; reasons, it should perhaps be said, which may be well taken, but which are beside the point so far as regards a scientific explanation of the changes under discussion.

The closing paragraphs of the section (91) dwell with a kindly insistence on the many elements of strength and beauty possessed by the form of household organisation handed down from the past generation to the present. The facts herewith recited by the author are, no doubt, of weight, and must be duly taken account of by any economist who ventures on a genetic discussion of the present situation and the changing fortunes of the received household. But Professor Schmoller has failed even to point out in what manner these elements of strength and beauty have in the recent past or may in the present and immediate future causally affect the fortunes of the institution. The failure to turn the material in question to scientific account becomes almost culpable in Professor Schmoller, since there are few, if any, who are in so favorable a position to outline the argument which a theoretical account of the situation at this point must take. Plainly, as shown by Professor Schmoller's argument, economic exigencies are working an incessant cumulative change in the form of organisation of the modern household; but he has done little towards pointing out in what manner and with what effect these exigencies come into play. Neither has he gone at all into the converse question, equally grave as a question of economic theory, of how the persistence, even though qualified, of the patriarchal family has modified and is modifying economic structure and function at other points and qualifying or accentuating the very exigencies themselves to which the changes wrought in the institution are to be traced.

Plainly, too, the strength and beauty of the traditionally received form of the household — that is to say, the habits of life and of complacency which are bound up with this household — are elements of importance in the modern situation as affects the degree of persistence and the direction of change which this institution shows under modern circumstances. They are psychological facts, facts of habit and propensity and spiritual fitness, the efficiency of which as live forces making for survival or variation is in this connection probably second to that of no other factors that could be named. We had, therefore, almost a right to expect that Professor Schmoller's profound and comprehensive erudition in the fields of psychology and cultural growth should turn these facts to better ends than a preachment concerning an intrinsically desirable consummation.

Regarding the present visible disintegration of the family, and the closely related " woman question," Professor Schmoller's observations are of much the same texture. He notes the growing disinclination to the old-fashioned family life on the part of the working population, and shows that there are certain economic causes for this growth or deterioration of sentiment. What he has to offer is made up of the commonplaces of latter-day social-economic discussion, and is charged with a strong undertone of deprecation. What the trend of the causes at work to alter or fortify this body of sentiment may be, counts for very little in what he says on the present movement or on the immediate future of the institution. The best he has to offer on the " woman question " is an off-hand reference of the ground of sentiment on which it rests to a recrudescence of the eighteenth century spirit of *égalité*. This notion of the equality of the sexes he refutes in graceful and affecting terms, and he pleads for the unbroken preservation of woman's sphere and man's primacy; as if the matter of superiority or inferiority between the sexes could conceivably be anything more than a conventional outcome of the habits of life imposed upon the community by the circumstances under which they live. How it has come to pass that under the economic exigencies of the past the physical and temperamental diversity between the sexes has been conventionally construed into a superiority of the man and an inferiority of the woman,— on this head he has no more to say or to suggest than on the correlate question of why this conventional interpretation of the facts has latterly not been holding its ancient ground. The discussion of the family and of the relation of the sexes, in modern culture, is marked throughout by unwillingness or inability to penetrate behind the barrier of conventional finality.

The discussion of the family just cited occupies the opening chapter of Book II. For a further instance of Professor Schmoller's handling of a modern economic problem, reference may be had to the closing chapter of Book I., on the " Development of Technological Expedients and its Economic Significance," but more particularly the sections (84–86) on the modern machine industry (pp. 211–228). In this discussion, also, the point of interest is the attention given to the latter-day phenomena of machine industry, and the author's method and animus in dealing with them. There is (pp. 211–218) a condensed and competent presentation of the main characteristics of the modern " machine age," followed (pp. 218–228) by a critical discussion of its cultural value. The customary eulogy, but with more than the customary discrimination, is given to the advantages of the régime of the machine in point of economy, creature comforts, and intellectual sweep; and it is pointed out how the régime of the machine has brought about a redistribution of wealth and of population and a reorganisation and redistribution of social and economic structures and functions. It is pointed out (p. 223) that the gravest social effect of the machine industry has been the creation of a large class of wage laborers. The material circumstances into which this class has been thrown, particularly in point of physical comfort, are dealt with in a sober and discriminating way; and it is shown (p. 224) that in the days of its fuller development the machine's régime has evolved a class of trained laborers who not only live in comfort, but are sound and strong in mind and body. But with the citation of these facts the pursuit of the chain of cause and effect in this modern machine situation comes to an end. The remainder of the space given to the subject is occupied with extremely sane and well-advised criticism, moral and æsthetic, and indications of what the proper ideals and ends of endeavor should be.

Professor Schmoller misses the opportunity he here has of dealing with this material in a scientific spirit and with some valuable results for economic theory. He could, it is not too bold to assume, have sketched for us an effective method and line of research to be pursued, for instance, in following up the scientific question of what may be the cultural, spiritual effects of the machine's régime upon this large body of trained workmen, and what this body of trained workmen in its turn counts for as a factor in shaping the institutional growth of the present and the economic and cultural situation of to-morrow. Work of this kind, there is reason to believe, Professor Schmoller could have done with better effect than any of his colleagues in the science; for he is, as already noticed above, possessed of the necessary qualifications in the way of psychological training, broad knowledge of the play of

cause and effect in cultural growth, and an ability to take
a scientific point of view.   Instead of this he harks back
again to the dreary homiletical waste of the traditional
Historismus.   It seems as if a topic which he deals with
as an objective matter so long as it lies outside the sphere
of every-day humanitarian and social solicitude, becomes
a matter to be passed upon by conventional standards of
taste, dignity, morality, and the like, so soon as it comes
within the sweep of latter-day German sentiment.

This habit of treating a given problem from these various and shifting points of view at times gives a kaleido-
scopic effect that is not without interest.   So in the matter of the technically trained working population in the
machine industry, to which reference has already been
made, something of an odd confusion appears when expressions taken from diverse phases of the discussion are
brought side by side.   He speaks of this class at one point
(p. 224) as " sound, strong, spiritually and morally advancing," superior in all these virtues to the working
classes of other times and places.   At another point (pp.
250–253) he speaks of the same popular element, under
the designation of " socialists," as perverse, degenerate,
and reactionary.   This latter characterisation may be
substantially correct, but it proceeds on grounds of taste
and predilection, not on grounds of scientifically determinable cause and effect.   And the two characterisations
apply to the same elements of population; for the substantial core and tone-giving factor of the radical socialistic element in the German community is, notoriously,
just this technically trained population of the industrial
towns where the discipline of the machine industry has
been at work with least mitigation.   The only other fairly
isolable element of a radical socialistic complexion is
found among the students of modern science.   Now, fur-

ther, in his speculations on the relation of technological knowledge to the advance of culture, Professor Schmoller points out (*e.g.,* p. 226) that a high degree of culture connotes, on the whole, a high degree of technological efficiency, and conversely. In this connection he makes use of the terms *Halbkulturvölker* and *Ganzkulturvölker* to designate different degrees of cultural maturity. It is curious to reflect, in the light of what he has to say on these several heads, that if the socialistically affected, technically trained population of the industrial towns, together with the radical-socialistic men of science, were abstracted from the German population, leaving substantially the peasantry, the slums, and the aristocracy great and small, the resulting German community would unquestionably have to be classed as a *Halbkulturvolk* in Professor Schmoller's scheme. Whereas the elements abstracted, if taken by themselves, would as unquestionably be classed among the *Ganzkulturvölker.*

In conclusion, one may turn to the concluding chapter (Book II., Chapter vii.) of the present volume for a final illustration of Professor Schmoller's method and animus in handling a modern economic problem. All the more so as this chapter on business enterprise better sustains that scientific attitude which the introductory outline leads the reader to look for throughout. It shows how modern business enterprise is in the main an outgrowth of commercial activity, as also that it has retained the commercial spirit down to the present. The motive force of business enterprise is the self-seeking quest of dividends; but Professor Schmoller shows, with more dispassionate insight than many economists, that this self-seeking motive is hemmed in and guided at all points in the course of its development by considerations and conventions that are not of a primarily self-seeking kind. He is not content to point to the beneficent working of a harmony of interests, but sketches the play of forces whereby a self-seeking business traffic has come to serve the interests of the community. Business enterprise has gradually emerged and come into its present central and dominant position in the community's industry as a concomitant of the growth of individual ownership and pecuniary discretion in modern life. It is therefore a phase of the modern cultural situation; and its survival and the direction of its further growth are therefore conditioned by the exigencies of the modern cultural situation. What this modern cultural situation is and what are the forces, essentially psychological, which shape the further growth of the situation, no one is better fitted to discuss than Professor Schmoller; and he has also given valuable indications (pp. 428–457) of what these factors are and how the inquiry into their working must be conducted. But even here, where a dispassionate tracing-out of the sequence of cause and effect should be easier to undertake, because less readily blurred with sentiment, than in the case, *e.g.,* of the family, the work of tracing the developmental sequence tapers off into advice and admonition proceeding on the assumption that the stage now reached is, or at least should be, final. The attention in the later pages diverges from the process of growth and its conditioning circumstances, to the desirability of maintaining the good results attained and to the ways and means of holding fast that which is good in the outcome already achieved. The question to which an answer is sought in discussing the present phase of the development is not a question as to what is taking place as respects the institution of business enterprise, but rather a question as to what form should be given to an optimistic policy of fostering business enterprise and turning it to account for the common good. At this point, as elsewhere, though perhaps in a less degree than elsewhere, the existing form of the institution is accepted as a finality. All this is disappointing in view of the fact that at no other point do modern economic institutions bear less of an air of finality than in the forms and conventions of business organisations and relations. As Professor Schmoller remarks (p. 455), the scope and character of business undertakings necessarily conform to the circumstances of the time, not to any logical scheme of development from small to great or from simple to complex. So also, one might be tempted to say, the expediency and the chance of ultimate survival of business enterprise is itself an open question, to be answered by a scrutiny of the forces that make for its survival or alteration, not by advice as to the best method of sustaining and controlling it.

What has here been said in criticism of Professor Schmoller's work, particularly as regards his departure from the path of scientific research in dealing with present-day phenomena, may, of course, have to be qualified, if not entirely set aside, when his work is completed with the promised genetic survey of modern institutions to be set forth in the concluding fourth book. Perhaps it may even be said that there is fair hope, on general grounds, of such a consummation; but the present volume does not afford ground for a confident expectation of this kind. It is perhaps needless, perhaps gratuitous, to add that the strictures offered indicate, after all, but relatively slight shortcomings in a work of the first magnitude.

INDUSTRIAL AND PECUNIARY EMPLOYMENTS [1]

FOR purposes of economic theory, the various activities of men and things about which economists busy themselves were classified by the early writers according to a scheme which has remained substantially unchanged, if not unquestioned, since their time. This scheme is the classical three-fold division of the factors of production under Land, Labor, and Capital. The theoretical aim of the economists in discussing these factors and the activities for which they stand has not remained the same throughout the course of economic discussion, and the three-fold division has not always lent itself with facility to new points of view and new purposes of theory, but the writers who have shaped later theory have, on the whole, not laid violent hands on the sacred formula. These facts must inspire the utmost reserve and circumspection in any one who is moved to propose even a subsidiary distinction of another kind between economic activities or agents. The terminology and the conceptual furniture of economics are complex and parti-colored enough without gratuitous innovation.

It is accordingly not the aim of this paper to set aside the time-honored classification of factors, or even to formulate an iconoclastic amendment, but rather to indicate how and why this classification has proved inadequate for certain purposes of theory which were not contemplated by the men who elaborated it. To this end a

[1] Reprinted by permission from *Publications of the American Economic Association.* series 3, Vol. II.

bit of preface may be in place as regards the aims which led to its formulation and the uses which the three-fold classification originally served.

The economists of the late eighteenth and early nineteenth centuries were believers in a Providential order, or an order of Nature. How they came by this belief need not occupy us here; neither need we raise a question as to whether their conviction of its truth was well or ill grounded. The Providential order or order of Nature is conceived to work in an effective and just way toward the end to which it tends; and in the economic field this objective end is the material welfare of mankind. The science of that time set itself the task of interpreting the facts with which it dealt, in terms of this natural order. The material circumstances which condition men's life fall within the scope of this natural order of the universe, and as members of the universal scheme of things men fall under the constraining guidance of the laws of Nature, who does all things well. As regards their purely theoretical work, the early economists are occupied with bringing the facts of economic life under natural laws conceived somewhat after the manner indicated; and when the facts handled have been fully interpreted in the light of this fundamental postulate the theoretical work of the scientist is felt to have been successfully done.

The economic laws aimed at and formulated under the guidance of this preconception are laws of what takes place "naturally" or "normally," and it is of the essence of things so conceived that in the natural or normal course there is no wasted or misdirected effort. The standpoint is given by the material interest of mankind, or, more concretely, of the community or "society" in

which the economist is placed; the resulting economic
theory is formulated as an analysis of the "natural"
course of the life of the community, the ultimate theoretical postulate of which might, not unfairly, be stated
as in some sort a law of the conservation of economic
energy. When the course of things runs off naturally
or normally, in accord with the exigencies of human
welfare and the constraining laws of nature, economic
income and outgo balance one another. The natural
forces at play in the economic field may increase indefinitely through accretions brought in under man's dominion and through the natural increase of mankind, and,
indeed, it is of the nature of things that an orderly
progress of this kind should take place; but within the
economic organism, as within the larger organism of
the universe, there prevails an equivalence of expenditure and returns, an equilibrium of flux and reflux,
which is not broken over in the normal course of things.
So it is, by implication, assumed that the product which
results from any given industrial process or operation is,
in some sense or in some unspecified respect, the equivalent of the expenditure of forces, or of the effort, or what
not, that has gone into the process out of which the
product emerges.

This theorem of equivalence is the postulate which
lies at the root of the classical theory of distribution,
but it manifestly does not admit of proof — or of disproof
either, for that matter; since neither the economic forces
which go into the process nor the product which emerges
are, in the economic respect, of such a tangible character as to admit of quantitative determination. They are
in fact incommensurable magnitudes. To this last remark the answer may conceivably present itself that the
equivalence in question is an equivalence in utility or in

exchange value, and that the quantitative determination of the various items in terms of exchange value or of utility is, theoretically, not impossible; but when it is called to mind that the forces or factors which go to the production of a given product take their utility or exchange value from that of the product, it will easily be seen that the expedient will not serve. The equivalence between the aggregate factors of production in any given case and their product remains a dogmatic postulate whose validity cannot be demonstrated in any terms that will not reduce the whole proposition to an aimless fatuity, or to metaphysical grounds which have now been given up.

The point of view from which the early, and even the later classical, economists discussed economic life was that of " the society " taken as a collective whole and conceived as an organic unit. Economic theory sought out and formulated the laws of the normal life of the social organism, as it is conceived to work out in that natural course whereby the material welfare of society is attained. The details of economic life are construed, for purposes of general theory, in terms of their subservience to the aims imputed to the collective life process. Those features of detail which will bear construction as links in the process whereby the collective welfare is furthered, are magnified and brought into the foreground, while such features as will not bear this construction are treated as minor disturbances. Such a procedure is manifestly legitimate and expedient in a theoretical inquiry whose aim is to determine the laws of health of the social organism and the normal functions of this organism in a state of health. The social organism is, in this theory, handled as an individual endowed with a consistent life purpose and something of an intelligent apprehension of what means will serve the ends which it seeks. With these collective ends the interests of the individual members are conceived to be fundamentally at one; and, while men may not see that their own individual interests coincide with those of the social organism, yet, since men are members of the comprehensive organism of nature and consequently subject to beneficent natural law, the ulterior trend of unrestrained individual action is, on the whole, in the right direction.

The details of individual economic conduct and its consequences are of interest to such a general theory chiefly as they further or disturb the beneficent " natural " course. But if the aims and methods of individual conduct were of minor importance in such an economic theory, that is not the case as regards individual rights. The early political economy was not simply a formulation of the natural course of economic phenomena, but it embodied an insistence on what is called " natural liberty." Whether this insistence on natural liberty is to be traced to utilitarianism or to a less specific faith in natural rights, the outcome for the purpose in hand is substantially the same. To avoid going too far afield, it may serve the turn to say that the law of economic equivalence, or conservation of economic energy, was, in early economics, backed by this second corollary of the order of nature, the closely related postulate of natural rights. The classical doctrine of distribution rests on both of these, and it is consequently not only a doctrine of what must normally take place as regards the course of life of society at large, but it also formulates what ought of right to take place as regards the remuneration for work and the distribution of wealth among men.

Under the resulting natural-economic law of equiva-

lence and equity, it is held that the several participants
or factors in the economic process severally get the
equivalent of the productive force which they expend.
They severally get as much as they produce; and conversely, in the normal case they severally produce as
much as they get. In the earlier formulations, as, for
example, in the authoritative formulation of Adam
Smith, there is no clear or consistent pronouncement as
regards the terms in which this equivalence between
production and remuneration runs. With the later,
classical economists, who had the benefit of a developed
utilitarian philosophy, it seems to be somewhat consistently conceived in terms of an ill-defined serviceability.
With some later writers it is an equivalence of exchange
values; but as this latter reduces itself to tautology, it
need scarcely be taken seriously. When we are told in
the later political economy that the several agents or
factors in production normally earn what they get, it is
perhaps fairly to be construed as a claim that the economic service rendered the community by any one of
the agents in production equals the service received by
the agent in return. In terms of serviceability, then, if
not in terms of productive force,[2] the individual agent,
or at least the class or group of agents to which the individual belongs, normally gets as much as he contributes and contributes as much as he gets. This applies
to all those employments or occupations which are ordinarily carried on in any community, throughout the
aggregate of men's dealings with the material means of
life. All activity which touches industry comes in under
this law of equivalence and equity.

[2] Some late writers, as, *e.g.*, J. B. Clark, apparently must be
held to conceive the equivalence in terms of productive force rather
than of serviceability; or, perhaps, in terms of serviceability on
one side of the equation and productive force on the other.

Now, to a theorist whose aim is to find the laws governing the economic life of a social organism, and who for this purpose conceives the economic community as a unit, the features of economic life which are of particular consequence are those which show the correlation of efforts and the solidarity of interests. For this purpose, such activities and such interests as do not fit into the scheme of solidarity contemplated are of minor importance, and are rather to be explained away or construed into subservience to the scheme of solidarity than to be incorporated at their face value into the theoretical structure. Of this nature are what are here to be spoken of under the term " pecuniary employments," and the fortune which these pecuniary employments have met at the hands of classical economic theory is such as is outlined in the last sentence.

In a theory proceeding on the premise of economic solidarity, the important bearing of any activity that is taken up and accounted for, is its bearing upon the furtherance of the collective life process. Viewed from the standpoint of the collective interest, the economic process is rated primarily as a process for the provision of the aggregate material means of life. As a late representative of the classical school expresses it: " Production, in fact, embraces every economic operation except consumption." [3] It is this aggregate productivity, and the bearing of all details upon the aggregate productivity, that constantly occupies the attention of the classical economists. What partially diverts their attention from this central and ubiquitous interest, is their persistent lapse into natural-rights morality.

The result is that acquisition is treated as a sub-head

[3] J. B. Clark, *The Distribution of Wealth,* p. 20.

under production, and effort directed to acquisition is construed in terms of production. The pecuniary activities of men, efforts directed to acquisition and operations incident to the acquisition or tenure of wealth, are treated as incidental to the distribution to each of his particular proportion in the production of goods. Pecuniary activities, in short, are handled as incidental features of the process of social production and consumption, as details incident to the method whereby the social interests are served, instead of being dealt with as the controlling factor about which the modern economic process turns.

Apart from the metaphysical tenets indicated above as influencing them, there are, of course, reasons of economic history for the procedure of the early economists in so relegating the pecuniary activities to the background of economic theory. In the days of Adam Smith, for instance, economic life still bore much of the character of what Professor Schmoller calls *Stadtwirtschaft*. This was the case to some extent in practice, but still more decidedly in tradition. To a greater extent than has since been the case, households produced goods for their own consumption, without the intervention of sale; and handicraftsmen still produced for consumption by their customers, without the intervention of a market. In a considerable measure, the conditions which the Austrian marginal-utility theory supposes, of a producing seller and a consuming buyer, actually prevailed. It may not be true that in Adam Smith's time the business operations, the bargain and sale of goods, were, in general, obviously subservient to their production and consumption, but it comes nearer being true at that time than at any time since then. And the tradition having once been put into form and authenticated by

Adam Smith, that such was the place of pecuniary transactions in economic theory, this tradition has lasted on in the face of later and further changes. Under the shadow of this tradition the pecuniary employments are still dealt with as auxiliary to the process of production, and the gains from such employments are still explained as being due to a productive effect imputed to them.

According to ancient prescription, then, all normal, legitimate economic activities carried on in a well regulated community serve a materially useful end, and so far as they are lucrative they are so by virtue of and in proportion to a productive effect imputed to them. But in the situation as it exists at any time there are activities and classes of persons which are indispensable to the community, or which are at least unavoidably present in modern economic life, and which draw some income from the aggregate product, at the same time that these activities are not patently productive of goods and can not well be classed as industrial, in any but a highly sophisticated sense. Some of these activities, which are concerned with economic matters but are not patently of an industrial character, are integral features of modern economic life, and must therefore be classed as normal; for the existing situation, apart from a few minor discrepancies, is particularly normal in the apprehension of present-day economists. Now, the law of economic equivalence and equity says that those who normally receive in income must perforce serve some productive end; and, since the existing organization of society is conceived to be eminently normal, it becomes imperative to find some ground on which to impute industrial productivity to those classes and employments which do not at first view appear to be industrial at all. Hence there is commonly visible in the classical political economy, ancient and modern, a strong inclination to make the schedule of industrially productive employments very comprehensive; so that a good deal of ingenuity has been spent in economically justifying their presence by specifying the productive effect of such non-industrial factors as the courts, the army, the police, the clergy, the schoolmaster, the physician, the opera singer.

But these non-economic employments are not so much to the point in the present inquiry; the point being employments which are unmistakably economic, but not industrial in the naïve sense of the word industry, and which yield an income.

Adam Smith analysed the process of industry in which he found the community of his time engaged, and found the three classes of agents or factors: Land, Labor, and Capital (stock). The productive factors engaged being thus determined, the norm of natural-economic equivalence and equity already referred to above, indicated what would be the natural sharers in the product. Later economists have shown great reserve about departing from this three-fold division of factors, with its correlated three-fold division of sharers of remuneration; apparently because they have retained an instinctive, indefeasible trust in the law of economic equivalence which underlies it. But circumstances have compelled the tentative intrusion of a fourth class of agent and income. The undertaker and his income presently came to be so large and ubiquitous figures in economic life that their presence could not be overlooked by the most normalising economist. The undertaker's activity has been interpolated in the scheme of productive factors, as a peculiar and fundamentally distinctive kind of labor, with the function of coördinating and directing industrial processes. Similarly, his income has been interpolated in the scheme of distribution, as a peculiar kind of wages, proportioned to the heightened productivity given the industrial process by his work.[4] His work is discussed in expositions of the theory of production. In discussions of his functions and his income the point of the argument is, how and in what degree does his activity increase the output of goods, or how and in what degree does it save wealth to the community. Beyond his effect in enhancing the effective volume of the aggregate wealth the undertaker receives but scant attention, apparently for the reason that so soon as that point has been disposed of the presence of the undertaker and his income has been reconciled with the tacitly accepted natural law of equivalence between productive service and remuneration. The normal balance has been established, and the undertaker's function has been justified and subsumed under the ancient law that Nature does all things well and equitably.

This holds true of the political economy of our grandfathers. But this aim and method of handling the phenomena of life for theoretical ends, of course, did not go out of vogue abruptly in the days of our grandfathers.[5] There is a large sufficiency of the like aim and animus in the theoretical discussions of a later time; but specifically to cite and analyse the evidence of its presence

[4] The undertaker gets an income; therefore he must produce goods. But human activity directed to the production of goods is labor; therefore the undertaker is a particular kind of laborer. There is, of course, some dissent from this position.

[5] The change which has supervened as regards the habitual resort to a natural law of equivalence is in large part a change with respect to the degree of immediacy and "reality" imputed to this law, and to a still greater extent a change in the degree of overtness with which it is avowed.

would be laborious, nor would it conduce to the general peace of mind.

Some motion towards a further revision of the scheme is to be seen in the attention which has latterly been given to the function and the profits of that peculiar class of undertakers whom we call speculators. But even on this head the argument is apt to turn on the question of how the services which the speculator is conceived to render the community are to be construed into an equivalent of his gains.[6] The difficulty of interpretation encountered at this point is considerable, partly because it is not quite plain whether the speculators as a class come out of their transactions with a net gain or with a net loss. A systematic net loss, or a no-profits balance, would, on the theory of equivalence, mean that the class which gets this loss or doubtful gain is of no service to the community; yet we are, out of the past, committed to the view that the speculator is useful — indeed economically indispensable — and shall therefore have his reward. In the discussions given to the speculator and his function some thought is commonly given to the question of the " legitimacy " of the speculator's traffic. The legitimate speculator is held to earn his gain by services of an economic kind rendered the community. The recourse to this epithet, " legitimate," is chiefly of interest as showing that the tacit postulate of a natural order is still in force. Legitimate are such speculative dealings as are, by the theorist, conceived to serve the ends of the community, while

[6] See, *e.g.*, a paper by H. C. Emery in the *Papers and Proceedings of the Twelfth Annual Meeting* of the American Economic Association, on " The Place of the Speculator in the Theory of Distribution," and more particularly the discussion following the paper.

illegitimate speculation is that which is conceived to be disserviceable to the community.

The theoretical difficulty about the speculator and his gains (or losses) is that the speculator *ex professo* is quite without interest in or connection with any given industrial enterprise or any industrial plant. He is, industrially speaking, without visible means of support. He may stake his risks on the gain or on the loss of the community with equal chances of success, and he may shift from one side to the other without winking.

The speculator may be treated as an extreme case of undertaker, who deals exclusively with the business side of economic life rather than with the industrial side. But he differs in this respect from the common run of business men in degree rather than in kind. His traffic is a pecuniary traffic, and it touches industry only remotely and uncertainly; while the business man as commonly conceived is more or less immediately interested in the successful operation of some concrete industrial plant. But since the undertaker first broke into economic theory, some change has also taken place as regards the immediacy of the relations of the common run of undertakers to the mechanical facts of the industries in which they are interested. Half a century ago it was still possible to construe the average business manager in industry as an agent occupied with the superintendence of the mechanical processes involved in the production of goods or services. But in the later development the connection between the business manager and the mechanical processes has, on an average, grown more remote; so much so, that his superintendence of the plant or of the processes is frequently visible only to the scientific imagination. That activity by virtue of which the undertaker is classed as such makes him a businessman, not a mechanic or foreman of the shop. His super-intendence is a superintendence of the pecuniary affairs of the concern, rather than of the industrial plant; especially is this true in the higher development of the modern captain of industry. As regards the nature of the employment which characterises the undertaker, it is possible to distinguish him from the men who are mechanically engaged in the production of goods, and to say that his employment is of a business or pecuniary kind, while theirs is of an industrial or mechanical kind. It is not possible to draw a similar distinction between the undertaker who is in charge of a given industrial concern, and the business man who is in business but is not interested in the production of goods or services. As regards the character of employment, then, the line falls not between legitimate and illegitimate pecuniary transactions, but between business and industry.

The distinction between business and industry has, of course, been possible from the beginning of economic theory, and, indeed, the distinction has from time to time temporarily been made in the contrast frequently pointed out between the proximate interest of the business man and the ulterior interest of society at large. What appears to have hindered the reception of the distinction into economic doctrine, is the constraining presence of a belief in an order of Nature and the habit of conceiving the economic community as an organism. The point of view given by these postulates has made such a distinction between employments not only useless, but even disserviceable for the ends to which theory has been directed. But the fact has come to be gradually more and more patent that there are constantly, normally present in modern economic life an important range of activities and classes of persons who work for an income but of whom it cannot be said that they, either proximately or remotely, apply themselves to the production of goods. Their services, proximate or remote, to society are often of quite a problematical character. They are ubiquitous, and it will scarcely do to say that they are anomalous, for they are of ancient prescription, they are within the law and within the pale of popular morals.

Of these strictly economic activities that are lucrative without necessarily being serviceable to the community, the greater part are to be classed as " business." Perhaps the largest and most obvious illustration of these legitimate business employments is afforded by the speculators in securities. By way of further illustration may be mentioned the extensive and varied business of real-estate men (land-agents) engaged in the purchase and sale of property for speculative gain or for a commission; so, also, the closely related business of promoters and boomers of other than real-estate ventures; as also attorneys, brokers, bankers, and the like, although the work performed by these latter will more obviously bear interpretation in terms of social serviceability. The traffic of these business men shades off insensibly from that of the *bona fide* speculator who has no ulterior end of industrial efficiency to serve, to that of the captain of industry or entrepreneur as conventionally set forth in the economic manuals.

The characteristic in which these business employments resemble one another, and in which they differ from the mechanical occupations as well as from other non-economic employments, is that they are concerned primarily with the phenomena of value — with exchange or market values and with purchase and sale — and only indirectly and secondarily, if at all, with mechanical processes. What holds the interest and guides and shifts the attention of men within these employments is the main chance. These activities begin and end within what may broadly be called " the higgling of the market." Of the industrial employments, in the stricter sense, it may be said, on the other hand, that they begin and end outside the higgling of the market. Their proximate aim and effect is the shaping and guiding of material things and processes. Broadly, they may be said to be primarily occupied with the phenomena of material serviceability, rather than with those of exchange value. They are taken up with phenomena which make the subject matter of Physics and the other material sciences.

The business man enters the economic life process from the pecuniary side, and so far as he works an effect in industry he works it through the pecuniary dispositions which he makes. He takes thought most immediately of men's convictions regarding market values; and his efforts as a business man are directed to the apprehension, and commonly also to the influencing of men's beliefs regarding market values. The objective point of business is the diversion of purchase and sale into some particular channel, commonly involving a diversion from other channels. The laborer and the man engaged in directing industrial processes, on the other hand, enter the economic process from the material side; in their characteristic work they take thought most immediately of mechanical effects, and their attention is directed to turning men and things to account for the compassing of some material end. The ulterior aim, and the ulterior effect, of these industrial employments may be some pecuniary result; work of this class commonly results in an enhancement, or at least an alteration, of market values. Conversely, business activity may, and in a majority of cases it perhaps does, effect an enhancement of the aggregate material wealth of the community, or the aggregate serviceability of the means at hand; but such an industrial outcome is by no means bound to follow from the nature of the business man's work.

From what has just been said it appears that, if we retain the classical division of economic theory into Production, Distribution, and Consumption, the pecuniary employments do not properly fall under the first of these divisions, Production, if that term is to retain the meaning commonly assigned to it. In an earlier and less specialised organisation of economic life, particularly, the undertaker frequently performs the work of a foreman or a technological expert, as well as the work of business management. Hence in most discussions of his work and his theoretical relations his occupation is treated as a composite one. The technological side of his composite occupation has even given a name to his gains (wages of superintendence), as if the undertaker were primarily a master-workman. The distinction at this point has been drawn between classes of persons instead of between classes of employments; with the result that the evident necessity of discussing his technological employment under production has given countenance to the endeavor to dispose of the undertaker's business activity under the same head. This endeavor has, of course, not wholly succeeded.

In the later development, the specialisation of work in the economic field has at this point progressed so far, and the undertaker now in many cases comes so near being occupied with business affairs alone, to the exclusion of technological direction and supervision, that, with this object lesson before us, we no longer have the same difficulty in drawing a distinction between business and industrial employments. And even in the earlier days of the doctrines, when the aim was to dispose of the undertaker's work under the theoretical head of Production, the business side of his work persistently obtruded itself for discussion in the books and chapters given to Distribution and Exchange. The course taken by the later theoretical discussion of the entrepreneur, leaves no question but that the characteristic fact about his work is that he is a business man, occupied with pecuniary affairs.

Such pecuniary employments, of which the purely fiscal or financiering forms of business are typical, are nearly all and nearly throughout, conditioned by the institution of property or ownership — an institution which, as John Stuart Mill remarks, belongs entirely within the theoretical realm of Distribution. Ownership, no doubt, has its effect upon productive industry, and, indeed, its effect upon industry is very large, both in scope and range, even if we should not be prepared to go the length of saying that it fundamentally conditions all industry; but ownership is not itself primarily or immediately a contrivance for production. Ownership directly touches the results of industry, and only indirectly the methods and processes of industry. If the institution of property be compared with such another feature of our culture, for instance, as the domestication of plants or the smelting of iron, the meaning of what has just been said may seem clearer.

So much then of the business man's activity as is conditioned by the institution of property, is not to be classed, in economic theory, as productive or industrial activity at all. Its objective point is an alteration of the distribution of wealth. His business is, essentially, to sell and buy — sell in order to buy cheaper, buy in order to sell dearer.[7] It may or may not, indirectly, and in a sense incidentally, result in enhanced production. The business man may be equally successful in his enterprise, and he may be equally well remunerated, whether his activity does or does not enrich the community. Immediately and directly, so long as it is confined to the pecuniary or business sphere, his activity is incapable of enriching or impoverishing the community as a whole except, after the fashion conceived by the mercantilists, through his dealings with men of other communities. The circulation and distribution of goods incidental to the business man's traffic is commonly, though not always or in the nature of the case, serviceable to the community; but the distribution of goods is a mechanical, not a pecuniary transaction, and it is not the objective point of business nor its invariable outcome. From the point of view of business, the distribution or circulation of goods is a means of gain, not an end sought.

It is true, industry is closely conditioned by business. In a modern community, the business man finally decides what may be done in industry, or at least in the greater number and the more conspicuous branches of industry. This is particularly true of those branches that are currently thought of as peculiarly modern. Under existing circumstances of ownership, the discretion in economic matters, industrial or otherwise, ultimately rests in the hands of the business men. It is their business to have to do with property, and property means the discretionary control of wealth. In point of character, scope and growth, industrial processes and plants adapt themselves to the exigencies of the market, wherever there is a developed market, and the exigencies of the market are pecuniary exigencies. The business man, through his

[7] Cf. *e.g.,* Marx, *Capital,* especially bk. I, ch. IV.

pecuniary dispositions, enforces his choice of what industrial processes shall be in use. He can, of course, not create or initiate methods or aims for industry; if he does so he steps out of the business sphere into the material domain of industry. But he can decide whether and which of the known processes and industrial arts shall be practiced, and to what extent. Industry must be conducted to suit the business man in his quest for gain; which is not the same as saying that it must be conducted to suit the needs or the convenience of the community at large. Ever since the institution of property was definitely installed, and in proportion as purchase and sale has been practiced, some approach has been made to a comprehensive system of control of industry by pecuniary transactions and for pecuniary ends, and the industrial organisation is nearer such a consummation now than it ever has been. For the great body of modern industry the final term of the sequence is not the production of the goods but their sale; the endeavor is not so much to fit the goods for use as for sale. It is well known that there are many lines of industry in which the cost of marketing the goods equals the cost of making and transporting them.

Any industrial venture which falls short in meeting the pecuniary exigencies of the market declines and yields ground to others that meet them with better effect. Hence shrewd business management is a requisite to success in any industry that is carried on within the scope of the market. Pecuniary failure carries with it industrial failure, whatever may be the cause to which the pecuniary failure is due — whether it be inferiority of the goods produced, lack of salesmanlike tact, popular prejudice, scanty or ill-devised advertising, excessive truthfulness, or what not. In this way industrial results are closely dependent upon the presence of business ability; but the cause of this dependence of industry upon business in a given case is to be sought in the fact that other rival ventures have the backing of shrewd business management, rather than in any help which business management in the aggregate affords to the aggregate industry of the community.   Shrewd and farsighted business management is a requisite of survival in the competitive pecuniary struggle in which the several industrial concerns are engaged, because shrewd and farsighted business management abounds and is employed by all the competitors.   The ground of survival in the selective process is fitness for pecuniary gain, not fitness for serviceability at large.   Pecuniary management is of an emulative character and gives, primarily, relative success only.   If the change were equitably distributed, an increase or decrease of the aggregate or average business ability in the community need not immediately affect the industrial efficiency or the material welfare of the community.   The like can not be said with respect to the aggregate or average industrial capacity of the men at work.   The latter are, on the whole, occupied with production of goods; the business men, on the other hand, are occupied with the acquisition of them.

Theoreticians who are given to looking beneath the facts and to contemplating the profounder philosophical meaning of life speak of the function of the undertaker as being the guidance and coördination of industrial processes with a view to economies of production.   No doubt, the remoter effect of business transactions often is such coördination and economy, and, no doubt also, the undertaker has such economy in view and is stimulated to his maneuvers of combination by the knowledge that certain economies of this kind are feasible and will inure to his gain if the proper business arrangements can be effected. But it is practicable to class even this indirect furthering of industry by the undertaker as a permissive guidance only. The men in industry must first create the mechanical possibility of such new and more economical methods and arrangements, before the undertaker sees the chance, makes the necessary business arrangements, and gives directions that the more effective working arrangements be adopted.

It is notorious, and it is a matter upon which men dilate, that the wide and comprehensive consolidations and coördinations of industry, which often add so greatly to its effectiveness, take place at the initiative of the business men who are in control. It should be added that the fact of their being in control precludes such coördination from being effected except by their advice and consent. And it should also be added, in order to a passably complete account of the undertaker's function, that he not only can and does effect economising co-ordinations of a large scope, but he also can and does at times inhibit the process of consolidation and coördination. It happens so frequently that it might fairly be said to be the common run that business interests and undertaker's maneuvers delay consolidation, combination, coördination, for some appreciable time after they have become patently advisable on industrial grounds. The industrial advisability or practicability is not the decisive point. Industrial advisability must wait on the eventual convergence of jarring pecuniary interests and on the strategical moves of business men playing for position.

Which of these two offices of the business man in modern industry, the furthering or the inhibitory, has the more serious or more far-reaching consequences is, on the whole, somewhat problematical. The furtherance of coördination by the modern captain of industry bulks large in our vision, in great part because the process of widening coördination is of a cumulative character. After a given step in coördination and combination has been taken, the next step takes place on the basis of the resulting situation. Industry, that is to say the working force engaged in industry, has a chance to develop new and larger possibilities to be taken further advantage of. In this way each successive move in the enhancement of the efficiency of industrial processes, or in the widening of coördination in industrial processes, pushes the captain of industry to a further concession, making possible a still farther industrial growth. But as regards the undertaker's inhibitory dealings with industrial coördination the visible outcome is not so striking. The visible outcome is simply that nothing of the kind then takes place in the premises. The potential cumulative sequence is cut off at the start, and so it does not figure in our appraisement of the disadvantage incurred. The loss does not commonly take the more obtrusive form of an absolute retreat, but only that of a failure to advance where the industrial situation admits of an advance.

It is, of course, impracticable to foot up and compare gain and loss in such a case, where the losses, being of the nature of inhibited growth, cannot be ascertained. But since the industrial serviceability of the captain of industry is, on the whole, of a problematical complexion, it should be advisable for a cautious economic theory not to rest its discussion of him on his serviceability.[8]

[8] It is not hereby intended to depreciate the services rendered the community by the captain of industry in his management of business. Such services are no doubt rendered and are also no

It appears, then, as all economists are no doubt aware, that there is in modern society a considerable range of

doubt of substantial value. Still less is it the intention to decry the pecuniary incentive as a motive to thrift and diligence. It may well be that the pecuniary traffic which we call business is the most effective method of conducting the industrial policy of the community; not only the most effective that has been contrived, but perhaps the best that can be contrived. But that is a matter of surmise and opinion. In a matter of opinion on a point that can not be verified, a reasonable course is to say that the majority are presumably in the right. But all that is beside the point. However probable or reasonable such a view may be, it can find no lodgment in modern scientific theory, except as a corollary of secondary importance. Nor can scientific theory build upon the ground it may be conceived to afford. Policy may so build, but science can not. Scientific theory is a formulation of the laws of phenomena in terms of the efficient forces at work in the sequence of phenomena. So long as (under the old dispensation of the order of nature) the animistically conceived natural laws, with their God-given objective end, were considered to exercise a constraining guidance over the course of events whereof they were claimed to be laws, so long it was legitimate scientific procedure for economists to formulate their theory in terms of these laws of the natural course; because so long they were speaking in terms of what was, to them, the efficient forces at work. But so soon as these natural laws were reduced to the plane of colorless empirical generalization as to what commonly happens, while the efficient forces at work are conceived to be of quite another cast, so soon must theory abandon the ground of the natural course, sterile for modern scientific purposes, and shift to the ground of the causal sequence, where alone it will have to do with the forces at work as they are conceived in our time. The generalisations regarding the normal course, as " normal" has been defined in economics since J. S. Mill, are not of the nature of theory, but only rule-of-thumb. And the talk about the " function" of this and that factor of production, etc., in terms of the collective life purpose, goes to the same limbo; since the collective life purpose is no longer avowedly conceived to cut any figure in the every-day guidance of economic activities or the shaping of economic results.

The doctrine of the social-economic function of the undertaker may for the present purpose be illustrated by a supposititious parallel from Physics. It is an easy generalisation, which will scarcely be questioned, that, in practice, pendulums commonly activities, which are not only normally present, but which constitute the vital core of our economic system; which are not directly concerned with production, but which are nevertheless lucrative. Indeed, the group comprises most of the highly remunerative employments in modern economic life. The gains from these employments must plainly be accounted for on other grounds than their productivity, since they need have no productivity.

But it is not only as regards the pecuniary employments that productivity and remuneration are constitutionally out of touch. It seems plain, from what has already been said, that the like is true for the remuneration gained in the industrial employments. Most wages, particularly those paid in the industrial employments proper, as contrasted with those paid for domestic or personal service, are paid on account of pecuniary serviceability to the employer, not on grounds of material serviceability to mankind at large. The product is valued, sought and paid for on account of and in some proportion to its vendibility, not for more recondite rea-

vibrate in a plane approximately parallel with the nearest wall of the clock-case in which they are placed. The normality of this parallelism is fortified by the further observation that the vibrations are also commonly in a plane parallel with the nearest wall of the room; and when it is further called to mind that the balance which serves the purpose of a pendulum in watches similarly vibrates in a plane parallel with the walls of its case, the absolute normality of the whole arrangement is placed beyond question. It is true, the parallelism is not claimed to be related to the working of the pendulum, except as a matter of fortuitous convenience; but it should be manifest from the generality of the occurrence that in the normal case, in the absence of disturbing causes, and in the long run, all pendulums will "naturally" tend to swing in a plane faultlessly parallel with the nearest wall. The use which has been made of the "organic concept," in economics and in social science at large, is fairly comparable with this supposititious argument concerning the pendulum.

sons of ulterior human welfare at large. It results that there is no warrant, in general theory, for claiming that the work of highly paid persons (more particularly that of highly paid business men) is of greater substantial use to the community than that of the less highly paid. At the same time, the reverse could, of course, also not be claimed. Wages, resting on a pecuniary basis, afford no consistent indication of the relative productivity of the recipients, except in comparisons between persons or classes whose products are identical except in amount, — that is to say, where a resort to wages as an index of productivity would be of no use anyway.[9]

A result of the acceptance of the theoretical distinction here attempted between industrial and pecuniary employments and an effective recognition of the pecuniary basis of the modern economic organisation would be to dissociate the two ideas of productivity and remuneration. In mathematical language, remuneration could no longer be conceived and handled as a " function " of productivity,— unless productivity be taken to mean pecuniary serviceability to the person who pays the remuneration. In modern life remuneration is, in the last analysis, uniformly obtained by virtue of an agreement between individuals who commonly proceed on their own interest in point of pecuniary gain. The remuneration may, therefore, be said to be a " function " of the pecuniary service rendered the person who grants

[9] Since the ground of payment of wages is the vendibility of the product, and since the ground of a difference in wages is the different vendibility of the product acquired through the purchase of the labor for which the wages are paid, it follows that wherever the difference in vendibility rests on a difference in the magnitude of the product alone, there wages should be somewhat in proportion to the magnitude of the product.

the remuneration; but what is pecuniarily serviceable to the individual who exercises the discretion in the matter need not be productive of material gain to the community as a whole. Nor does the algebraic sum of individual pecuniary gains measure the aggregate serviceability of the activities for which the gains are got.

In a community organized, as modern communities are, on a pecuniary basis, the discretion in economic matters rests with the individuals, in severalty; and the aggregate of discrete individual interests nowise expresses the collective interest. Expressions constantly recur in economic discussions which imply that the transactions discussed are carried out for the sake of the collective good or at the initiative of the social organism, or that "society" rewards so and so for their services. Such expressions are commonly of the nature of figures of speech and are serviceable for homiletical rather than for scientific use. They serve to express their user's faith in a beneficent order of nature, rather than to convey or to formulate information in regard to facts.

Of course, it is still possible consistently to hold that there is a natural equivalence between work and its reward, that remuneration is naturally, or normally, or in the long run, proportioned to the material service rendered the community by the recipient; but that proposition will hold true only if "natural" or "normal" be taken in such a sense as to admit of our saying that the natural does not coincide with the actual; and it must be recognised that such a doctrine of the "natural" apportionment of wealth or of income disregards the efficient facts of the case. Apart from effects of this kind in the way of equitable arrangements traceable to grounds of sentiment, the only recourse which modern science would afford the champion of a doctrine of natural distribution, in the sense indicated, would be a doctrine of natural selection; according to which all disserviceable or unproductive, wasteful employments would, perforce, be weeded out as being incompatible with the continued life of any community that tolerated them. But such a selective elimination of unserviceable or wasteful employments would presume the following two conditions, neither of which need prevail: (1) It must be assumed that the disposable margin between the aggregate productivity of industry and the aggregate necessary consumption is so narrow as to admit of no appreciable waste of energy or of goods; (2) it must be assumed that no deterioration of the condition of society in the economic respect does or can "naturally" take place. As to the former of these two assumptions, it is to be said that in a very poor community, and under exceptionally hard economic circumstances, the margin of production may be as narrow as the theory would require. Something approaching this state of things may be found, for instance, among some Eskimo tribes. But in a modern industrial community — where the margin of admissible waste probably always exceeds fifty per cent. of the output of goods — the facts make no approach to the hypothesis. The second assumed condition is, of course, the old-fashioned assumption of a beneficent, providential order or meliorative trend in human affairs. As such, it needs no argument at this day. Instances are not far to seek of communities in which economic deterioration has taken place while the system of distribution, both of income and of accumulated wealth, has remained on a pecuniary basis.

To return to the main drift of the argument. The

pecuniary employments have to do with wealth in point of ownership, with market values, with transactions of exchange, purchase and sale, bargaining for the purpose of pecuniary gain. These employments make up the characteristic occupations of business men, and the gains of business are derived from successful endeavors of the pecuniary kind. These business employments are the characteristic activity (constitute the " function ") of what are in theory called undertakers. The dispositions which undertakers, *qua* business men, make are pecuniary dispositions — whatever industrial sequel they may or may not have — and are carried out with a view to pecuniary gain. The wealth of which they have the discretionary disposal may or may not be in the form of " production goods "; but in whatever form the wealth in question is conceived to exist, it is handled by the undertakers in terms of values and is disposed of by them in the pecuniary respect. When, as may happen, the undertaker steps down from the pecuniary plane and directs the mechanical handling and functioning of " production goods," he becomes for the time a foreman. The undertaker, if his business venture is of the industrial kind, of course takes cognizance of the aptness of a given industrial method or process for his purpose, and he has to choose between different industrial processes in which to invest his values; but his work as undertaker, simply, is the investment and shifting of the values under his hand from the less to the more gainful point of investment. When the investment takes the form of material means of industry, or industrial plant, the sequel of a given business transaction is commonly some particular use of such means; and when such industrial use follows, it commonly takes place at the hands of other men than the undertaker, although it takes place within limits imposed by the pecuniary exigencies of which the undertaker takes cognizance. Wealth turned to account in the way of investment or business management may or may not, in consequence, be turned to account, materially, for industrial effect. Wealth, values, so employed for pecuniary ends is capital in the business sense of the word.[10] Wealth, material means of industry, physically employed for industrial ends is capital in the industrial sense. Theory, therefore, would require that care be taken to distinguish between capital as a pecuniary category, and capital as an industrial category, if the term capital is retained to cover the two concepts.[11] The distinction here made substantially coincides with a distinction which many late writers have arrived at from a different point of approach and have, with varying success, made use of under different terms.[12]

A further corollary touching capital may be pointed out. The gains derived from the handling of capital in the pecuniary respect have no immediate relation, stand in no necessary relation of proportion, to the productive effect compassed by the industrial use of the material

[10] All wealth so used is capital, but it does not follow that all pecuniary capital is social wealth.

[11] In current theory the term capital is used in these two senses; while in business usage it is employed pretty consistently in the former sense alone. The current ambiguity in the term capital has often been adverted to by economists, and there may be need of a revision of the terminology at this point; but this paper is not concerned with that question.

[12] Professor Fetter, in a recent paper (*Quarterly Journal of Economics,* November, 1900) is, perhaps, the writer who has gone the farthest in this direction in the definition of the capital concept. Professor Fetter wishes to confine the term capital to pecuniary capital, or rather to such pecuniary capital as is based on the ownership of material goods. The wisdom of such a terminological expedient is, of course, not in question here.

means over which the undertaker may dispose; although the gains have a relation of dependence to the effects achieved in point of vendibility. But vendibility need not, even approximately, coincide with serviceability, except serviceability be construed in terms of marginal utility or some related conception, in which case the outcome is a tautology. Where, as in the case commonly assumed by economists as typical, the investing undertaker seeks his gain through the production and sale of some useful article, it is commonly also assumed that his effort is directed to the most economical production of as large and serviceable a product as may be, or at least it is assumed that such production is the outcome of his endeavors in the natural course of things. This account of the aim and outcome of business enterprise may be natural, but it does not describe the facts. The facts being, of course, that the undertaker in such a case seeks to produce economically as vendible a product as may be. In the common run vendibility depends in great part on the serviceability of the goods, but it depends also on several other circumstances; and to that highly variable, but nearly always considerable extent to which vendibility depends on other circumstances than the material serviceability of the goods, the pecuniary management of capital must be held not to serve the ends of production. Neither immediately, in his purely pecuniary traffic, nor indirectly, in the business guidance of industry through his pecuniary traffic, therefore, can the undertaker's dealings with his pecuniary capital be accounted a productive occupation, nor can the gains of capital be taken to mark or to measure the productivity due to the investment. The " cost of production " of goods in the case contemplated is to an appreciable, but indeterminable, extent a cost of production of vendibility — an outcome which is often of doubtful service to the body of consumers, and which often counts in the aggregate as waste. The material serviceability of the means employed in industry, that is to say the functioning of industrial capital in the service of the community at large, stands in no necessary or consistent relation to the gainfulness of capital in the pecuniary respect. Productivity can accordingly not be predicated of pecuniary capital. It follows that productivity theories of interest should be as difficult to maintain as productivity theories of the gains of the pecuniary employments, the two resting on the same grounds.

It is, further, to be remarked that pecuniary capital and industrial capital do not coincide in respect of the concrete things comprised under each. From this and from the considerations already indicated above, it follows that the magnitude of pecuniary capital may vary independently of variations in the magnitude of industrial capital — not indefinitely, perhaps, but within a range which, in its nature, is indeterminate. Pecuniary capital is a matter of market values, while industrial capital is, in the last analysis, a matter of mechanical efficiency, or rather of mechanical effects not reducible to a common measure or a collective magnitude. So far as the latter may be spoken of as a homogenous aggregate — itself a doubtful point at best — the two categories of capital are disparate magnitudes, which can be mediated only through a process of valuation conditioned by other circumstances besides the mechanical efficiency of the material means valued. Market values being a psychological outcome, it follows that pecuniary capital, an aggregate of market values, may vary in magnitude with a freedom which gives the whole an air of caprice,— such as psychological phenomena, particularly the psychological phenomena of crowds, frequently present, and such as becomes strikingly noticeable in times of panic or of speculative inflation. On the other hand, industrial capital, being a matter of mechanical contrivances and adaptation, cannot similarly vary through a revision of valuations. If it is taken as an aggregate, it is a physical magnitude, and as such it does not alter its complexion or its mechanical efficiency in response to the greater or less degree of appreciation with which it is viewed. Capital pecuniarily considered rests on a basis of subjective value; capital industrially considered rests on material circumstances reducible to objective terms of mechanical, chemical and physiological effect.

The point has frequently been noted that it is impossible to get at the aggregate social (industrial) capital by adding up the several items of individual (pecuniary) capital. A reason for this, apart from variations in the market values of given material means of production, is that pecuniary capital comprises not only material things but also conventional facts, psychological phenomena not related in any rigid way to material means of production,— as *e. g.*, good will, fashions, customs, prestige, effrontery, personal credit. Whatever ownership touches, and whatever affords ground for pecuniary discretion, may be turned to account for pecuniary gain and may therefore be comprised in the aggregate of pecuniary capital. Ownership, the basis of pecuniary capital, being itself a conventional fact, that is to say a matter of habits of thought, it is intelligible that phenomena of convention and opinion should figure in an inventory of pecuniary capital; whereas, industrial capital being of a mechanical character, conventional circumstances do not affect it — except as the future production of material means to replace the existing outfit may be guided by convention — and items having but a conventional existence are, therefore, not comprised in its aggregate. The disparity between pecuniary and industrial capital, therefore, is something more than a matter of an arbitrarily chosen point of view, as some recent discussions of the capital concept would have us believe; just as the difference between the pecuniary and the industrial employments, which are occupied with the one or the other category of capital, means something more than the same thing under different aspects.

But the distinction here attempted has a farther bearing, beyond the possible correction of a given point in the theory of distribution. Modern economic science is to an increasing extent concerning itself with the question of what men do and how and why they do it, as contrasted with the older question of how Nature, working through human nature, maintains a favorable balance in the output of goods. Neither the practical questions of our generation, nor the pressing theoretical questions of the science, run on the adequacy or equity of the share that goes to any class in the normal case. The questions are rather such realistic ones as these: Why do we, now and again, have hard times and unemployment in the midst of excellent resources, high efficiency and plenty of unmet wants? Why is one-half our consumable product contrived for consumption that yields no material benefit? Why are large coördinations of industry, which greatly reduce cost of production, a cause of perplexity and alarm? Why is the family disintegrating among the industrial classes, at the same time that the wherewithal to maintain it is easier to compass? Why are large and increasing portions of the community penniless in spite of a scale of remuneration which is very appreciably above the subsistence minimum? Why is there a widespread disaffection among the intelligent workmen who ought to know better? These and the like questions, being questions of fact, are not to be answered on the grounds of normal equivalence. Perhaps it might better be said that they have so often been answered on those grounds, without any approach to disposing of them, that the outlook for help in that direction has ceased to have a serious meaning. These are, to borrow Professor Clark's phrase, questions to be answered on dynamic, not on static grounds. They are questions of conduct and sentiment, and so far as their solution is looked for at the hands of economists it must be looked for along the line of the bearing which economic life has upon the growth of sentiment and canons of conduct. That is to say, they are questions of the bearing of economic life upon the cultural changes that are going forward.

For the present it is the vogue to hold that economic life, broadly, conditions the rest of social organization or the constitution of society. This vogue of the proposition will serve as excuse from going into an examination of the grounds on which it may be justified, as it is scarcely necessary to persuade any economist that it has substantial merits even if he may not accept it in an unqualified form. What the Marxists have named the " Materialistic Conception of History " is assented to with less and less qualification by those who make the growth of culture their subject of inquiry. This materialistic conception says that institutions are shaped by economic conditions; but, as it left the hands of the

Marxists, and as it still functions in the hands of many who knew not Marx, it has very little to say regarding the efficient force, the channels, or the methods by which the economic situation is conceived to have its effect upon institutions. What answer the early Marxists gave to this question, of how the economic situation shapes institutions, was to the effect that the causal connection lies through a selfish, calculating class interest. But, while class interest may count for much in the outcome, this answer is plainly not a competent one, since, for one thing, institutions by no means change with the alacrity which the sole efficiency of a reasoned class interest would require.

Without discrediting the claim that class interest counts for something in the shaping of institutions, and to avoid getting entangled in preliminaries, it may be said that institutions are of the nature of prevalent habits of thought, and that therefore the force which shapes institutions is the force or forces which shape the habits of thought prevalent in the community. But habits of thought are the outcome of habits of life. Whether it is intentionally directed to the education of the individual or not, the discipline of daily life acts to alter or reënforce the received habits of thought, and so acts to alter or fortify the received institutions under which men live. And the direction in which, on the whole, the alteration proceeds is conditioned by the trend of the discipline of daily life. The point here immediately at issue is the divergent trend of this discipline in those occupations which are prevailingly of an industrial character, as contrasted with those which are prevailingly of a pecuniary character. So far as regards the different cultural outcome to be looked for on the basis of the present economic situation as contrasted with the past, therefore, the question immediately in hand is as to the greater or less degree in which occupations are differentiated into industrial and pecuniary in the present as compared with the past.

The characteristic feature which is currently held to differentiate the existing economic situation from that out of which the present has developed, or out of which it is emerging, is the prevalence of the machine industry with the consequent larger and more highly specialised organisation of the market and of the industrial force and plant.   As has been pointed out above, and as is well enough known from the current discussions of the economists, industrial life is organised on a pecuniary basis and managed from the pecuniary side.   This, of course, is true in a degree both of the present and of the nearer past, back at least as far as the Middle Ages. But the larger scope of organisations in modern industry means that the pecuniary management has been gradually passing into the hands of a relatively decreasing class, whose contact with the industrial classes proper grows continually less immediate.   The distinction between employments above spoken of is in an increasing degree coming to coincide with a differentiation of occupations and of economic classes.   Some degree of such specialisation and differentiation there has, of course, been, one might almost say, always.   But in our time, in many branches of industry, the specialisation has been carried so far that large bodies of the working population have but an incidental contact with the business side of the enterprise, while a minority have little if any other concern with the enterprise than its pecuniary management.   This was not true, *e. g.,* at the time when the undertaker was still salesman, purchasing agent, business manager, foreman of the shop, and master work-

man. Still less was it true in the days of the self-suf-
ficing manor or household, or in the days of the closed
town industry. Neither is it true in our time of what
we call the backward or old-fashioned industries. These
latter have not been and are not organised on a large
scale, with a consistent division of labor between the
owners and business managers on the one side and the
operative employees on the other. Our standing illustrations of this less highly organised class of industries are
the surviving handicrafts and the common run of farming as carried on by relatively small proprietors. In that
earlier phase of economic life, out of which the modern
situation has gradually grown, all the men engaged had
to be constantly on their guard, in a pecuniary sense, and
were constantly disciplined in the husbanding of their
means and in the driving of bargains,— as is still true,
e. g., of the American farmer. The like was formerly
true also of the consumer, in his purchases, to a greater
extent than at present. A good share of the daily attention of those who were engaged in the handicrafts was
still perforce given to the pecuniary or business side of
their trade. But for that great body of industry which
is conventionally recognised as eminently modern, specialisation of function has gone so far as, in great measure,
to exempt the operative employees from taking thought
of pecuniary matters.

Now, as to the bearing of all this upon cultural changes
that are in progress or in the outlook. Leaving the
" backward," relatively unspecialised, industries on one
side, as being of an equivocal character for the point in
hand and as not differing characteristically from the corresponding industries in the past so far as regards their
disciplinary value ; modern occupations may, for the sake
of the argument, be broadly distinguished, as economic

employments have been distinguished above, into business and industrial. The modern industrial and the modern business occupations are fairly comparable as regards the degree of intelligence required in both, if it be borne in mind that the former occupations comprise the highly trained technological experts and engineers as well as the highly skilled mechanics. The two classes of occupations differ in that the men in the pecuniary occupations work within the lines and under the guidance of the great institution of ownership, with its ramifications of custom, perogative, and legal right; whereas those in the industrial occupations are, in their work, relatively free from the constraint of this conventional norm of truth and validity. It is, of course, not true that the work of the latter class lies outside the reach of the institution of ownership; but it is true that, in the heat and strain of the work, when the agent's powers and attention are fully taken up with the work which he has in hand, that of which he has perforce to take cognisance is not conventional law, but the conditions impersonally imposed by the nature of material things. This is the meaning of the current commonplace that the required close and continuous application of the operative in mechanical industry bars him out of all chance for an all-around development of the cultural graces and amenities. It is the periods of close attention and hard work that seem to count for most in the formation of habits of thought.

An *a priori* argument as to what cultural effects should naturally follow from such a difference in discipline between occupations, past and present, would probably not be convincing, as *a priori* arguments from half-authenticated premises commonly are not. And the experiments along this line which later economic developments have so far exhibited have been neither neat enough, comprehensive enough, nor long continued enough to give definite results. Still, there is something to be said under this latter head, even if this something may turn out to be somewhat familiar.

It is, *e. g.,* a commonplace of current vulgar discussions of existing economic questions, that the classes engaged in the modern mechanical or factory industries are improvident and apparently incompetent to take care of the pecuniary details of their own life. In this indictment may well be included not only factory hands, but the general class of highly skilled mechanics, inventors, technological experts. The rule does not hold in any hard and fast way, but there seems to be a substantial ground of truth in the indictment in this general form. This will be evident on comparison of the present factory population with the class of handicraftsmen of the older culture whom they have displaced, as also on comparison with the farming population of the present time, especially the small proprietors of this and other countries. The inferiority which is currently conceded to the modern industrial classes in this respect is not due to scantier opportunities for saving, whether they are compared with the earlier handicraftsmen or with the modern farmer or peasant. This phenomenon is commonly discussed in terms which impute to the improvident industrial classes something in the way of total depravity, and there is much preaching of thrift and steady habits. But the preaching of thrift and self-help, unremitting as it is, is not producing an appreciable effect. The trouble seems to run deeper than exhortation can reach. It seems to be of the nature of habit rather than of reasoned conviction. Other causes may be present and may be competent partially to explain the improvidence of these classes; but the inquiry is at least a pertinent one; how far the absence of property and thrift among them may be traceable to the relative absence of pecuniary training in the discipline of their daily life. If, as the general lie of the subject would indicate, this peculiar pecuniary situation of the industrial classes is in any degree due to comprehensive disciplinary causes, there is material in it for an interesting economic inquiry.

The surmise that the trouble with the industrial class is something of this character is strengthened by another feature of modern vulgar life, to which attention is directed as a further, and, for the present, a concluding illustration of the character of the questions that are touched by the distinction here spoken for. The most insidious and most alarming malady, as well as the most perplexing and unprecedented, that threatens the modern social and political structure is what is vaguely called socialism. The point of danger to the social structure, and at the same time the substantial core of the socialistic disaffection, is a growing disloyalty to the institution of property, aided and abetted as it is by a similarly growing lack of deference and affection for other conventional features of social structure. The classes affected by socialistic vagaries are not consistently averse to a competent organisation and control of society, particularly not in the economic respect, but they are averse to organisation and control on conventional lines. The sense of solidarity does not seem to be either defective or in abeyance, but the ground of solidarity is new and unexpected. What their constructive ideals may be need not concern nor detain us; they are vague and inconsistent and for the most part negative. Their disaffection has been set down to discontent with their lot by comparison with others, and to a mistaken view of their own interests; and much and futile effort has been spent in showing them the error of their ways of thinking. But what the experience of the past suggests that we should expect under the guidance of such motives and reasoning as these would be a demand for a redistribution of property, a reconstitution of the conventions of ownership on such new lines as the apprehended interests of these classes would seem to dictate. But such is not the trend of socialistic thinking, which contemplates rather the elimination of the institution of property. To the socialists property or ownership does not seem inevitable or inherent in the nature of things; to those who criticise and admonish them it commonly does.

Compare them in this respect with other classes who have been moved by hardship or discontent, whether well or ill advised, to put forth denunciations and demands for radical economic changes; as *e. g.,* the American farmers in their several movements, of grangerism, populism, and the like. These have been loud enough in their denunciations and complaints, and they have been accused of being socialistic in their demand for a virtual redistribution of property. They have not felt the justice of the accusation, however, and it is to be noted that their demands have consistently run on a rehabilitation of property on some new basis of distribution, and have been uniformly put forth with the avowed purpose of bettering the claimants in point of ownership. Ownership, property " honestly " acquired, has been sacred to the rural malcontents, here and elsewhere; what they have aspired to do has been to remedy what they have conceived to be certain abuses under the institution, without questioning the institution itself.

Not so with the socialists, either in this country or elsewhere. Now, the spread of socialistic sentiment shows a curious tendency to affect those classes particularly who are habitually employed in the specialised industrial occupations, and are thereby in great part exempt from the intellectual discipline of pecuniary management. Among these men, who by the circumstances of their daily life are brought to do their serious and habitual thinking in other than pecuniary terms, it looks as if the ownership preconception were becoming obsolescent through disuse. It is the industrial population, in the modern sense, and particularly the more intelligent and skilled men employed in the mechanical industries, that are most seriously and widely affected. With exceptions both ways, but with a generality that is not to be denied, the socialistic disaffection spreads through the industrial towns, chiefly and most potently among the better classes of operatives in the mechanical employments; whereas the relatively indigent and unintelligent regions and classes, which the differentiation between pecuniary and industrial occupations has not reached, are relatively free from it. In like manner the upper and middle classes, whose employments are of a pecuniary character, if any, are also not seriously affected; and when avowed socialistic sentiment is met with among these upper and middle classes it commonly turns out to be merely a humanitarian aspiration for a more "equitable" redistribution of wealth — a readjustment of ownership under some new and improved method of control — not a contemplation of the traceless disappearance of ownership.

Socialism, in the sense in which the word connotes a subversion of the economic foundations of modern culture, appears to be found only sporadically and uncertainly outside the limits, in time and space, of the discipline exercised by the modern mechanical, non-pecuniary occupations. This state of the case need of course not be due solely to the disciplinary effects of the industrial employments, nor even solely to effects traceable to those employments whether in the way of disciplinary results, selective development, or what not. Other factors, particularly factors of an ethnic character, seem to coöperate to the result indicated; but, so far as evidence bearing on the point is yet in hand and has been analysed, it indicates that this differentiation of occupations is a necessary requisite to the growth of a consistent 'body of socialistic sentiment; and the indication is also that wherever this differentiation prevails in such a degree of accentuation and affects such considerable and compact bodies of people as to afford ground for a consistent growth of common sentiment, a result is some form of iconoclastic socialism. The differentiation may of course have a selective as well as a disciplinary effect upon the population affected, and an off-hand separation of these two modes of influence can of course not be made. In any case, the two modes of influence seem to converge to the outcome indicated; and, for the present purpose of illustration simply, the tracing out of the two strands of sequence in the case neither can nor need be undertaken. By force of this differentiation, in one way and another, the industrial classes are learning to think in terms of material cause and effect, to the neglect of prescription and conventional grounds of validity; just as, in a faintly incipient way, the economists are also learning to do in their discussion of the life of these classes. The resulting decay of the popular sense of conventional validity of course extends to other matters than the pecuniary conventions alone, with the outcome that the socialistically affected industrial classes are pretty uniformly affected with an effortless iconoclasm in other directions as well. For the discipline to which their work and habits of life subject them gives not so much a training away from the pecuniary conventions, specifically, as a positive and somewhat unmitigated training in methods of observation and inference proceeding on grounds alien to all conventional validity.   But the practical experiment going on in the specialisation of discipline, in the respect contemplated, appears still to be near its beginning, and the growth of aberrant views and habits of thought due to the peculiar disciplinary trend of this late and unprecedented specialisation of occupations has not yet had time to work itself clear.

The effects of the like one-sided discipline are similarly visible in the highly irregular, conventionally indefensible attitude of the industrial classes in the current labor and wage disputes, not of an avowedly socialistic aim.   So also as regards the departure from the ancient norm in such non-economic, or secondarily economic matters as the family relation and responsibility, where the disintegration of conventionalities in the industrial towns is said to threaten the foundations of domestic life and morality; and again as regards the growing inability of men trained to materialistic, industrial habits of thought to appreciate, or even to apprehend, the meaning of religious appeals and consolations that proceed on the old-fashioned conventional or metaphysical grounds of validity.   But these and other like directions in which the cultural effects of the modern specialisation of occupations, whether in industry or in business, may be traceable can not be followed up here.

ON THE NATURE OF CAPITAL [1]

I. The Productivity of Capital Goods

It has been usual in expositions of economic theory to speak of capital as an array of "productive goods." What is immediately had in mind in this expression, as well as in the equivalent "capital goods," is the industrial equipment, primarily the mechanical appliances employed in the processes of industry. When the productive efficiency of these and of other subsidiary classes of capital goods is subjected to further analysis, it is not unusual to trace it back to the productive labor of the workmen, the labor of the individual workman being the ultimate productive factor in the commonly accepted systems of theory. The current theories of production, as also those of distribution, are drawn in individualistic terms, particularly when these theories are based on hedonistic premises, as they commonly are.

Now, whatever may or may not be true for human conduct in some other bearing, in the economic respect man has never lived an isolated, self-sufficient life as an individual, either actually or potentially. Humanly speaking, such a thing is impossible. Neither an individual person nor a single household, nor a single line of descent, can maintain its life in isolation. Economically speaking, this is the characteristic trait of humanity that separates mankind from the other animals. The life-history of the race has been a life-history of human com-

[1] Reprinted by permission from *The Quarterly Journal of Economics,* Vol. XXII, Aug., 1908.

munities, of more or less considerable size, with more or less of group solidarity, and with more or less of cultural continuity over successive generations. The phenomena of human life occur only in this form.

This continuity, congruity, or coherence of the group, is of an immaterial character. It is a matter of knowledge, usage, habits of life and habits of thought, not a matter of mechanical continuity or contact, or even of consanguinity. Wherever a human community is met with, as, *e.g.*, among any of the peoples of the lower cultures, it is found in possession of something in the way of a body of technological knowledge,— knowledge serviceable and requisite to the quest of a livelihood, comprising at least such elementary acquirements as language, the use of fire, of a cutting edge, of a pointed stick, of some tool for piercing, of some form of cord, thong, or fiber, together with some skill in the making of knots and lashings. Coördinate with this knowledge of ways and means, there is also uniformly present some matter-of-fact knowledge of the physical behavior of the materials with which men have to deal in the quest of a livelihood, beyond what any one individual has learned or can learn by his own experience alone. This information and proficiency in the ways and means of life vests in the group at large; and, apart from accretions borrowed from other groups, it is the product of the given group, though not produced by any single generation. It may be called the immaterial equipment, or, by a license of speech, the intangible assets [2] of the community; and, in the early days at least, this is far and away the most important and

[2] " Assets " is, of course, not to be taken literally in this connection. The term properly covers a pecuniary concept, not an industrial (technological) one, and it connotes ownership as well as value; and it will be used in this literal sense when, in a later article, ownership and investment come into the discussion. In

consequential category of the community's assets or equipment. Without access to such a common stock of immaterial equipment no individual and no fraction of the community can make a living, much less make an advance. Such a stock of knowledge and practice is perhaps held loosely and informally; but it is held as a common stock, pervasively, by the group as a body, in its corporate capacity, as one might say; and it is transmitted and augmented in and by the group, however loose and haphazard the transmission may be conceived to be, not by individuals and in single lines of inheritance.

The requisite knowledge and proficiency of ways and means is a product, perhaps a by-product, of the life of the community at large; and it can also be maintained and retained only by the community at large. Whatever may be true for the unsearchable prehistoric phases of the life-history of the race, it appears to be true for the most primitive human groups and phases of which there is available information that the mass of technological knowledge possessed by any community, and necessary to its maintenance and to the maintenance of each of its members or subgroups, is too large a burden for any one individual or any single line of descent to carry. This holds true, of course, all the more rigorously and consistently, the more advanced the " state of the industrial arts " may be. But it seems to hold true with a generality that is fairly startling, that whenever a given cultural community is broken up or suffers a serious diminution of numbers, its technological heritage deteriorates and dwindles, even though it may have been apparently meager enough before. On the other hand, it seems to hold

the present connection it is used figuratively, for want of a better term, to convey the connotation of value and serviceability without thereby implying ownership.

true with a similar uniformity that, when an individual member or a fraction of a community on what we call a lower stage of economic development is drawn away and trained and instructed in the ways of a larger and more efficient technology, and is then thrown back into his home community, such an individual or fraction proves unable to make head against the technological bent of the community at large or even to create a serious diversion. Slight, perhaps transient, and gradually effective technological consequences may result from such an experiment; but they become effective by diffusion and assimilation through the body of the community, not in any marked degree in the way of an exceptional efficiency on the part of the individual or fraction which has been subjected to exceptional training. And inheritance in technological matters runs not in the channels of consanguinity, but in those of tradition and habituation, which are necessarily as wide as the scheme of life of the community. Even in a relatively small and primitive community the mass of detail comprised in its knowledge and practice of ways and means is large,— too large for any one individual or household to become competently expert in it all; and its ramifications are extensive and diverse, at the same time that all these ramifications bear, directly or indirectly, on the life and work of each member of the community. Neither the standard and routine of living nor the daily work of any individual in the community would remain the same after the introduction of an appreciable change, for good or ill, in any branch of the community's equipment of technological expedients. If the community grows larger, to the dimensions of a modern civilised people, and this immaterial equipment grows proportionately great and various, then it will become increasingly difficult to trace the connection between any given change in technological detail and the fortunes of any given obscure member of the community. But it is at least safe to say that an increase in the volume and complexity of the body of technological knowledge and practice does not progressively emancipate the life and work of the individual from its dominion.

The complement of technological knowledge so held, used, and transmitted in the life of the community is, of course, made up out of the experience of individuals. Experience, experimentation, habit, knowledge, initiative, are phenomena of individual life, and it is necessarily from this source that the community's common stock is all derived. The possibility of its growth lies in the feasibility of accumulating knowledge gained by individual experience and initiative, and therefore it lies in the feasibility of one individual's learning from the experience of another. But the initiative and technological enterprise of individuals, such, *e.g.,* as shows itself in inventions and discoveries of more and better ways and means, proceeds on and enlarges the accumulated wisdom of the past. Individual initiative has no chance except on the ground afforded by the common stock, and the achievements of such initiative are of no effect except as accretions to the common stock. And the invention or discovery so achieved always embodies so much of what is already given that the creative contribution of the inventor or discoverer is trivial by comparison.

In any known phase of culture this common stock of intangible, technological equipment is relatively large and complex,— *i.e.,* relatively to the capacity of any individual member to create or to use it; and the history of its growth and use is the history of the development of material civilisation. It is a knowledge of ways and means, and is embodied in the material contrivances and processes by means of which the members of the community make their living. Only by such means does technological efficiency go into effect. These "material contrivances" ("capital goods," material equipment) are such things as tools, vessels, vehicles, raw materials, buildings, ditches, and the like, including the land in use; but they include also, and through the greater part of the early development chiefly, the useful minerals, plants, and animals. To say that these minerals, plants, and animals are useful — in other words, that they are economic goods — means that they have been brought within the sweep of the community's knowledge of ways and means.

In the relatively early stages of primitive culture the useful plants and minerals are, no doubt, made use of in a wild state, as, *e.g.,* fish and timber have continued to be used. Yet in so far as they are useful they are unmistakably to be counted in among the material equipment ("tangible assets") of the community. The case is well illustrated by the relation of the Plains Indians to the buffalo, and by the northwest coast Indians to the salmon, on the one hand, and by the use of a wild flora by such communities as the Coahuilla Indians,[3] the Australian blacks, or the Andamanese, on the other hand.

But with the current of time, experience, and initiative, domesticated (that is to say improved) plants and animals come to take the first place. We have then such "technological expedients" in the first rank as the many species and varieties of domestic animals, and more particularly still the various grains, fruits, root-crops, and the like, virtually all of which were created by man for human use; or perhaps a more scrupulously veracious account would say that they were in the main created by the women, through long ages of workmanlike selection and

[3] Barrows.

cultivation. These things, of course, are useful because men have learned their use, and their use, so far as it has been learned, has been learned by protracted and voluminous experience and experimentation, proceeding at each step on the accumulated achievements of the past. Other things, which may in time come to exceed these in usefulness are still useless, economically non-existent, on the early levels of culture, because of what men in that time have not yet learned.

While this immaterial equipment of industry, the intangible assets of the community, have apparently always been relatively very considerable and are always mainly in the keeping of the community at large, the material equipment, the tangible assets, on the other hand, have, in the early stages (say the earlier 90 per cent.) of the life-history of human culture, been relatively slight, and have apparently been held somewhat loosely by individuals or household groups. This material equipment is relatively very slight in the earlier phases of technological development, and the tenure by which it is held is apparently vague and uncertain. At a relatively primitive phase of the development, and under ordinary conditions of climate and surroundings, the possession of the concrete articles ("capital goods") needed to turn the commonplace knowledge of ways and means to account is a matter of slight consequence,— contrary to the view commonly spoken for by the economists of the classical line. Given the commonplace technological knowledge and the commonplace training,— and these are given by common notoriety and the habituation of daily life,— the acquisition, construction, or usufruct of the slender material equipment needed arranges itself almost as a matter of course, more particularly where this material equipment does not include a stock of domestic animals or a plantation of domesticated trees and vegetables. Under given circumstances a relatively primitive technological scheme may involve some large items of material equipment, as the buffalo pens (*piskun*) of the Blackfoot Indians or the salmon weirs of the river Indians of the northwest coast. Such items of material equipment are then likely to be held and worked collectively, either by the community at large or by subgroups of a considerable size. Under ordinary, more generally prevalent conditions, it appears that even after a relatively great advance has been made in the cultivation of crops the requisite industrial equipment is not a matter of serious concern, particularly so aside from the tilled ground and the cultivated trees, as is indicated by the singularly loose and inconsequential notions of ownership prevalent among peoples occupying such a stage of culture. A primitive stage of communism is not known.

But as the common stock of technological knowledge increases in volume, range, and efficiency, the material equipment whereby this knowledge of ways and means is put into effect grows greater, more considerable relatively to the capacity of the individual. And so soon, or in so far, as the technological development falls into such shape as to require a relatively large unit of material equipment for the effective pursuit of industry, or such as otherwise to make the possession of the requisite material equipment a matter of consequence, so as seriously to handicap the individuals who are without these material means, and to place the current possessors of such equipment at a marked advantage, then the strong arm intervenes, property rights apparently begin to fall into definite shape, the principles of ownership gather force and consistency, and men begin to accumulate capital goods and take measures to make them secure.

An appreciable advance in the industrial arts is commonly followed or accompanied by an increase of population. The difficulty of procuring a livelihood may be no greater after such an increase; it may even be less; but there results a relative curtailment of the available area and raw materials, and commonly also an increased accessibility of the several portions of the community. A wide-reaching control becomes easier. At the same time a larger unit of material equipment is needed for the effective pursuit of industry. As this situation develops, it becomes worth while — this is to say, it becomes feasible — for the individual with the strong arm to engross, or " corner," the usufruct of the commonplace knowledge of ways and means by taking over such of the requisite material as may be relatively scarce and relatively indispensable for procuring a livelihood under the current state of the industrial arts.[4] Circumstances of space and numbers prevent escape from the new technological situation. The commonplace knowledge of ways and means cannot be turned to account, under the new conditions, without a material equipment adapted to the then current state of the industrial arts; and such a suitable material equipment is no longer a slight matter, to be compassed by workmanlike initiative and application. *Beati possidentes.*

The emphasis of the technological situation, as one

[4] Motives of exploit and emulation, no doubt, play a serious part in bringing on the practice of ownership and in establishing the principles on which it rests; but this play of motives and the concomitant growth of institutions cannot be taken up here. *Cf. The Theory of the Leisure Class,* chaps. i, ii, iii.

might say, may fall now on one line of material items, now on another, according as the exigencies of climate, topography, flora and fauna, density of population, and the like, may decide. So also, under the rule of the same exigencies, the early growth of property rights and of the principles (habits of thought) of ownership may settle on one or another line of material items, according as one or another affords the strategic advantage for engrossing the current technological efficiency of the community.

Should the technological situation, the state of the industrial arts, be such as to throw the strategic emphasis on manual labor, on workmanlike skill and application, and if at the same time the growth of population has made land relatively scarce, or hostile contact with other communities has made it impracticable for members of the community to range freely over outlying tracts, then it would be expected that the growth of ownership should take the direction primarily of slavery, or of some equivalent form of servitude, so effecting a naïve and direct monopolistic control of the current knowledge of ways and means.[5] Whereas if the development has taken such a turn, and the community is so placed as to make the quest of a livelihood a matter of the natural increase of flocks and herds, then it should reasonably be expected that these items of equipment will be the chief and primary subject of property rights. In point of fact, it appears that a pastoral culture commonly involves also some degree of servitude, along with the ownership of flocks and herds.

Under different circumstances the mechanical appliances of industry, or the tillable land, might come into the position of strategic advantage, and might come in for the

[5] *Cf.* H. Nieboer, *Slavery as an Industrial System,* chap. iv, sect. 12.

foremost place in men's consideration as objects of ownership. The evidence afforded by the known (relatively) primitive cultures and communities seems to indicate that slaves and cattle have in this way come into the primacy as objects of ownership at an earlier period in the growth of material civilisation than land or the mechanical appliances. And it seems similarly evident — more so, indeed — that land has on the whole preceded the mechanical equipment as the stronghold of ownership and the means of engrossing the community's industrial efficiency.

It is not until a late period in the life-history of material civilisation that ownership of the industrial equipment, in the narrower sense in which that phrase is commonly employed, comes to be the dominant and typical method of engrossing the immaterial equipment. Indeed, it is a consummation which has been reached only a very few times even partially, and only once with such a degree of finality as to leave the fact indisputable. If it may be said, loosely, that mastery through the ownership of slaves, cattle, or land comes on in force only after the economic development has run through some nine-tenths of its course hitherto, then it may be said likewise that some ninety-nine one-hundredths of this course of development had been completed before the ownership of the mechanical equipment came into undisputed primacy as the basis of pecuniary dominion. So late an innovation, indeed, is this modern institution of " capitalism,"— the predominant ownership of industrial capital as we know it,— and yet so intimate a fact is it in our familiar scheme of life, that we have some difficulty in seeing it in perspective at all, and we find ourselves hesitating between denying its existence, on the one hand, and affirming it to be a fact of nature antecedent to all human institutions, on the other hand.

In so speaking of the ownership of industrial equipment as being an institution for cornering the community's intangible assets, there is conveyed an unavoidably implied, though unintended, note of condemnation. Such an implication of merit or demerit is an untoward circumstance in any theoretical inquiry. Any sentimental bias, whether of approval or disapproval, aroused by such an implied censure, must unavoidably hamper the dispassionate pursuit of the argument. To mitigate the effect of this jarring note as far as may be, therefore, it will be expedient to turn back for a moment to other, more primitive and remoter forms of the institution,— as slavery and landed wealth,— and so reach the modern facts of industrial capital by a roundabout and gradual approach.

These ancient institutions of ownership, slavery and landed wealth, are matters of history. Considered as dominant factors in the community's scheme of life, their record is completed; and it needs no argument to enforce the proposition that it is a record of economic dominion by the owners of the slaves or the land, as the case may be. The effect of slavery in its best day, and of landed wealth in mediæval and early modern times, was to make the community's industrial efficiency serve the needs of the slave-owners in the one case and of the land-owners in the other. The effect of these institutions in this respect is not questioned now, except in such sporadic and apologetical fashion as need not detain the argument.

But the fact that such was the direct and immediate effect of these institutions of ownership in their time by no means involves the instant condemnation of the institutions in question. It is quite possible to argue that slavery and landed wealth, each in its due time and due cultural setting, have served the amelioration of the lot of man and the advance of human culture. What these arguments may be that aim to show the merits of slavery and landed wealth as a means of cultural advance does not concern the present inquiry, neither do the merits of the case in which the arguments are offered. The matter is referred to here to call to mind that any similar theoretical outcome of an analysis of the productivity of " capital goods " need not be admitted to touch the merits of the case in controversy between the socialistic critics of capitalism and the spokesmen of law and order.

The nature of landed wealth, in point of economic theory, especially as regards its productivity, has been sifted with the most jealous precautions and the most tenacious logic during the past century; and any economic student can easily review the course of the argument whereby that line of economic theory has been run to earth. It is only necessary here to shift the point of view slightly to bring the whole argument concerning the rent of land to bear on the present question. Rent is of the nature of a differential gain, resting on a differential advantage in point of productivity of the industry employed upon or about it. This differential advantage attaching to a given parcel of land may be a differential as against another parcel or as against industry applied apart from land. The differential advantage attaching to agricultural land — *e.g.,* as against industry at large — rests on certain broad peculiarities of the technological situation. Among them are such peculiarities as these: the human species, or the fraction of it concerned in the case, is numerous, relatively to the extent of its habitat; the methods of getting a living, as hitherto elaborated, the ways and means of life, make use of certain crop-plants and certain domestic animals. Apart from such conditions, taken for granted in arguments concerning agricultural rent, there could manifestly be no differential advantage attaching to land, and no production of rent. With increased command of methods of transportation, the agricultural lands of England, *e.g.,* and of Europe at large, declined in value, not because these lands became less fertile, but because an equivalent result could more advantageously be got by a new method. So, again, the flint- and amber-bearing regions that are now Danish and Swedish territory about the waters at the entrance to the Baltic were in the neolithic culture of northern Europe the most favored and valuable lands within that cultural region. But, with the coming of the metals and the relative decline of the amber trade, they began to fall behind in the scale of productivity and preference. So also in later time, with the rise of " industry " and the growth of the technology of communication, urban property has gained, as contrasted with rural property, and land placed in an advantageous position relatively to shipping and railroads has acquired a value and a " productiveness " which could not be claimed for it apart from these modern technological expedients.

The argument of the single-tax advocates and other economists as to the " unearned increment " is sufficiently familiar, but its ulterior implications have not commonly been recognised. The unearned increment, it is held, is produced by the growth of the community in numbers and in the industrial arts. The contention seems to be sound, and is commonly accepted; but it has commonly been overlooked that the argument involves the ulterior conclusion that all land values and land productivity, including the " original and indestructible powers of the soil," are a function of the " state of the industrial art." It is only within the given technological situation, the current scheme of ways and means, that any parcel of land has such productive powers as it has. It is, in other words, useful only because, and in so far, and in such manner, as men have learned to make use of it. This is what brings it into the category of "land," economically speaking. And the preferential position of the landlord as a claimant of the "net product" consists in his legal right to decide whether, how far, and on what terms men shall put this technological scheme into effect in those features of it which involve the use of his parcel of land.

All this argument concerning the unearned increment may be carried over, with scarcely a change of phrase, to the case of "capital goods." The Danish flint supply was of first-rate economic consequence, for a thousand years or so, during the stone age; and the polished-flint utensils of that time were then "capital goods" of inestimable importance to civilisation, and were possessed of a "productivity" so serious that the life of mankind in that world may be said to have been balanced on the fine-ground edge of those magnificent polished-flint axes. All that lasted through its technological era. The flint supply and the mechanical expedients and "capital goods," whereby it was turned to account, were valuable and productive then, but neither before nor after that time. Under a changed technological situation the capital goods of that time have become museum exhibits, and their place in human economy has been taken by technological expedients which embody another "state of the industrial arts," the outcome of later and different phases of human experience. Like the polished-flint ax, the metal utensils which gradually displaced it and its like in the economy of the Occidental culture were the product of long experience and the gradual learning of ways and means. The steel ax, as well as the flint ax, embodies the same ancient technological expedient of a cutting edge, as well as the use of a helve and the efficiency due to the weight of the tool. And in the case of the one or the other, when seen in historical perspective and looked at from the point of view of the community at large, the knowledge of ways and means embodied in the utensils was the serious and consequential matter. The construction or acquisition of the concrete " capital goods " was simply an easy consequence. It " cost nothing but labor," as Thomas Mun would say.

Yet it might be argued that each concrete article of " capital goods " was the product of some one man's labor, and, as such, its productivity, when put to use, was but the indirect, ulterior, deferred productiveness of the maker's labor. But the maker's productivity in the case was but a function of the immaterial technological equipment at his command, and that in its turn was the slow spiritual distillate of the community's time-long experience and initiative. To the individual producer or owner, to whom the community's accumulated stock of immaterial equipment was open by common notoriety, the cost of the concrete material goods would be the effort involved in making or getting them and in making good his claim to them. To his neighbor who had made or acquired no such parcel of " productive goods," but to whom the resources of the community, material and immaterial, were open on the same easy terms, the matter would look very much the same. He would have no grievance, nor would he have occasion to seek one. Yet, as a resource in the maintenance of the community's life and a factor in the advance of material civilisation, the whole matter would have a different meaning.

So long, or rather in so far, as the " capital goods " required to meet the technological demands of the time were slight enough to be compassed by the common man

with reasonable diligence and proficiency, so long the
draft upon the common stock of immaterial assets by any
one would be no hindrance to any other, and no differential advantage or disadvantage would emerge. The economic situation would answer passably to the classical
theory of a free competitive system,— " the obvious and
simple system of natural liberty," which rests on the presumption of equal opportunity. In a roughly approximate way, such a situation supervened in the industrial
life of western Europe on the transition from mediæval
to modern times, when handicraft and " industrial " enterprise superseded landed wealth as the chief economic
factor. Within the " industrial system," as distinct from
the privileged non-industrial classes, a man with a modicum of diligence, initiative, and thrift might make his
way in a tolerable fashion without special advantages in
the way of prescriptive right or accumulated means. The
principle of equal opportunity was, no doubt, met only in
a very rough and dubious fashion ; but so favorable became the conditions in this respect that men came to persuade themselves in the course of the eighteenth century
that a substantially equitable allotment of opportunities
would result from the abrogation of all prerogatives other
than the ownership of goods. But so precarious and
transient was this approximation to a technologically
feasible system of equal opportunity that, while the liberal
movement which converged upon this great economic reform was still gathering head, the technological situation
was already outgrowing the possibility of such a scheme
of reform. After the Industrial Revolution came on, it
was no longer true, even in the roughly approximate way
in which it might have been true some time earlier, that
equality before the law, barring property rights, would
mean equal opportunity. In the leading, aggressive in-

dustries which were beginning to set the pace for all that economic system that centered about the market, the unit of industrial equipment, as required by the new technological era, was larger than one man could compass by his own efforts with the free use of the commonplace knowledge of ways and means. And the growth of business enterprise progressively made the position of the small, old-fashioned producer more precarious. But the speculative theoreticians of that time still saw the phenomena of current economic life in the light of the handicraft traditions and of the preconceptions of natural rights associated with that system, and still looked to the ideal of "natural liberty" as the goal of economic development and the end of economic reform. They were ruled by the principles (habits of thought) which had arisen out of an earlier situation, so effectually as not to see that the rule of equal opportunity which they aimed to establish was already technologically obsolete.[6]

During the hundred years and more of this ascendancy of the natural-rights theories in economic science, the growth of technological knowledge has unremittingly gone forward, and concomitantly the large-scale industry has grown great and progressively dominated the field. This large-scale industrial régime is what the socialists, and some others, call "capitalism." "Capitalism," as so used, is not a neat and rigid technical term, but it is definite enough to be useful for many purposes. On its technological side the characteristic trait of this capitalism is that the current pursuit of industry requires a larger unit of material equipment than one individual can com-

[6] For a more extended discussion of this point see the *Quarterly Journal of Economics,* July, 1899, "The Preconceptions of Economic Science"; also *The Theory of Business Enterprise,* chap. iv, especially pp. 70–82.

pass by his own labor, and larger than one person can make use of alone.

So soon as the capitalist régime, in this sense, comes in, it ceases to be true that the owner of the industrial equipment (or the controller of it) in any given case is or may be the producer of it, in any naïve sense of "production." He is under the necessity of acquiring its ownership or control by some other expedient than that of industrially productive work. The pursuit of industry requires an accumulation of wealth, and, barring force, fraud, and inheritance, the method of acquiring such an accumulation of wealth is necessarily some form of bargaining; that is to say, some form of business enterprise. Wealth is accumulated, within the industrial field, from the gains of business; that is to say, from the gains of advantageous bargaining.[7] Taking the situation by and large, looking to the body of business enterprise as a whole, the advantageous bargaining from which gains accrue and from which, therefore, accumulations of capital are derived, is necessarily, in the last analysis, a bargaining between those who own (or control) industrial wealth and those whose work turns this wealth to account in productive industry. This bargaining for hire — commonly a wage agreement — is conducted under the rule of free contract, and is concluded according to the play of demand and supply, as has been well set forth by many writers.

On this technological view of capital, as here spoken

[7] Marx holds that the "primitive accumulation" from which capitalism takes its rise is a matter of force and fraud (*Capital*, Book I, chap. xxiv.). Sombart holds the source to have been landed wealth (*Moderne Kapitalismus*, Book II, Part II, especially chap. xii). Ehrenberg and other critics of Sombart incline to the view that the most important source was usury and the petty trade (*Zeitalter der Fugger*, chaps. i, ii).

for, the relations between the two parties to the bargain, the capitalist-employer and the working class, stand as follows. More or less rigorously, the technological situation enforces a certain scale and method in the various lines of industry.[8] The industry can, in effect, be carried on only by recourse to the technologically requisite scale and method, and this requires a material equipment of a certain (large) magnitude; while material equipment of this required magnitude is held exclusively by the capitalist-employer, and is *de facto* beyond the reach of the common man.

A corresponding body of immaterial equipment — knowledge and practice of ways and means — is likewise requisite, under the rule of the same technological exigencies. This immaterial equipment is in part drawn on in the making of the material equipment held by the capitalist-employers, in part in the use to be made of this material equipment in the further processes of industry. This body of immaterial equipment so drawn on in any line of industry is, relatively, still larger, being, on any exhaustive analysis, virtually the whole body of industrial experience accumulated by the community up to date. A free draft on this common stock of technological wisdom must be had both in the construction and in the subsequent use of the material equipment; although no one person can master, or himself employ, more than an inconsiderable fraction of the immaterial equipment so

[8] The phrase "more or less" covers a certain margin of tolerance in respect of scale and method, which may be very appreciably wider in some lines of industry than in others, and which cannot be more adequately defined or described here within such space as could reasonably be allowed. The requirement of scale and method is enforced by competition. The force and reach of this competitive adjustment can also not be dealt with here, but the familiar current acceptance of the fact will dispense with details.

drawn on for the installation or operation of any given block of the material equipment.

The owner of the material equipment, the capitalist-employer, is, in the typical case, not possessed of any appreciable fraction of the immaterial equipment necessarily drawn on in the construction and subsequent use of the material equipment owned (controlled) by him. His knowledge and training, so far as it enters into the question, is a knowledge of business, not of industry.[9] The slight technological proficiency which he has or needs for his business ends is of a general character, wholly superficial and impracticable in point of workmanlike efficiency; nor is it turned to account in actual workmanship. He therefore " needs in his business " the service of persons who have a competent working mastery of this immaterial technological equipment, and it is with such persons that his bargains for hire are made. By and large, the measure of their serviceability for his ends is the measure of their technological competency. No workman not possessed of some fractional mastery of the technological requirements is employed,— imbeciles are useless in proportion to their imbecility; and even unskilled and " unintelligent " workmen, so called, are of relatively little use, although they may be possessed of a proficiency in the commonplace industrial details such as would bulk large in absolute magnitude. The " common laborer " is, in fact, a highly trained and widely proficient workman when contrasted with the conceivable human blank supposed to have drawn on the community for nothing but his physique.

In the hands of these workmen — the industrial community, the bearers of the immaterial, technological equipment — the capital goods owned by the capitalist become

[9] *Cf. Theory of Business Enterprise,* chap. iii.

a " means of production." Without them, or in the hands of men who do not know their use, the goods in question would be simply raw materials, somewhat deranged and impaired through having been given the form which now makes them " capital goods." The more proficient the workmen in their mastery of the technological expedients involved, and the greater the facility with which they are able to put these expedients into effect, the more productive will be the processes in which the workmen turn the employer's capital goods to account. So, also, the more competent the work of " superintendence," the foreman-like oversight and correlation of the work in respect of kind, speed, volume, the more will it count in the aggregate of productive efficiency. But this work of correlation is a function of the foreman's mastery of the technological situation at large and his facility in proportioning one process of industry to the requirements and effects of another. Without this due and sagacious correlation of the processes of industry, and their current adaptation to the demands of the industrial situation at large, the material equipment engaged would have but slight efficiency and would count for but little in the way of capital goods. The efficiency of the control exercised by the master-workman, engineer, superintendent, or whatever term may be used to designate the technological expert who controls and correlates the productive processes,— this workmanlike efficiency determines how far the given material equipment is effectually to be rated as " capital goods."

Through all this functioning of the workman and the foreman the capitalist's business ends are ever in the background, and the degree of success that attends his business endeavors depends, other things equal, on the efficiency with which these technologists carry on the processes of industry in which he has invested. His working arrangements with these workmen, the bearers of the immaterial equipment engaged, enables the capitalist to turn the processes for which his capital goods are adapted to account for his own profit, but at the cost of such a deduction from the aggregate product of these processes as the workmen may be able to demand in return for their work. The amount of this deduction is determined by the competitive bidding of other capitalists who may have use for the same lines of technological efficiency, in the manner set forth by writers on wages.

With the conceivable consolidation of all material assets under one business management, so as to eliminate competitive bidding between employers, it is plain that the resulting business concern would command the undivided forces of the technological situation, with such deduction as is involved in the livelihood of the working population. This livelihood would in such a case be reduced to the most economical footing, as seen from the standpoint of the employer. And the employer (capitalist) would be the *de facto* owner of the community's aggregate knowledge of ways and means, except so far as this body of immaterial equipment serves also the housekeeping routine of the working population. How nearly the current economic situation may approach to this finished state is a matter of opinion. There is also place for a broad question whether the conditions are more or less favorable to the working population under the existing business régime, involving competitive bidding between the several business concerns, than they would be in case a comprehensive business consolidation had eliminated competition and placed the ownership of the material assets on a footing of unqualified monopoly. Nothing but vague surmises can apparently be offered in answer to these questions.

But as bearing on the question of monopoly and the use of the community's immaterial equipment it is to be kept in mind that the technological situation as it stands to-day does not admit of a complete monopolisation of the community's technological expedients, even if a complete monopolisation of the existing aggregate of material property were effected. There is still current a large body of industrial processes to which the large-scale methods do not apply and which do not presume such a large unit of material equipment or involve such rigorous correlation with the large-scale industry as to take them out of the range of discretionary use by persons not possessed of appreciable material wealth. Typical of such lines of work, hitherto not amenable to monopolisation, are the details of housekeeping routine alluded to above. It is, in fact, still possible for an appreciable fraction of the population to " pick up a living," more or less precarious, without recourse to the large-scale processes that are controlled by the owners of the material assets. This somewhat precarious margin of free recourse to the commonplace knowledge of ways and means appears to be what stands in the way of a neater adjustment of wages to the " minimum of subsistence " and the virtual ownership of the immaterial equipment by the owners of the material equipment.

It follows from what has been said that all tangible [10] assets owe their productivity and their value to the immaterial industrial expedients which they embody or which

[10] " Tangible assets " is here taken to signify serviceable capital goods considered as valuable possessions yielding income to their owner.

their ownership enables their owner to engross. These immaterial industrial expedients are necessarily a product of the community, the immaterial residue of the community's experience, past and present; which has no existence apart from the community's life, and can be transmitted only in the keeping of the community at large. It may be objected by those who make much of the productivity of capital that tangible capital goods on hand are themselves of value and have a specific productive efficiency, if not apart from the industrial processes in which they serve, then at least as a prerequisite to these processes, and therefore a material condition-precedent standing in a causal relation to the industrial product. But these material goods are themselves a product of the past exercise of technological knowledge, and so back to the beginning. What there is involved in the material equipment, which is not of this immaterial, spiritual nature, and so what is not an immaterial residue of the community's experience, is the raw material out of which the industrial appliances are constructed, with the stress falling wholly on the " raw."

The point is illustrated by what happens to a mechanical contrivance which goes out of date because of a technological advance and is displaced by a new contrivance embodying a new process. Such a contrivance " goes to the junk-heap," as the phrase has it. The specific technological expedient which it embodies ceases to be effective in industry, in competition with " improved methods." It ceases to be an immaterial asset. When it is in this way eliminated, the material repository of it ceases to have value as capital. It ceases to be a material asset. " The original and indestructible powers " of the material constituents of capital goods, to adapt Ricardo's phrase, do not make these constituents capital goods; nor, indeed, do these original and indestructible powers of themselves bring the objects in question into the category of economic goods at all. The raw materials — land, minerals, and the like — may, of course, be valuable property, and may be counted among the assets of a business. But the value which they so have is a function of the anticipated use to which they may be put, and that is a function of the technological situation under which it is anticipated that they will be useful.

All this may seem to undervalue or perhaps to overlook the physical facts of industry and the physical nature of commodities. There is, of course, no call to understate the importance of material goods or of manual labor. The goods about which this inquiry turns are the products of trained labor working on the available materials; but the labor has to be trained, in the large sense, in order to be labor, and the materials have to be available in order to be materials of industry. And both the trained efficiency of the labor and the availability of the material objects engaged are a function of the " state of the industrial arts."

Yet the state of the industrial arts is dependent on the traits of human nature, physical, intellectual, and spiritual, and on the character of the material environment. It is out of these elements that the human technology is made up; and this technology is efficient only as it meets with the suitable material conditions and is worked out, practically, in the material forces required. The brute forces of the human animal are an indispensable factor in industry, as are likewise the physical characteristics of the material objects with which industry deals. And it seems bootless to ask how much of the products of industry or of its productivity is to be imputed to these brute forces, human and non-human, as contrasted with the specifically human factors that make technological efficiency. Nor is it necessary to go into questions of that import here, since the inquiry here turns on the productive relation of capital to industry; that is to say, the relation of the material equipment and its ownership to men's dealings with the physical environment in which the race is placed. The question of capital goods (including that of their ownership and therefore including the question of investment) is a question of how mankind as a species of intelligent animals deals with the brute forces at its disposal. It is a question of how the human agent deals with his means of life, not of how the forces of the environment deal with man. Questions of the latter class belong under the head of Ecology, a branch of the biological sciences dealing with the adaptive variability of plants and animals. Economic inquiry would belong under that category if the human response to the forces of the environment were instinctive and variational only, including nothing in the way of a technology. But in that case there would be no question of capital goods, or of capital, or of labor. Such questions do not arise in relation to the non-human animals.

In an inquiry into the productivity of labor some perplexity might be met with as to the share or the place of the brute forces of the human organism in the theory of production; but in relation to capital that question does not arise, except so far as these forces are involved in the production of the capital goods. As a parenthesis, more or less germane to the present inquiry into capital, it may be remarked that an analysis of the productive powers of labor would apparently take account of the brute energies of mankind (nervous and muscular energies) as material forces placed at the disposal of man by circumstances largely beyond human control, and in great part not theoretically dissimilar to the like nervous and muscular forces afforded by the domestic animals.

ON THE NATURE OF CAPITAL

II. Investment, Intangible Assets, and the Pecuniary Magnate

WHAT has been said in the earlier section of this paper [2] applies to " capital goods," so called, and it is intended to apply to these in their character of " productive goods " rather than in their character of " capital "; that is to say, what is had in mind is the industrial, or technological, efficiency and subservience of the material means of production, rather than the pecuniary use and effect of invested wealth. The inquiry has dealt with the industrial equipment as " plant " rather than as " assets." In the course of this inquiry it has appeared that out of the profitable engrossing of the community's industrial efficiency through control of the material equipment there arises the practice of investment, which has further consequences that merit more detailed attention.

Investment is a pecuniary transaction, and its aim is pecuniary gain,— gain in terms of value and ownership. Invested wealth is capital, a pecuniary magnitude, measured in terms of value and determined in respect of its magnitude by a valuation which proceeds on an appraisement of the gain expected from the ownership of this invested wealth. In modern business practice, capital is distinguished into two coördinate categories of assets,

[1]Reprinted by permission from *The Quarterly Journal of Economics,* Vol. XXIII, Nov., 1908.
[2] See this Journal for August, 1908.

tangible and intangible. " Tangible assets " is here taken to designate pecuniarily serviceable capital goods, considered as a valuable possession yielding an income to their owner. Such goods, material items of wealth, are " assets" to the amount of their capitalisable value, which may be more or less closely related to their industrial serviceability as productive goods. " Intangible assets " are immaterial items of wealth, immaterial facts owned, valued, and capitalised on an appraisement of the gain to be derived from their possession. These are also assets to the amount of their capitalisable value, which has commonly little, if any, relation to the industrial serviceability of these items of wealth considered as factors of production.

Before going into the matter of intangible assets, it is necessary to speak further of the consequences which investment — and hence capitalisation — has for the use and serviceability of (material) capital goods. It has commonly been assumed by economists, without much scrutiny, that the gains which accrue from invested wealth are derived from and (roughly) measured by the productivity of the industrial process in which the items of wealth so invested are employed, productivity being counted in some terms of material serviceability to the community, conduciveness to the livelihood, comfort, or consumptive needs of the community. In the course of the present inquiry it has appeared that the gainfulness of such invested wealth (tangible assets) is due to a more or less extensive engrossing of the community's industrial efficiency. The aggregate gains of the aggregate material capital accrue from the community's industrial activity, and bear some relation to the productive capacity of the industrial traffic so engrossed. But it will be noted that there is no warrant in the analysis of these phenomena as here set forth for alleging that the gains of investment bear a relation of equality or proportion to the material serviceability of the capital goods, as rated in terms of effectual usefulness to the community. Given capital goods, tangible assets, may owe their pecuniary serviceability to their owner, and so their value, to other things than their serviceability to the community; although the gains of investment in the aggregate are drawn from the aggregate material productivity of the community's industry.

The ownership of the material equipment gives the owner not only the right of use over the community's immaterial equipment, but also the right of abuse and of neglect or inhibition. This power of inhibition may be made to afford an income, as well as the power to serve; and whatever will yield an income may be capitalised and become an item of wealth to its possessor. Under modern conditions of investment it happens not infrequently that it becomes pecuniarily expedient for the owner of the material equipment to curtail or retard the processes of industry,—" restraint of trade." The motive in all such cases of retardation is the pecuniary expediency of the measure for the owner (controller) of capital, — expediency in terms of income from investment, not expediency in terms of serviceability to the community at large or to any fraction of the community except the owner (manager). Except for the exigencies of investment, *i.e.,* exigencies of pecuniary gain to the investor, phenomena of this character would have no place in the industrial system. They invariably come of the endeavors of business men to secure a pecuniary gain or to avoid a pecuniary loss. More frequently, perhaps, manœuvers of inhibition — advised idleness of plant — in industry aim to effect a saving or avoid a waste than to procure an increase of gain; but the saving to be effected and the waste to be avoided are always pecuniary saving to the owner and pecuniary waste in the matter of ownership, not a saving of goods to the community or a prevention of wasteful consumption or wasteful expenditure of effort and resources on the part of the community. Pecuniary — that is to say, differential — advantage to the capitalist-manager has, under the régime of investment, taken precedence of economic advantage to the community; or rather, the differential advantage of ownership is alone regarded in the conduct of industry under this system.

Business practices which inhibit industrial efficiency and curtail the industrial output are too well known to need particular enumeration. Nor is it necessary to cite evidence to show that such inhibition and curtailment are resorted to from motives of pecuniary expediency. But an illustrative example or two will make the theoretical point clearer, and perhaps more plainly bring out the wholly pecuniary grounds of such business procedure. The most comprehensive principle involved in this class of business management is that of raising prices, and so increasing the net gains of business, by limiting the supply, or " charging what the traffic will bear." Of a similar effect, for the point here in question, are the obstructive tactics designed to hinder the full efficiency of a business rival. These phenomena lie along the line of division between tangible and intangible assets. Successful strategy of this kind may, by force of custom, legislation, or the " freezing-out " of rival concerns, pass into settled conditions of differential advantage for the given business concern, which so may be capitalised as an item of intangible assets and take their place in the business community as articles of invested wealth.

But, aside from such capitalisation of inefficiency, it is at least an equally consequential fact that the processes of productive industry are governed in detail by the exigencies of investment, and therefore by the quest of gain as counted in terms of price, which leads to the dependence of production on the course of prices. So that, under the régime of capital, the community is unable to turn its knowledge of ways and means to account for a livelihood except at such seasons and in so far as the course of prices affords a differential advantage to the owners of the material equipment. The question of advantageous — which commonly means rising — prices for the owners (managers) of the capital goods is made to decide the question of livelihood for the rest of the community. The recurrence of hard times, unemployment, and the rest of that familiar range of phenomena, goes to show how effectual is the inhibition of industry exercised by the ownership of capital under the price system.[3]

So also as regards the discretionary abuse of the community's industrial efficiency vested in the owner of the material equipment. Disserviceability may be capitalised as readily as serviceability, and the ownership of the capital goods affords a discretionary power of misdirecting the industrial processes and perverting[4] industrial efficiency, as well as of inhibiting or curtailing industrial processes and their output, while the outcome may still be profitable to the owner of the capital goods. There is a large volume of capital goods whose value lies in their turning the technological inheritance to the injury of man-

---

[3] For the connection between prices and prosperity, hard times, unemployment, etc., see *The Theory of Business Enterprise,* chap. vii (pp. 185–252, especially 196–212).

[4] By " perversion " is here meant such disposition of the industrial forces as entails a net waste or detriment to the community's livelihood.

kind. Such are, *e.g.*, naval and military establishments, together with the docks, arsenals, schools, and manufactories of arms, ammunition, and naval and military stores, that supplement and supply such establishments. These armaments and the like are, of course, public and quasi-public enterprises, under the current régime, with somewhat disputable relations to the system of current business enterprise. But it is no far-fetched interpretation to say that they are, in great part, a material equipment for the maintenance of law and order, and so enable the owners of capital goods with immunity to inhibit or pervert the industrial processes when the exigencies of business profits make it expedient; that they are, further, a means — more or less ineffectual, it is true — for extending and protecting trade, and so serve the differential advantage of business men at the cost of the community; and that they are also in large part a material equipment set apart for the diversion of a livelihood from the community at large to the military, naval, diplomatic, and other official classes. These establishments may in any case be taken as illustrating how items of material equipment may be devoted to and may be valued for the use of the technological expedients for the damage and discomfort of mankind, without sensible offset or abatement.

Typical of a class of investments which derive profits from capital goods devoted to uses that are altogether dubious, with a large presumption of net detriment, are such establishments as race-tracks, saloons, gambling-houses, and houses of prostitution.[5] Some spokesmen of

---

[5] Should the connection at this point with the main argument of the paper as set forth in the earlier section seem doubtful or obscure, it may be called to mind that these dubious enterprises in dissipation are cases of investment for a profit, and that the "capital goods" engaged are invested wealth yielding an income, but that they yield an income only on the fulfillment of two

the " non-Christian tribes " might wish to include churches under the same category, but the consensus of opinion in modern communities inclines to look on churches as serviceable, on the whole; and it may be as well not to attempt to assign them a specific place in the scheme of serviceable and disserviceable use of invested wealth.

There is, further, a large field of business, employing much capital goods and many technological processes, whose profits come from products in which serviceability and disserviceability are mingled with waste in the most varying proportions. Such are the production of goods of fashion, disingenuous proprietary articles, sophisticated household supplies, newspapers and advertising enterprise. In the degree in which business of this class draws its profits from wasteful practices, spurious goods, illusions and delusions, skilled mendacity, and the like, the capital goods engaged must be said to owe their capitalisable value to a perverse use of the technological expedients employed.

These wasteful or disserviceable uses of capital goods have been cited, not as implying that the technological proficiency embodied in these goods or brought into ef-

conditions: (*a*) the possession and employment of these capital goods enables their holder to turn to account the common stock of technological proficiency, in those bearings in which it may be of use in his enterprise; and (*b*) the limited amount of wealth available for the purpose enables their holder to "engross" the usufruct of such a fraction of the common stock of technological proficiency, in the degree determined by this limitation of the amount available. In so far, these enterprises are like any other industrial enterprise; but beyond this they have the peculiarity that they do not, or need not, even ostensibly, turn the current knowledge and use of ways and means to "productive" account for the community at large, but simply take their stand on the (institutionally sacred) "accomplished fact" of invested wealth. They have less of the fog of apology about them than the common run of business enterprise.

fect in their use, intrinsically has a disserviceable bearing, nor that investment in these things, and business enterprise in the management of them, need aim at disserviceability, but only to bring out certain minor points of theory, obvious but commonly overlooked: (*a*) technological proficiency is not of itself and intrinsically serviceable or disserviceable to mankind,— it is only a means of efficiency for good or ill; (*b*) the enterprising use of capital goods by their businesslike owner aims not at serviceability to the community, but only at serviceability to the owner; (*c*) under the price system — under the rule of pecuniary standards and management — circumstances make it advisable for the business man at times to mismanage the processes of industry, in the sense that it is expedient for his pecuniary gain to inhibit, curtail, or misdirect industry, and so turn the community's technological proficiency to the community's detriment. These somewhat commonplace points of theory are of no great weight in themselves, but they are of consequence for any theory of business or of life under the rules of the price system, and they have an immediate bearing here on the question of intangible assets.

At the risk of some tedium it is necessary to the theory of intangible assets to pursue this analysis and piecing together of commonplaces somewhat farther. As has already been remarked, " assets " is a pecuniary concept, not a technological one; a concept of business, not of industry. Assets are capital, and tangible assets are items of material equipment and the like, considered as available for capitalisation. The tangibility of tangible assets is a matter of the materiality of the items of wealth of which they are made up, while they are assets to the amount of their value. Capital goods, which typically make up the category of tangible assets, are capital goods by virtue of their technological serviceability, but they are capital in the measure, not of their technological serviceability, but in the measure of the income which they may yield to their owner. The like is, of course, true of intangible assets, which are likewise capital, or assets, in the measure of their income-yielding capacity. Their intangibility is a matter of the immateriality of the items of wealth — objects of ownership — of which they are made up, but their character and magnitude as assets is a matter of the gainfulness to their owner of the processes which their ownership enables him to engross. The facts so engrossed, in the case of intangible assets, are not of a technological or industrial character; and herein lies the substantial disparity between tangible and intangible assets.

Mankind has other dealings with the material means of life, besides those covered by the community's technological proficiency. These other dealings have to do with the use, distribution, and consumption of the goods procured by the employment of the community's technological proficiency, and are carried out under working arrangements of an institutional character,— use and wont, law and custom. The principles and practice of the distribution of wealth vary with the changes in technology and with the other cultural changes that are going forward; but it is probably safe to assume that the principles of apportionment,— that is to say, the consensus of habitual opinion as to what is right and good in the distribution of the product,— these principles and the concomitant methods of carrying them out in practice have always been such as to give one person or group or class something of a settled preference above another. Something of this kind, something in the way of a conventionally arranged differential advantage in the apportionment of the common livelihood, is to be found in all cultures and communities that have been observed at all carefully; and it is perhaps needless to remark that in the higher cultures such economic preferences, privileges, prerogatives, differential advantages and disadvantages, are numerous and varied, and that they make up an intricate fabric of economic institutions. Indeed, peculiarities of class difference in some such respect are among the most striking and decisive features that distinguish one cultural era from another. In all phases of material civilisation these preferential advantages are sought and valued. Classes or groups which are in a position to make good a claim to such differential advantages commonly come, in due course, to put forward such claims; as, *e.g.,* the priesthood, the princely and ruling class, the men as contrasted with the women, the adults as against minors, the able-bodied as against the infirm. Principles (habits of thought) countenancing some form of class or personal preference in the distribution of income are to be found incorporated in the moral code of all known civilisations and embodied in some form of institution. Such items of immaterial wealth are of a differential character, in that the advantage of those who secure the preference is the disadvantage of those who do not; and it may be mentioned in passing, that such a differential advantage inuring to any one class or person commonly carries a more than equal disadvantage to some other class or person or to the community at large.[6]

[6] This statement may not seem clear without indicating in a more concrete manner some terms in which to measure the relative differential advantage and disadvantage which so emerge in such a case of prerogative or privilege. Where, as in the earlier, non-pecuniary phases of culture, no price test is applicable, the statement in the text may be taken to mean that the differential

When property rights fall into definite shape and the price system comes in, and more particularly when the practice of investment arises and business enterprise comes into vogue, such differential advantages take on something of the character of intangible assets. They come to have a pecuniary value and rating, whether they are transferable or not; and if they are transferable, if they can be sold and delivered, they become assets in a fairly clear and full sense of that term. Such immaterial wealth, preferential benefits of the nature of intangible assets, may be a matter of usage simply, as the vogue of a given public house, or of a given tradesman, or of a given brand of consumable goods; or may be a matter of arrogation, as the King's Customs in early times, or the once notorious Sound Dues, or the closing of public highways by large land-owners; or of contractual concession, as the freedom of a city or a guild, or a franchise in the Hanseatic League or in the Associated Press; or of government concession, whether on the basis of a bargain or otherwise, as the many trade monopolies of early modern times, or a corporation charter, or a railway franchise, or letters of marque, or letters patent; or of statutory creation, as trade protection by import, export, or excise duties or navigation laws; or of conventionalised superstitious punctilio, as the creation of a demand for wax by the devoutly obligatory consumption of consecrated tapers, or the similar devout consumption of and demand for fish during Lent.

Under the régime of investment and business enterprise these and the like differential benefits may turn to the business advantage of a given class, group, or con-

disadvantage at the cost of which the differential benefit in question is gained is greater than the beneficiary would be willing to undergo in order to procure this benefit.

cern, and in such an event the resulting differential business advantage in the pursuit of gain becomes an asset, capitalised on the basis of its income-yielding capacity, and possibly vendible under the cover of a corporation security (as, *e.g.,* common stock), or even under the usual form of private sale (as, *e.g.,* the appraised good-will of a business concern).

But the régime of business enterprise has not only taken over various forms of institutional privileges and prerogatives out of the past: it also gives rise to new kinds of differential advantage and capitalises them into intangible assets. These are all (or virtually all) of one kind, in that their common aim and common basis of value and capitalisation is a preferentially advantageous sale. Naturally so, since the end of all business endeavor, in the last analysis, is an advantageous sale. The commonest and typical kind of such intangible assets is " goodwill," so called,— a term which has come to cover a great variety of differential business advantages, but which in the original business usage of it meant the customary resort of a clientèle to the concern so possessed of the goodwill. It seems originally to have implied a kindly sentiment of trust and esteem on the part of a customer, but as the term is now used it has lost this sentimental content. In the broad and loose sense in which it is now currently employed it is extended to cover such special advantages as inure to a monopoly or a combination of business concerns through its power to limit or engross the supply of a given line of goods or services. So long as such a special advantage is not specifically protected by special legislation or by a due legal instrument,— as in the case of a franchise or a patent right,— it is likely to be spoken of loosely as " good-will."

The results of the analysis may be summed up to show

the degree of coincidence and the distinctions between the two categories of assets: (*a*) the value (that is to say, the amount) of given assets, whether tangible or intangible, is the capitalised (or capitalisable) value of the given articles of wealth, rated on the basis of their income-yielding capacity to their owner; (*b*) in the case of tangible assets there is a presumption that the objects of wealth involved have some (at least potential) serviceability at large, since they serve a materially productive work, and there is therefore a presumption, more or less well founded, that their value represents, though it by no means measures, an item of serviceability at large; (*c*) in the case of intangible assets there is no presumption that the objects of wealth involved have any serviceability at large, since they serve no materially productive work, but only a differential advantage to the owner in the distribution of the industrial product;[7] (*d*) given tangible

[7] A doubt has been offered as to the applicability of this characterization to such intangible assets as a patent right and other items of the same class. The doubt seems to arise from a misapprehension of the analysis and of its intention. It should be remarked that there is no intention to condemn or disapprove any of the items here spoken of as intangible assets. The patent right may be justifiable or it may not: there is no call to discuss that question here. Other intangible assets are in the same case in this respect.

Further, as to the character of a patent right considered as an asset. The invention or innovation covered by the patent right is a contribution to the common stock of technological proficiency. It may be (immediately) serviceable to the community at large, or it may not; — *e.g.,* a cash register, a bank-check punch, a streetcar fare register; a burglar-proof safe, and the like are of no immediate service to the community at large, but serve only a pecuniary use to their users. But, whether the innovation is useful or not, the patent right, as an asset, has no (immediate) usefulness at large, since its essence is the restriction of the usufruct of the innovation to the patentee. Immediately and directly the patent right must be considered a detriment to the community at large, since its purport is to prevent the community from making

assets may be disserviceable to the community,— a given
material equipment may owe its value as capital to a disserviceable use, though in the aggregate or on an average
the body of tangible assets are (presumptively) serviceable; (*e*) given intangible assets may be indifferent in respect of serviceability at large, though in the aggregate,
or on an average, intangible assets are (presumably) disserviceable to the community.

On this showing it would appear that the substantial
difference between tangible and intangible assets lies in
the different character of the immaterial facts which are
turned to pecuniary account in the one case and in the
other. The former, in effect, capitalise such fraction of
the technological proficiency of the community as the
ownership of the capital goods involved enables the owner
to engross. The latter capitalise such habits of life, of a
non-technological character,— settled by usage, convention, arrogation, legislative action, or what not,— as will
effect a differential advantage to the concern to which
the assets in question appertain. The former owe their
existence and magnitude to the usufruct of technological
expedients involved in the industrial process proper;
while the latter are in like manner due to the usufruct
of what may be called the interstitial correlations and adjustments both within the industrial system and between
industry proper and the market, in so far as these relations are of a pecuniary rather than a technological character. Much the same distinction may be put in other
words, so as to bring the expression nearer the current
popular apprehension of the matter, by saying that tangible assets, commonly so called, capitalise the processes
of production, while intangible assets, so called, capitalise

use of the patented innovation, whatever may be its ulterior bene-
ficial effects or its ethical justification.

certain expedients and processes of acquisition, not productive of wealth, but affecting only its distribution.
Formulated in either way, the distinction seems not to
be an altogether hard-and-fast one, as will immediately
appear if it is called to mind that intangible assets may be
converted into tangible assets, and conversely, as the
exigencies of business may decide. Yet, while the two
categories of assets stand in such close relation to one
another as this state of things presumes, it is still evident
from the same state of things that they are not to be confounded with one another.

Taking " good-will " as typical of the category of " intangible assets," as being the most widely prevalent and
at the same time the farthest removed in its characteristics from the range of " tangible assets," some slight further discussion of it may serve to bring out the difference
between the two categories of assets and at the same
time to enforce their essential congruity as assets as well
as the substantial connection between them. In the
earlier days of the concept, in the period of growth to
which it owes its name, when good-will was coming into
recognition as a factor affecting assets, it was apparently
looked on habitually as an adventitious differential advantage accruing spontaneously to the business concern
to which it appertained; an immaterial by-product of the
concern's conduct of business,— commonly presumed to
be an adventitious blessing incident to an upright and
humane course of business life. Poor Richard would
express this sense of the matter in the saying that " hon-
esty is the best policy." But presently, no doubt, some
thought would be taken of the acquirement of good-will,
and some effort would be expended by the wise business
man in that behalf. Goods would be given a more ele-
gant finish for the sake of a readier sale, beyond what

would conduce to their brute serviceability simply; smooth-spoken and obsequious salesmen and solicitors, gifted with a tactful effrontery, have come to be preferred to others, who, without these merits, may be possessed of all the diligence, dexterity, and muscular force required in their trade; something is expended on convincing, not to say vain-glorious, show-windows that shall promise something more than one would like to commit one's self to in words; itinerant agents, and the like, are employed at some expense to secure a clientèle; much thought and substance is spent on advertising of many kinds.

This last-named item may be taken as typical of the present stage of growth in the production or generation of good-will, and therefore in the creation of intangible assets. Advertising has come to be an important branch of business enterprise by itself, and it employs a large and varied array of material appliances and processes (tangible assets). Investment is made in certain material items (productive goods), such as printed matter, billboards, and the like, with a view to creating a certain body of good-will. The precise magnitude of the product may not be foreseen, but, if sagaciously made, such investment rarely fails of the effect aimed at — unless a business rival with even greater sagacity should outmanœuver and offset these endeavors with a superior array of appliances (productive goods) and workmen for the generation of good-will. The product aimed at, commonly with effect, is good-will,— an intangible asset,— which may be considered to have been generated by converting certain tangible assets into this intangible; or it may be considered as an industrial product, the output of certain industrial processes in which the given items of material equipment are employed and give effect to the requisite technological proficiency. Whichever view be taken of the causal relation between the material equipment and processes employed, on the one hand, and the output of good-will, on the other hand, the result is substantially the same for the purpose in hand.

The ulterior end of the advertising is, it may be said, the sale of an increased quantity of the advertised articles, at an increased net gain; which would mean an increased value of the material items offered for sale; which, in turn, is the same as saying an increase of tangible assets. It may be assumed without debate that the end of business endeavor is a gain in final terms of tangible values. But this ulterior end is, in the case of advertising enterprise, to be gained only by the intermediate step of a production of an immaterial item of good-will, an intangible asset.

So the case in illustration shows not only the conversion of tangible assets (material capital goods, such as printed matter) into intangible wealth, or, if that formula be preferred, the production of immaterial wealth by the productive use of material wealth, but also, conversely, in the second step of the process, it shows the conversion of intangible assets into tangible wealth (enhanced value of vendible goods), or, if the expression seems preferable, the production of tangible assets by the use of intangible wealth.

This creation of tangible wealth out of intangible assets is seen perhaps at its neatest in the enhancement of land values by the endeavors of interested parties. Real estate is, of course, a tangible asset of the most authentic tangibility, and it is an asset to the amount of its value, which is determined, say, by the figures at which the real estate in question is currently bought and sold. This is the current value of the real estate, and therefore its current actual magnitude as a tangible asset. The value of the real estate might also be computed by capitalising its rental value; but, where the current market value does not coincide with the capitalised rental value, the former must, according to business conceptions, be accepted as the actual value. In many parts of this country, perhaps in most, but particularly in the Western States and in the neighborhood of flourishing towns, these two methods of rating the pecuniary magnitude of real estate will habitually not coincide. Due allowance, often very considerable, being made, the capitalised rental value of the land may be taken as measuring its current serviceability as an item of material equipment; while the amount by which the market value of the land exceeds its capitalised rental value may be taken as the product, the tangible residue, of an intangible asset of the nature of good-will, turned to account, or " productively employed," in behalf of this parcel of land.[8]

Some of the lands of California may be taken as a very good, though perhaps not an extreme, example of such a creation of real estate by spiritual instrumentalities. It is probably well within the mark to say that some of these lands owe not more than one-half their current market value to their current serviceability as an instrument of production or use. The excess may be attributable to illusions touching the chances of future sale, to anticipation of a prospective enhanced usefulness, and the like;

[8] Neither as a physical magnitude ("land") nor as a pecuniary magnitude ("real estate") is the capitalised land in question an item of "good-will"; but its value as real estate — *i.e.,* its magnitude as an asset — is in part a product of the "good-will" (illusions and the like) worked up in its behalf and turned to account, by the land agent. The real estate is a tangible asset, an item of material wealth, while the "good-will" to which in part it owes its magnitude as an item of wealth is an intangible asset, an item of immaterial wealth.

but all these are immaterial factors, of the nature of good-will. Like other assets, these lands are capitalised on the basis of the anticipated income from them, part of which income is anticipated from profitable sales to persons who, it is hoped, will be persuaded to take a very sanguine view of the land situation, while part of it may be due to over-sanguine anticipations of usefulness generated by the advertising matter and the efforts of the land agents directed to what is called " developing the country."

To any one preoccupied with the conceit that " capital " means " capital goods " such a conversion of intangible into tangible goods, or such a generation of intangible assets by the productive use of tangible assets, might be something of a puzzle. If " assets " were a physical concept, covering a range of physical things, instead of a pecuniary concept, such conversion of tangible into intangible assets, and conversely, would be a case of transubstantiation. But there is nothing miraculous in the matter. " Assets " are a pecuniary magnitude, and belong among the facts of investment. Except in relation to investment the items of wealth involved are not assets. In other words, assets are a matter of capitalisation, which is a special case of valuation; and the question of tangibility or intangibility as regards a given parcel of assets is a question of what article or class of articles the valuation shall attach to or be imputed to. If, *e.g.,* the fact to which value is imputed in the valuation is the habitual demand for a given article of merchandise, or the habitual resort of a given group of customers to a particular shop or merchant, or a monopolistic control or limitation of price and supply, then the resulting item of assets will be " intangible," since the object to which the capitalised value in question is imputed is an immaterial object. If the fact which is by imputation made the bearer of the capitalised value is a material object, as, *e.g.*, the merchantable goods of which the supply is arbitrarily limited or the price arbitrarily fixed, or if it is the material means of supplying such goods, then the capitalised value in question is a case of tangible assets. The value involved is, like all value, a matter of imputation, and as assets it is a matter of capitalisation; but capitalisation is an appraisement of a pecuniary " income-stream " in terms of the vendible objects to the ownership of which the income is assumed to inure. To what object the capitalised value of the " income-stream " shall be imputed is a question of what object of ownership secures to the owner an effectual claim on this " income-stream "; that is to say, it is a question of what object of ownership the strategic advantages is assumed to attach to, which is a question of the play of business exigencies in the given case.

The " income-stream " in question is a pecuniary income-stream, and is in the last resort traceable to transactions of sale. Within the confines of business — and therefore within the scope of capital, investments, assets, and the like business concepts — transactions of purchase and sale are the final terms of any analysis. But beyond these confines, comprehending and conditioning the business system, lie the material facts of the community's work and livelihood. In the final transaction of sale the merchantable goods are valued by the consumer, not as assets, but as livelihood;[9] and in the last analysis and long run it is to some such transaction that all business imputations of value and capitalistic appraisement of

[9] " Livelihood " is, of course, here taken in a loose sense, not as denoting the means of subsistence simply or even the means of physical comfort, but as signifying that the purchases in question are made with a view to the consumptive use of the goods rather than with a view to their use for a profit.

assets must have regard and by which they must finally be checked. Dissociated from the facts of work and livelihood, therefore, assets cease to be assets; but this does not preclude their relation to these facts of work and livelihood being at times somewhat remote and loose.

Without recourse, immediately or remotely, to certain material facts of industrial process and equipment, assets would not yield earnings; that is to say, wholly disjoined from these material facts, they would in effect not be assets. This is true for both tangible and intangible assets, although the relation of the assets to the material facts of industry is not the same in the two cases. The case of tangible assets needs no argument. Intangible assets, such as patent right or monopolistic control, are likewise of no effect except in effectual contact with industrial facts. The patent right becomes effective for the purpose only in the material working of the innovation covered by it; and monopolistic control is a source of gain only in so far as it effectually modifies or divides the supply of goods.

In the light of these considerations it seems feasible to indicate both the congruence and the distinction between the two categories of assets a little more narrowly than was done above. Both are assets,— that is to say, both are values determined by a capitalisation of anticipated income-yielding capacity; both depend for their income-yielding capacity on the preferential use of certain immaterial factors; both depend for their efficiency on the use of certain material objects; both may increase or decrease, as assets, apart from any increase or decrease of the material objects involved. The tangible assets capitalise the preferential use of technological, industrial expedients,— expedients of production, dealing with the facts of brute nature under the laws of physical cause and

effect,— this preferential use being secured by the ownership of material articles employed in the processes in
which these expedients are put into effect. The intangible assets capitalise the preferential use of certain facts
of human nature — habits, propensities, beliefs, aspirations, necessities — to be dealt with under the psychological laws of human motivation; this preferential use being secured by custom, as in the case of old-fashioned
good-will, by legal assignment, as in patent or copyright,
by ownership of the instruments of production, as in the
case of industrial monopolies.[10]

Intangible assets are capital as well as tangible assets;
that is to say, they are items of capitalised wealth. Both
categories of assets, therefore, represent expected " income-streams " which are of such definite character as to
admit of their being rated in set terms per cent. per time
unit; although the expected income need not therefore be
anticipated to come in an even flow or to be distributed
in any equable manner over a period of time. The income-streams to be so rated and capitalised are associated
in such a manner with some external fact (impersonal to
their claimant), whether material or immaterial, as to
permit their being traced or attributed to an income-yielding capacity on the part of this external fact, to which
their valuation as a whole may be imputed and which
may then be capitalised as an item of wealth yielding
this income-stream. Income-streams which do not meet

[10] The instruments of production so monopolised are, of course,
tangible assets, but the ownership of such means of production in
amount sufficient to enable the owner to monopolise or control
the market, whether for purchase (as of materials or labor) or
for sale (as of marketable goods or services), gives rise to a
differential business advantage which is to be classed as intangible assets.

these requirements do not give rise to assets in the accepted sense of the term, and so do not swell the volume of capitalised wealth.

There are income-streams which do not meet the necessary specifications of capitalisable wealth; and in modern business traffic, particularly, there are large and secure sources of income that are in this way not capitalisable and yet yield a legitimate business income. Such are, indeed, to be rated among the most consequential factors in the current business situation. Under the guidance of traditions carried over from a more primitive business situation, it has been usual to speak of income-streams derived in such a manner as " wages of superintendence," or " undertaker's wages," or " entrepreneur's profits," or, latterly, as " profits " simply and specifically. Such phenomena of this class as are of consequence in business are commonly accounted for, theoretically, under this head; and the effort so to account for them is to be taken as, at least, a laudable endeavor to avoid an undue multiplication of technical terms and categories.[11] Yet the most striking phenomena of this class, and the most consequential for modern business and industry, both in respect of their magnitude and in respect of the pecuniary dominion and discretion which they represent, cannot well be accounted undertaker's gains, in the ordinary sense of that term. The great gains of the great industrial financiers or of the great " interests," *e.g.*, do not answer the description of undertaker's gains, in that they do not accrue to the captain of industry on the basis of

[11] One writer even goes so far in the endeavor to bring the facts within the scope of the staple concepts of theory at this point as to rate the persons concerned in such a case as " capital," after having satisfied himself that such income-streams are traceable to a personal source.— See Fisher, *Nature of Capital and Income,* chap. v.

his "managerial ability" alone, apart from his wealth or out of relation to his wealth; and yet it is not safe to say that such gains (which are over and above ordinary returns on his investments) accrue on the ground of the requisite amount of wealth alone, apart from the exercise of a large business direction on the part of the owner of such wealth, or on the part of his agent to whom discretion has been delegated. Administrative, or strategic, discretion and activity must necessarily be present in the case: otherwise, the income in question would rightly be rated as income from capital simply.

The captain of industry, the pecuniary magnate, is normally in receipt of income in excess of the ordinary rate per cent. on investment; but apart from his large holdings he is not in a position to get these large gains. Dissociated from his large holdings, he is not a large captain of industry; but it is not the size of his holdings alone that determines what the gains of the pecuniary magnate in modern industry shall be. Gains of the kind and magnitude that currently come to this class of business men come only on condition that the owner (or his agent) shall exercise a similarly large discretion and control in the affairs of the business community; but the magnitude of the gains, as well as of the discretion and control exercised, is somewhat definitely conditioned by the magnitude of the wealth which gives effect to this discretion.

The disposition of pecuniary forces in such matters may be well seen in the work and remuneration of any coalition of "interests," such as the modern business community has become familiar with. The "interests" in such a case are of a personal character,— they are "interested parties,"— and the sagacity, experience, and animus of these various interested parties counts in the outcome, both as regards the aggregate gains of the coalition and as regards the distribution of these gains among the several parties in interest; but the weight of any given "interest" in a coalition or "system" is more nearly proportioned to the wealth controlled by the given "interest," and to the strategic position of such wealth, than to any personal talents or proficiency of the "interested party." The talents and proficiency involved are not the main facts. Indeed, the movements of such a "system," and of the several component "interests," are largely a matter of artless routine, in which the greatest ingenuity and initiative engaged in the premises are commonly exercised by the legal counsel working for a fee.

A dispassionate student of the current business traffic, who is not overawed by round numbers, will be more impressed by the ease and simplicity of the manœuvers that lead to large pecuniary results in the higher business finance than by any evidence of preëminent sagacity and initiative among the pecuniary magnates. One need only call to mind the simple and obvious way in which the promoters of the Steel Corporation were magnificently checkmated by the financiers of the Carnegie "interest," when that great and reluctant corporation was floated, or the pettyfogging tactics of Standard Oil in its later career. In extenuation of their visible lack of initiative and insight it may not be ungraceful to call to mind that many of the discretionary heads of the great "interests" are men of advanced years, and that in the nature of the case the pecuniary magnates of the present generation must commonly be men of a somewhat advanced age; and it is only during the present generation that the existing situation has arisen, with its characteristic opportunities and demands. To take their present foremost rank in the new business finance which is here under inquiry, they have had to accumulate the great wealth on which alone their discretionary control of business affairs rests, and their best vigor has been spent in this work of preparation; so that they have commonly attained the requisite strategic position only after they had outlived their " years of discretion."

But there is no intention here to depreciate the work of the pecuniary magnates or the spokesmen of the great " interests." The matter has been referred to only as it bears on this category of capitalistic income which accrues on other grounds than the " earning-capacity " of the assets involved, and which still cannot be imputed to the " earning-capacity " of these business men apart from these assets. The case is evidently not one of " wages of superintendence " or " undertaker's profits "; but it is as evidently not a case of the earning-capacity of the assets. The proof of the latter point is quite as easy as of the former. If the gains of the " system " or of its constituent " interests " and magnates were imputable to the earning-capacity of the assets involved,— in any accepted sense of " earnings,"— then it would immediately follow that these assets would be recapitalised on the basis of these extraordinary earnings, and that the income derived in this class of traffic should reappear as interest or dividends on the capital so increased to correspond with the increased earnings. But such recapitalisation takes place only to a relatively very limited extent, and the question then bears on the income which is not so accounted for in the recapitalisation.

The gains of this class of traffic are, of course, themselves capitalised,— for the most part they accrue in the capitalised form, as issues of securities and the like; but the sources of this income are not capitalised as such. The (large) accumulated wealth, or assets, which gives weight to the movements of the " interests " and magnates in question, and which affords the ground for the discretionary control of business affairs exercised by them, are, for the most part at least, invested in ordinary business ventures, in the form of corporation securities and the like, and are there earning dividends or interest at current rates; and these assets are valued in the market (and thereby capitalised) on the basis of their current earnings in the various enterprises in which they are so invested. But their being so invested in profitable business enterprises does not in the least hinder their usefulness in the hands of the magnates as a basis or means of carrying on the large and highly profitable transactions of the higher industrial finance. To impute these gains to these assets as "earnings," therefore, would be to count the assets twice as capital, or rather to count them over and over.

An additional perplexity in endeavoring to handle gains of this class theoretically as earnings, in the ordinary sense, arises from the fact that they stand in no definable time relation to their underlying assets. They have no definable "time-shape," as Mr. Fisher might put it.[12] Such gains are timeless, in the sense that the time relation does not count in any substantial manner or in any sensible degree in their determination.[13]

In a more painstaking statement of this point of theory it would be necessary to note that these gains are "timeless," in the sense indicated, in so far as the enterprise from which they accrue is dissociated from the technological circumstances and processes of industry, and only in

[12] *Cf.* Fisher, *Rate of Interest,* chap. vi.

[13] This conclusion is reached, *e.g.,* by Mr. G. P. Watkins (*The Growth of Large Fortunes,* chap. iii, sec. 10), although through a curious etymological misapprehension he rejects the term "timeless" as not available.

so far. Technological (industrial) procedure, being of the nature of physical causation, is subject to the time relation under which causal sequence runs. This is the basis of such discussions of capital and interest as those of Böhm-Bawerk, and of Fisher. But business traffic, as distinguished from the processes of industry, being not immediately concerned with the technological process, is also not immediately or uniformly subject to the time relation involved in the causal sequence of the technological process. Business traffic is subject to the time relation because and in so far as it depends upon and follows up the processes of production. The commonplace or old-fashioned business enterprise, the competitive system of investment in industrial business simply, commonly rests pretty directly on the due sequence of the industrial processes in which the investments of such enterprise are placed. Such enterprise, as conceived by the current theories of capital, does business at first hand in the industrial efficiency of the community, which is conditioned by the time relation of the causal sequence, and which is, indeed, in great measure a function of the time consumed in the technological processes. Therefore, the gains, as well as the transactions, of such enterprise are also commonly somewhat closely conditioned by the like time relation, and they typically emerge under the form of a per-cent. per time unit; that is to say, as a function of the lapse of time. Yet the business transactions themselves are not a matter of the lapse of time. Time is not of the essence of the case. The magnitude of a pecuniary transaction is not a function of the time consumed in concluding it, nor are the gains which accrue from the transaction. In business enterprise on the higher plane, which is here under inquiry, the relation of the transactions, and of their gains, to the consecution of the technological processes remotely underlying them is distant, loose, and uncertain, so that the time element here does not obtrude itself : rather, it somewhat obviously falls into abeyance, marking the degree of its remoteness. Yet this phase of business enterprise, like any other, of course takes place in time; and, it is also to be remarked, the volume of the traffic and the gains derived from it are, no doubt, somewhat closely conditioned in the long run by the time relation which dominates that technological (industrial) efficiency on which this enterprise, too, ultimately and indirectly rests and from which in the last resort its gains are finally drawn, however remotely and indirectly.

An analysis of these phenomena on lines similar to those which have been followed in the discussion of assets above is not without difficulty, nor can it fairly be expected to yield any but tentative and provisional results. The matter has received so little attention from economic theoreticians that even significant mistakes in this connection are of very rare occurrence.[14] The cause of this scant attention to these matters lies, no doubt, in the relative novelty of the facts in question. The facts may be roughly drawn together under the caption " Traffic in Vendible Capital "; although that term serves rather as a comprehensive designation of the class of business enterprise from which these gains accrue than as an adequate chararacterisation of the play of forces involved.[15] Traffic in vendible capital has not been un-

---

[14] Even Mr. Watkins (as cited above), *e.g.*, is led by a superficial generalisation to class these gains as " speculative," and so to excuse himself from a closer acquaintance with their character and with the bearings of the class of business enterprise out of which they arise.

[15] *Cf. Theory of Business Enterprise,* chap. v, pp. 119–130; chap. vi, pp. 162–174.

known in the past, but it is only recently that it has come into the foreground as the most important line of business enterprise. Such it now is, in that it is in this traffic that the ultimate initiative and discretion in business are now to be found. It is at the same time the most gainful of business enterprise, not only in absolute terms, but relatively to the magnitude of the assets involved as well. One reason for this superior gainfulness is the fact that the assets involved in this traffic are at the same time engaged as assets to their full extent in ordinary business, so that the peculiar gains of this traffic are of the nature of a bonus above the earnings of the invested wealth. " It is like finding money."

As was said above, the method, or the ways and means, characteristic of this superior business enterprise is a traffic in vendible capital. The wealth gained in this field is commonly in the capitalised form, and constitutes in each transaction, or " deal," a deduction or abstraction from the capitalised wealth of the business community in favor of the magnates or " interests " to whom the gains accrue. Its proximate aim is a transfer of capitalised wealth from other capitalists to those who so gain. This transfer or abstraction of capitalised wealth from the former owners is commonly effected by an augmentation of the nominal capital, based on a (transient) advantage inuring to the particular concerns whose capitalisation is so augmented.[16] Any such increase of the community's aggregate capitalisation, without a corresponding increase of the material wealth on which the capitalisation is based, involves, of course, in effect a redistribution of the aggregate capitalised wealth ; and in this redistribution the great financiers are in a position to gain. The gains in question, it will be seen, come out of the

[16] *Cf. Theory of Business Enterprise*, footnote on pp. 169–170.

business community, out of invested wealth, and only remotely and indirectly out of the community at large from which the business community draws its income. These gains, therefore, are a tax on commonplace business enterprise, in much the same manner and with much the like effects as the gains of commonplace business (ordinary profits and interest) are a tax on industry.[17]

In a manner analogous to the old-fashioned capitalist-employer's engrossing of the industrial community's technological efficiency does the modern pecuniary magnate engross the business community's capitalistic efficiency. This capitalistic efficiency lies in the capitalist-employer's ability — by force of the ownership of the material equipment — to induce the industrial community, through suitable bargaining, to turn over to the owner of the material equipment the excess of the product above the industrial community's livelihood. The fortunes of the capitalist-employer are closely dependent on the run of the market, — the conjunctures of advantageous purchase and sale; and it is his constant endeavor to create or gain for himself some peculiar degree of advantage in the market, in the way of monopoly, good-will, legalised privilege, and the like,— something in the way of intangible assets. But the pecuniary magnate, in the measure in which he truly answers to the concept, is superior to the market on which the capitalist-employer depends, and can make or

[17] As should be evident from the run of the argument in the earlier portions of this paper, the use of the words "tax," "deduction," "abstraction," in this connection, is not to be taken as implying approval or disapproval of the phenomena so characterised. The words are used for want of better terms to indicate the source of business gains, and objectively to characterise the relation of give-and-take between industry and ordinary capitalistic business, on the one hand, and between ordinary business and this business enterprise on the higher plane, on the other hand.

mar its conjunctures of advantageous purchase and sale of goods; that is to say, he is in a position to make or mar any peculiar advantage possessed by the given capitalist-employer who comes in his way. He does this by force of his large holdings of capital at large, the weight of which he can shift from one point of investment to another as the relative efficiency — earning-capacity — of one and another line of investment may make it expedient; and at each move of this kind, in so far as it is effective for his ends, he cuts into and assimilates a fraction of the invested wealth involved, in that he cuts into and sequesters a fraction of the capital's earning-capacity in the given line. That is to say, in the measure in which he is a pecuniary magnate, and not simply a capitalist-employer, he engrosses the capitalistic efficiency of invested wealth; he turns to his own account the capitalist-employer's effectual engrossing of the community's industrial efficiency. He engrosses the community's pecuniary initiative and proficiency. In the measure, therefore, in which this relatively new-found serviceability of extraordinarily large wealth is effective for its peculiar business function, the old-fashioned capitalist-employer loses his discretionary initiative and becomes a mediator, an instrumentality of extraction and transmission, a collector and conveyer of revenue from the community at large to the pecuniary magnate, who, in the ideal case, should leave him only such an allowance out of the gross earnings collected and transmitted as will induce him to continue in business.

To the community at large, whose industrial efficiency is already virtually engrossed by the capitalist-employer's ownership and control of the material equipment, this later step in the evolution of the economic situation should apparently not be a matter of substantial consequence or a matter for sentimental disturbance. On the face of it, it should appear to have little more than a speculative interest for those classes of the community who do not derive an income from investments; particularly not for the working classes, who own nothing to speak of and whose only dependence is their technological efficiency, which has virtually ceased to be their own. But such is not the current state of sentiment. This inchoate new phase of capitalism, this business enterprise on the higher plane, is in fact viewed with the most lively apprehension. In a maze of consternation and solicitude the boldest, wisest, most public-spirited, most illustrious gentlemen of our time are spending their manhood in an endeavor to make the hen continue sitting on the nest after the chickens are out of the shell. The modern community is imbued with business principles — of the old dispensation. By precept and example, men have learned that the business interests (of the authentic superannuated scale and kind) are the palladium of our civilisation, as Mr. Dooley would say; and it is felt that any disturbance of the existing pecuniary dominion of the capitalist-employer — as contrasted with the pecuniary magnate — would involve the well-being of the community in one common agony of desolation.

The merits of this perturbation, or of the remedies proposed for saving the pecuniary life of the old-fashioned capitalist-employer, of course do not concern the present inquiry; but the matter has been referred to here as evidence that the pecuniary magnate's work, and the dominion which his extraordinarily large wealth gives him, are, in effect, substantially a new phase of the economic development, and that these phenomena are distastefully unfamiliar and are felt to be consequential enough to threaten the received institutional structure.

That is to say, it is felt to be a new phase of business enterprise,— distasteful to those who stand to lose by it.

The basis of this business enterprise on the higher plane is capital-at-large, as distinguished from capital invested in a given line of industrial enterprise, and it becomes effective when wealth has accumulated in holdings sufficiently large to give the holder (or combination of holders, the " system ") a controlling weight in any group or ramification of business interests into which he may throw his weight by judicious investment (or by underwriting and the like). The pecuniary magnate must be able effectually to engross the pecuniary initiative and the business opportunities on which such a section or ramification of the business community depends for its ordinary gains. How large a proportion of the business community's capital is needed for such an effectual engrossing of its capitalistic efficiency, in any given bearing, is a question that cannot be answered in anything like absolute terms, or even in relative terms of a satisfactorily definite kind. It is, of course, evident that a relatively large disposable body of capital is needed for such a purpose; and it is also evident, from the current facts of business, that the body of capital so disposed of need not amount to a majority, or anything near a majority, of the investments involved,— at least not at the present relatively inchoate phase of this larger business enterprise. The larger the holdings of the magnate, the more effectual and expeditious will be his work of absorbing the holdings of the smaller capitalist-employer, and the more precipitately will the latter yield his assets to the new claimant.

Evidently, this work of the pecuniary magnate bears a great resemblance to the creation of intangible assets under the ordinary competitive system. This is, no doubt, the point of its nearest relation to the current capitalistic enterprise. But, as has already been indicated above, it cannot be said that the magnate's peculiar work is the creation of intangible, or other assets, although there is commonly some recapitalisation involved in his manœuvers, and although his gains commonly come as assets, *i.e.,* in the capitalised form. Nor can it, as has also been indicated above, be said that the wealth which serves him as the means of his peculiar enterprise stands in the relation of assets to this enterprise or to the gains in question, since this wealth already stands in an exhaustive relation as assets to some corporate enterprise in ordinary business and to the corresponding items of interest and dividends. It may, of course, be contended that the present state of things on this higher plane of enterprise is transient and transitional only, and that in the settled condition which may conceivably supervene, the magnate's relation to business at large will be capitalised in some form of intangible assets, after the manner in which the monopoly advantage of an ordinary " trust " is now capitalised. But this is at the best only a surmise, guided by inapplicable generalisations drawn from a past situation in which this higher enterprise has not engrossed the pecuniary initiative and played the ruling part.

SOME NEGLECTED POINTS IN THE
THEORY OF SOCIALISM [1]

THE immediate occasion for the writing of this paper
was given by the publication of Mr. Spencer's essay,
" From Freedom to Bondage "; [2] although it is not altogether a criticism of that essay. It is not my purpose
to controvert the position taken by Mr. Spencer as regards the present feasibility of any socialist scheme.
The paper is mainly a suggestion, offered in the spirit
of the disciple, with respect to a point not adequately
covered by Mr. Spencer's discussion, and which has
received but very scanty attention at the hands of any
other writer on either side of the socialist controversy.
This main point is as to an economic ground, as a matter
of fact, for the existing unrest that finds expression in
the demands of socialist agitators.

I quote from Mr. Spencer's essay a sentence which does
fair justice, so far as it goes, to the position taken by
agitators : " In presence of obvious improvements, joined
with that increase of longevity, which even alone yields
conclusive proof of general amelioration, it is proclaimed,
with increasing vehemence, that things are so bad that
society must be pulled to pieces and reorganised on another plan." The most obtrusive feature of the change
demanded by the advocates of socialism is governmental
control of the industrial activities of society — the na-

[1] Reprinted by permission from the *Annals of American
Academy of Political and Social Science,* Vol. II, 1892.

[2] Introductory paper of *A Plea for Liberty;* edited by Thomas
Mackay.

tionalisation of industry. There is also, just at present, a distinct movement in practice, towards a more extended control of industry by the government, as Mr. Spencer has pointed out. This movement strengthens the position of the advocates of a complete nationalisation of industry, by making it appear that the logic of events is on their side.

In America at least, this movement in the direction of a broader assertion of the paramount claims of the community, and an extension of corporate action on part of the community in industrial matters, has not generally been connected with or based on an adherence to socialistic dogmas. This is perhaps truer of the recent past than of the immediate present. The motive of the movement has been, in large part, the expediency of each particular step taken. Municipal supervision, and, possibly, complete municipal control, has come to be a necessity in the case of such industries — mostly of recent growth — as elementary education, street-lighting, water-supply, etc. Opinions differ widely as to how far the community should take into its own hands such industries as concern the common welfare, but the growth of sentiment may fairly be said to favor a wider scope of governmental control.

But the necessity of some supervision in the interest of the public extends to industries which are not simply of municipal importance. The modern development of industry and of the industrial organisation of society makes it increasingly necessary that certain industries — often spoken of as " natural monopolies "— should be treated as being of a semi-public character. And through the action of the same forces a constantly increasing number of occupations are developing into the form of " natural monopolies."

The motive of the movement towards corporate action on the part of the community — State control of industry — has been largely that of industrial expediency. But another motive has gone with this one, and has grown more prominent as the popular demands in this direction have gathered wider support and taken more definite form. The injustice, the inequality, of the existing system, so far as concerns these natural monopolies especially, are made much of. There is a distinct unrest abroad, a discontent with things as they are, and the cry of injustice is the expression of this more or less widely prevalent discontent. This discontent is the truly socialistic element in the situation.

It is easy to make too much of this popular unrest. The clamor of the agitators might be taken to indicate a wider prevalence and a greater acuteness of popular discontent than actually exists; but after all due allowance is made for exaggeration on the part of those interested in the agitation, there can still be no doubt of the presence of a chronic feeling of dissatisfaction with the working of the existing industrial system, and a growth of popular sentiment in favor of a leveling policy. The economic ground of this popular feeling must be found, if we wish to understand the significance, for our industrial system, of the movement to which it supplies the motive. If its causes shall appear to be of a transient character, there is little reason to apprehend a permanent or radical change of our industrial system as the outcome of the agitation; while if this popular sentiment is found to be the outgrowth of any of the essential features of the existing social system, the chances of its ultimately working a radical change in the system will be much greater.

The explanation offered by Mr. Spencer, that the popular unrest is due essentially to a feeling of *ennui* — to a desire for a change of posture on part of the social body, is assuredly not to be summarily rejected; but the analogy will hardly serve to explain the sentiment away. This may be a cause, but it can hardly be accepted as a sufficient cause.

Socialist agitators urge that the existing system is necessarily wasteful and industrially inefficient. That may be granted, but it does not serve to explain the popular discontent, because the popular opinion, in which the discontent resides, does notoriously not favor that view. They further urge that the existing system is unjust, in that it gives an advantage to one man over another. That contention may also be true, but it is in itself no explanation, for it is true only if it be granted that the institutions which make this advantage of one man over another possible are unjust, and that is begging the question. This last contention is, however, not so far out of line with popular sentiment. The advantage complained of lies, under modern conditions, in the possession of property, and there is a feeling abroad that the existing order of things affords an undue advantage to property, especially to owners of property whose possessions rise much above a certain rather indefinite average. This feeling of injured justice is not always distinguishable from envy; but it is, at any rate, a factor that works towards a leveling policy. With it goes a feeling of slighted manhood, which works in the same direction. Both these elements are to a great extent of a subjective origin. They express themselves in the general, objective form, but it is safe to say that on the average they spring from a consciousness of disadvantage and slight suffered by the person expressing them, and by persons whom he classes with himself. No flippancy is intended in saying that the rich are not so generally alive to the necessity of any leveling policy as are people of slender means. Any question as to the legitimacy of the dissatisfaction, on moral grounds, or even on grounds of expediency, is not very much to the point; the question is as to its scope and its chances of persistence.

The modern industrial system is based on the institution of private property under free competition, and it cannot be claimed that these institutions have heretofore worked to the detriment of the material interests of the average member of society. The ground of discontent cannot lie in a disadvantageous comparison of the present with the past, so far as material interests are concerned. It is notorious, and, practically, none of the agitators deny, that the system of industrial competition, based on private property, has brought about, or has at least co-existed with, the most rapid advance in average wealth and industrial efficiency that the world has seen. Especially can it fairly be claimed that the result of the last few decades of our industrial development has been to increase greatly the creature comforts within the reach of the average human being. And, decidedly, the result has been an amelioration of the lot of the less favored in a relatively greater degree than that of those economically more fortunate. The claim that the system of competition has proved itself an engine for making the rich richer and the poor poorer has the fascination of epigram; but if its meaning is that the lot of the average, of the masses of humanity in civilised life, is worse to-day, as measured in the means of livelihood, than it was twenty, or fifty, or a hundred years ago, then it is farcical. The cause of discontent must be sought elsewhere than in any increased difficulty in obtaining the means of subsistence or of comfort. But there is a sense in which the aphorism is true, and in it lies at least a partial explanation of the unrest which our conservative people so greatly deprecate. The existing system has not made, and does not tend to make, the industrious poor poorer as measured absolutely in means of livelihood; but it does tend to make them relatively poorer, in their own eyes, as measured in terms of comparative economic importance, and, curious as it may seem at first sight, that is what seems to count. It is not the abjectly poor that are oftenest heard protesting; and when a protest is heard in their behalf it is through spokesmen who are from outside their own class, and who are not delegated to speak for them. They are not a negligible element in the situation, but the unrest which is ground for solicitude does not owe its importance to them. The protest comes from those who do not habitually, or of necessity, suffer physical privation. The qualification " of necessity," is to be noticed. There is a not inconsiderable amount of physical privation suffered by many people in this country, which is not physically necessary. The cause is very often that what might be the means of comfort is diverted to the purpose of maintaining a decent appearance, or even a show of luxury.

Man as we find him to-day has much regard to his good fame — to his standing in the esteem of his fellowmen. This characteristic he always has had, and no doubt always will have. This regard for reputation may take the noble form of a striving after a good name; but the existing organisation of society does not in any way preëminently foster that line of development. Regard for one's reputation means, in the average of cases, emulation. It is a striving to be, and more immediately to be thought to be, better than one's neighbor. Now, modern society, the society in which competition without prescription is predominant, is preëminently an industrial, economic society, and it is industrial — economic — excellence that most readily attracts the approving regard of that society. Integrity and personal worth will, of course, count for something, now as always; but in the case of a person of moderate pretentions and opportunities, such as the average of us are, one's reputation for excellence in this direction does not penetrate far enough into the very wide environment to which a person is exposed in modern society to satisfy even a very modest craving for respectability. To sustain one's dignity — and to sustain one's self-respect — under the eyes of people who are not socially one's immediate neighbors, it is necessary to display the token of economic worth, which practically coincides pretty closely with economic success. A person may be well-born and virtuous, but those attributes will not bring respect to the bearer from people who are not aware of his possessing them, and these are ninety-nine out of every one hundred that one meets. Conversely, by the way, knavery and vulgarity in any person are not reprobated by people who know nothing of the person's shortcomings in those respects.

In our fundamentally industrial society a person should be economically successful, if he would enjoy the esteem of his fellowmen. When we say that a man is " worth " so many dollars, the expression does not convey the idea that moral or other personal excellence is to be measured in terms of money, but it does very distinctly convey the idea that the fact of his possessing many dollars is very much to his credit. And, except in cases of extraordinary excellence, efficiency in any direction which is not immediately of industrial importance, and does not redound to a person's economic benefit, is not of great value as a means of respectability. Economic success is in our day the most widely accepted as well as the most readily ascertainable measure of esteem. All this will hold with still greater force of a generation which is born into a world already encrusted with this habit of a mind.

But there is a further, secondary stage in the development of this economic emulation. It is not enough to possess the talisman of industrial success. In order that it may mend one's good fame efficiently, it is necessary to display it. One does not " make much of a showing " in the eyes of the large majority of the people whom one meets with, except by unremitting demonstration of ability to pay. That is practically the only means which the average of us have of impressing our respectability on the many to whom we are personally unknown, but whose transient good opinion we would so gladly enjoy. So it comes about that the appearance of success is very much to be desired, and is even in many cases preferred to the substance. We all know how nearly indispensable it is to afford whatever expenditure other people with whom we class ourselves can afford, and also that it is desirable to afford a little something more than others.

This element of human nature has much to do with the " standard of living." And it is of a very elastic nature, capable of an indefinite extension. After making proper allowance for individual exceptions and for the action of prudential restraints, it may be said, in a general way, that this emulation in expenditure stands ever ready to absorb any margin of income that remains after ordinary physical wants and comforts have been provided for, and, further, that it presently becomes as hard to give up that part of one's habitual " standard of living " which is due to the struggle for respectability, as it is to give up many physical comforts. In a general way, the need of expenditure in this direction grows as fast as the means of satisfying it, and, in the long run, a large expenditure comes no nearer satisfying the desire than a smaller one.

It comes about through the working of this principle that even the creature comforts, which are in themselves desirable, and, it may even be, requisite to a life on a passably satisfactory plane, acquire a value as a means of respectability quite independent of, and out of proportion to, their simple utility as a means of livelihood. As we are all aware, the chief element of value in many articles of apparel is not their efficiency for protecting the body, but for protecting the wearer's respectability; and that not only in the eyes of one's neighbors but even in one's own eyes. Indeed, it happens not very rarely that a person chooses to go ill-clad in order to be well dressed. Much more than half the value of what is worn by the American people may confidently be put down to the element of " dress," rather than to that of " clothing." And the chief motive of dress is emulation —" economic emulation." The like is true, though perhaps in a less degree, of what goes to food and shelter.

This misdirection of effort through the cravings of human vanity is of course not anything new, nor is " economic emulation " a modern fact. The modern system of industry has not invented emulation, nor has even this particular form of emulation originated under that system. But the system of free competition has accentuated this form of emulation, both by exalting the industrial activity of man above the rank which it held under more primitive forms of social organisation, and by in great measure cutting off other forms of emulation from the chance of efficiently ministering to the craving for a good fame. Speaking generally and from the standpoint of the average man, the modern industrial organization of society has practically narrowed the scope of emulation to this one line; and at the same time it has made the means of sustenance and comfort so much easier to obtain as very materially to widen the margin of human exertion that can be devoted to purposes of emulation. Further, by increasing the freedom of movement of the individual and widening the environment to which the individual is exposed — increasing the number of persons before whose eyes each one carries on his life, and, *pari passu,* decreasing the chances which such persons have of awarding their esteem on any other basis than that of immediate appearances, it has increased the relative efficiency of the economic means of winning respect through a show of expenditure for personal comforts.

It is not probable that further advance in the same direction will lead to a different result in the immediate future; and it is the *immediate* future we have to deal with. A further advance in the efficiency of our industry, and a further widening of the human environment to which the individual is exposed, should logically render emulation in this direction more intense. There are, indeed, certain considerations to be set off against this tendency, but they are mostly factors of slow action, and are hardly of sufficient consequence to reverse the general rule. On the whole, other things remaining the same, it must be admitted that, within wide limits, the easier the conditions of physical life for modern civilised man become, and the wider the horizon of each and the extent of the personal contact of each with his fellowmen, and the greater the opportunity of each to compare notes with his fellows, the greater will be the preponderance of economic success as a means of emulation, and the greater the straining after economic respectability. Inasmuch as the aim of emulation is not any absolute degree of comfort or of excellence, no advance in the average well-being of the community can end the struggle or lessen the strain. A general amelioration cannot quiet the unrest whose source is the craving of everybody to compare favorably with his neighbor.

Human nature being what it is, the struggle of each to possess more than his neighbor is inseparable from the institution of private property. And also, human nature being what it is, one who possesses less will, on the average, be jealous of the one who possesses more; and "more" means not more than the average share, but more than the share of the person who makes the comparison. The criterion of complacency is, largely, the *de facto* possession or enjoyment; and the present growth of sentiment among the body of the people — who possess less — favors, in a vague way, a readjustment adverse to the interests of those who possess more, and adverse to the possibility of legitimately possessing or enjoying "more"; that is to say, the growth of sentiment favors a socialistic movement. The outcome of modern industrial development has been, so far as concerns the present purpose, to intensify emulation and the jealousy that goes with emulation, and to focus the emulation and the jealousy on the possession and enjoyment of material goods. The ground of the unrest with which we are concerned is, very largely, jealousy,— envy, if you choose; and the ground of this particular form of jealousy, that makes for socialism, is to be found in the institution of private property. With private property, under modern conditions, this jealousy and unrest are unavoidable.

The corner-stone of the modern industrial system is the institution of private property. That institution is also the objective point of all attacks upon the existing system of competitive industry, whether open or covert, whether directed against the system as a whole or against any special feature of it. It is, moreover, the ultimate ground — and, under modern conditions, necessarily so — of the unrest and discontent whose proximate cause is the struggle for economic respectability. The inference seems to be that, human nature being what it is, there can be no peace from this — it must be admitted — ignoble form of emulation, or from the discontent that goes with it, this side of the abolition of private property. Whether a larger measure of peace is in store for us after that event shall have come to pass, is of course not a matter to be counted on, nor is the question immediately to the point.

This economic emulation is of course not the sole motive, nor the most important feature, of modern industrial life; although it is in the foreground, and it pervades the structure of modern society more thoroughly perhaps than any other equally powerful moral factor. It would be rash to predict that socialism will be the inevitable outcome of a continued development of this emulation and the discontent which it fosters, and it is by no means the purpose of this paper to insist on such an inference. The most that can be claimed is that this emulation is one of the causes, if not the chief cause, of the existing unrest and dissatisfaction with things as they are; that this unrest is inseparable from the existing system of industrial organisation; and that the growth of popular sentiment under the influence of these conditions is necessarily adverse to the institution of private property, and therefore adverse to the existing industrial system of free competition.

The emulation to which attention has been called in the preceding section of this paper is not only a fact of importance to an understanding of the unrest that is urging us towards an untried path in social development, but it has also a bearing on the question of the practicability of any scheme for the complete nationalisation of industry. Modern industry has developed to such a degree of efficiency as to make the struggle of subsistence alone, under average conditions, relatively easy, as compared with the state of the case a few generations ago. As I have labored to show, the modern competitive system has at the same time given the spirit of emulation such a direction that the attainment of subsistence and comfort no longer fixes, even approximately, the limit of the required aggregate labor on the part of the community. Under modern conditions the struggle for existence has, in a very appreciable degree, been transformed into a struggle to keep up appearances. The ultimate ground of this struggle to keep up appearance by otherwise unnecessary expenditure, is the institution of private property. Under a régime which should allow no inequality of acquisition or of income, this form of emulation, which is due to the possibility of such inequality, would also tend to become obsolete. With the abolition of private property, the characteristic of human nature which now finds its exercise in this form of emulation, should logically find exercise in other, perhaps nobler and socially more serviceable, activities; it is at any rate not easy to imagine it running into any line of action more futile or less worthy of human effort.

Supposing the standard of comfort of the community to remain approximately at its present average, the abolition of the struggle to keep up economic appearances would very considerably lessen the aggregate amount of labor required for the support of the community. How great a saving of labor might be effected is not easy to say. I believe it is within the mark to suppose that the struggle to keep up appearances is chargeable, directly and indirectly, with one-half the aggregate labor, and abstinence from labor — for the standard of respectability requires us to shun labor as well as to enjoy the fruits of it — on part of the American people. This does not mean that the same community, under a system not allowing private property, could make its way with half the labor we now put forth; but it means something more or less nearly approaching that. Any one who has not seen our modern social life from this point of view will find the claim absurdly extravagant, but the startling character of the proposition will wear off with longer and closer attention to this aspect of the facts of everyday life. But the question of the exact amount of waste due to this factor is immaterial. It will not be denied that is is a fact of considerable magnitude, and that is all that the argument requires.

It is accordingly competent for the advocates of the nationalisation of industry and property to claim that even if their scheme of organisation should prove less effective for production of goods than the present, as measured absolutely in terms of the aggregate output of our industry, yet the community might readily be maintained at the present average standard of comfort. The required aggregate output of the nation's industry would be considerably less than at present, and there would therefore be less necessity for that close and strenuous industrial organisation and discipline of the members of society under the new régime, whose evils unfriendly critics are apt to magnify. The chances of practicability for the scheme should logically be considerably increased by this lessening of the necessity for severe application. The less irksome and exacting the new régime, the less chance of a reversion to the earlier system.

Under such a social order, where common labor would no longer be a mark of peculiar economic necessity and consequent low economic rank on part of the laborer, it is even conceivable that labor might practically come to assume that character of nobility in the eyes of society at large, which it now sometimes assumes in the speculations of the well-to-do, in their complacent moods. Much has sometimes been made of this possibility by socialist speculators, but the inference has something of a utopian look, and no one, certainly, is entitled to build institutions for the coming social order on this dubious ground.

What there seems to be ground for claiming is that a society which has reached our present degree of industrial efficiency would not go into the Socialist or Nationalist state with as many chances of failure as a community whose industrial development is still at the stage at which strenuous labor on the part of nearly all members is barely sufficient to make both ends meet.

In Mr. Spencer's essay, in conformity with the line of argument of his " Principles of Sociology," it is pointed out that, as the result of constantly operative social forces, all social systems, as regards the form of organisation, fall into the one or the other of Sir Henry Maine's two classes — the system of status or the system of contract. In accordance with this generalisation it is concluded that whenever the modern system of contract or free competition shall be displaced, it will necessarily be replaced by the only other known system — that of status; the type of which is the military organisation, or, also, a hierarchy, or a bureaucracy. It is something after the fashion of the industrial organisation of ancient Peru that Mr. Spencer pictures as the inevitable sequel of the demise of the existing competitive system. Voluntary coöperation can be replaced only by compulsory coöperation, which is identified with the system of status and defined as the subjection of man to his fellow-man.

Now, at least as a matter of speculation, this is not the only alternative. These two systems, of status, or prescription, and of contract, or competition, have divided the field of social organisation between them in some proportion or other in the past. Mr. Spencer has shown that, very generally, where human progress in its advanced stages has worked towards the amelioration of the lot of the average member of society, the movement has been away from the system of status and towards the system of contract. But there is at least one, if not more than one exception to the rule, as concerns the recent past. The latest development of the industrial organisation among civilised nations — perhaps in an especial degree in the case of the American people — has not been entirely a continuation of the approach to a régime of free contract. It is also, to say the least, very doubtful if the movement has been towards a régime of status, in the sense in which Sir Henry Maine uses the term. This is especially evident in the case of the great industries which we call "natural monopolies"; and it is to be added that the present tendency is for a continually increasing proportion of the industrial activities of the community to fall into the category of "natural monopolies." No revolution has been achieved; the system of competition has not been discarded, but the course of industrial development is not in the direction of an extension of that system at all points; nor does the principle of status always replace that of competition wherever the latter fails.

The classification of methods of social organisation under the two heads of status or of contract, is not logically exhaustive. There is nothing in the meaning of the terms employed which will compel us to say that whenever man escapes from the control of his fellow man, under a system of status, he thereby falls into a system of free contract. There is a conceivable escape from the dilemma, and it is this conceivable, though perhaps impracticable, escape from both these systems that the socialist agitator wishes to effect. An acquaintance with the aims and position of the more advanced and consistent advocates of a new departure leaves no doubt but that the principles of contract and of status, both, are in substance familiar to their thoughts — though often in a vague and inadequate form — and that they distinctly repudiate both. This is perhaps less true of those who take the socialist position mainly on ethical grounds.

As bearing on this point it may be remarked that while the industrial system, in the case of all communities with whose history we are acquainted, has always in the past been organised according to a scheme of status or of contract, or of the two combined in some proportion, yet the social organisation has not in all cases developed along the same lines, so far as concerns such social functions as are not primarily industrial. Especially is this true of the later stages in the development of those communities whose institutions we are accustomed to contemplate with the most complacency, *e.g.,* the case of the English-speaking peoples. The whole system of modern constitutional government in its latest developed forms, in theory at least, and, in a measure, in practice, does not fall under the head of either contract or status. It is the analogy of modern constitutional government through an impersonal law and impersonal institutions, that comes nearest doing justice to the vague notions of our socialist propagandists. It is true, some of the most noted among them are fond of the analogy of the military organisation, as a striking illustration of one feature of the system they advocate, but that must after all be taken as an *obiter dictum.*

Further, as to the manner of the evolution of existing institutions and their relation to the two systems spoken of. So far as concerns the communities which have figured largely in the civilised world, the political organisation has had its origin in a military system of government. So, also, has the industrial organisation. But while the development of industry, during its gradual escape from the military system of status, has been, at least until lately, in the direction of a system of free contract, the development of the political organisation, so far as it has escaped from the régime of status, has not been in that direction. The system of status is a system of subjection to personal authority,— of prescription and class distinctions, and privileges and immunities; the system of constitutional government, especially as seen at its best among a people of democratic traditions and habits of mind, is a system of subjection to the will of the social organism, as expressed in an impersonal law. This difference between the system of status and the "constitutional system" expresses a large part of the meaning of the boasted free institutions of the English-speaking people. Here, subjection is not to the person of the public functionary, but to the powers vested in him. This has, of course, something of the ring of latter-day popular rhetoric, but it is after all felt to be true, not only speculatively, but in some measure also in practice.

The right of eminent domain and the power to tax, as interpreted under modern constitutional forms, indicate something of the direction of development of the political functions of society at a point where they touch the province of the industrial system. It is along the line indicated by these and kindred facts that the socialists are advancing; and it is along this line that the later developments made necessary by the exigencies of industry under modern conditions are also moving. The aim of the propagandists is to sink the industrial community in the political community; or perhaps better, to identify the two organisations; but always with insistence on the necessity of making the political organisation, in some further developed form, the ruling and only one in the outcome. Distinctly, the system of contract is to be done away with; and equally distinctly, no system of status is to take its place.

All this is pretty vague, and of a negative character, but it would quickly pass the limits of legitimate inference from the accepted doctrines of the socialists if it should attempt to be anything more. It does not have much to say as to the practicability of any socialist scheme. As a matter of speculation, there seems to be an escape from the dilemma insisted on by Mr. Spencer. We may conceivably have nationalism without status and without contract. In theory, both principles are entirely obnoxious to that system. The practical question, as to whether modern society affords the materials out of which an industrial structure can be erected on a system different from either of these, is a problem of constructive social engineering which calls for a consideration of details far too comprehensive to be entered on here. Still, in view of the past course of development of character and institutions on the part of the people to which we belong, it is perhaps not extravagant to claim that no form of organisation which should necessarily eventuate in a thoroughgoing system of status could endure among us. The inference from this proposition may be, either that a near approach to nationalisation of industry would involve a régime of status, a bureaucracy, which would be unendurable, and which would therefore drive us back to the present system before it had been entirely abandoned; or that the nationalisation would be achieved with such a measure of success, in conformity with the requirements of our type of character, as would make it preferable to what we had left behind. In either case the ground for alarm does not seem so serious as is sometimes imagined.

A reversion to the system of free competition, after it had been in large part discarded, would no doubt be a matter of great practical difficulty, and the experiment which should demonstrate the necessity of such a step might involve great waste and suffering, and might seriously retard the advance of the race toward something better than our present condition; but neither a permanent deterioration of human society, nor a huge catastrophe, is to be confidently counted on as the outcome of the movement toward nationalisation, even if it should prove necessary for society to retrace its steps.

It is conceivable that the application of what may be called the " constitutional method " to the organisation of industry — for that is essentially what the advocates of Nationalisation demand — would result in a course of development analogous to what has taken place in the case of the political organisation under modern constitutional forms. Modern constitutional government — the system of modern free institutions — is by no means an unqualified success, in the sense of securing to each the rights and immunities which in theory are guaranteed to him.

Our modern republics have hardly given us a foretaste of that political millennium whereof they proclaim the fruition. The average human nature is as yet by no means entirely fit for self-government according to the " constitutional method." Shortcomings are visible at every turn. These shortcomings are grave enough to furnish serious arguments against the practicability of our free institutions. On the continent of Europe the belief seems to be at present in the ascendant that man must yet, for a long time, remain under the tutelage of absolutism before he shall be fit to organise himself into an autonomous political body. The belief is not altogether irrational. Just how great must be the advance of society and just what must be the character of the advance, pre-liminary to its advantageously assuming the autonomous — republican — form of political organisation, must be admitted to be an open question. Whether we, or any people, have yet reached the required stage of the advance is also questioned by many. But the partial success which has attended the movement in this direction, among the English-speaking people for example, goes very far towards proving that the point in the development of human character at which the constitutional method may be advantageously adopted in the political field, lies far this side the point at which human nature shall have be-come completely adapted for that method. That is to say, it does not seem necessary, as regards the functions of society which we are accustomed to call political, to be entirely ready for nationalisation before entering upon it. How far the analogy of this will hold when applied to the industrial organisation of society is difficult to say, but some significance the analogy must be admitted to possess.

Certainly, the fact that constitutional government — the nationalisation of political functions — seems to have been a move in the right direction is not to be taken as proof of the advisability of forthwith nationalising the industrial functions. At the same time this fact does af-ford ground for the claim that a movement in this direc-tion may prove itself in some degree advantageous, even if it takes place at a stage in the development of human nature at which mankind is still far from being entirely fit for the duties which the new system shall impose. The question, therefore, is not whether we have reached the perfection of character which would be necessary in order to a perfect working of the scheme of nationalisation of industry, but whether we have reached such a degree of development as would make an imperfect working of the scheme possible.

THE SOCIALIST ECONOMICS OF KARL MARX AND HIS FOLLOWERS [1]

I. The Theories of Karl Marx

THE system of doctrines worked out by Marx is characterised by a certain boldness of conception and a great logical consistency. Taken in detail, the constituent elements of the system are neither novel nor iconoclastic, nor does Marx at any point claim to have discovered previously hidden facts or to have invented recondite formulations of facts already known; but the system as a whole has an air of originality and initiative such as is rarely met with among the sciences that deal with any phase of human culture. How much of this distinctive character the Marxian system owes to the personal traits of its creator is not easy to say, but what marks it off from all other systems of economic theory is not a matter of personal idiosyncrasy. It differs characteristically from all systems of theory that had preceded it, both in its premises and in its aims. The (hostile) critics of Marx have not sufficiently appreciated the radical character of his departure in both of these respects, and have, therefore, commonly lost themselves in a tangled scrutiny of supposedly abstruse details; whereas those writers who have been in sympathy with his teachings have too commonly been disciples bent on exegesis and on confirming their fellow-disciples in the faith.

[1] The substance of lectures before students in Harvard University in April, 1906. Reprinted by permission from *The Quarterly Journal of Economics*, Vol. XX, Aug., 1906

Except as a whole and except in the light of its postulates and aims, the Marxian system is not only not tenable, but it is not even intelligible. A discussion of a given isolated feature of the system (such as the theory of value) from the point of view of classical economics (such as that offered by Böhm-Bawerk) is as futile as a discussion of solids in terms of two dimensions.

Neither as regards his postulates and preconceptions nor as regards the aim of his inquiry is Marx's position an altogether single-minded one. In neither respect does his position come of a single line of antecedents. He is of no single school of philosophy, nor are his ideals those of any single group of speculators living before his time. For this reason he takes his place as an originator of a school of thought as well as the leader of a movement looking to a practical end.

As to the motives which drive him and the aspirations which guide him, in destructive criticism and in creative speculation alike, he is primarily a theoretician busied with the analysis of economic phenomena and their organisation into a consistent and faithful system of scientific knowledge; but he is, at the same time, consistently and tenaciously alert to the bearing which each step in the progress of his theoretical work has upon the propaganda. His work has, therefore, an air of bias, such as belongs to an advocate's argument; but it is not, therefore, to be assumed, nor indeed to be credited, that his propagandist aims have in any substantial way deflected his inquiry or his speculations from the faithful pursuit of scientific truth. His socialistic bias may color his polemics, but his logical grasp is too neat and firm to admit of any bias, other than that of his metaphysical preconceptions, affecting his theoretical work.

There is no system of economic theory more logical

than that of Marx. No member of the system, no single article of doctrine, is fairly to be understood, criticised, or defended except as an articulate member of the whole and in the light of the preconceptions and postulates which afford the point of departure and the controlling norm of the whole. As regards these preconceptions and postulates, Marx draws on two distinct lines of antecedents,— the Materialistic Hegelianism and the English system of Natural Rights. By his earlier training he is an adept in the Hegelian method of speculation and inoculated with the metaphysics of development underlying the Hegelian system. By his later training he is an expert in the system of Natural Rights and Natural Liberty, ingrained in his ideals of life and held inviolate throughout. He does not take a critical attitude toward the underlying principles of Natural Rights. Even his Hegelian preconceptions of development never carry him the length of questioning the fundamental principles of that system. He is only more ruthlessly consistent in working out their content than his natural-rights antagonists in the liberal-classical school. His polemics run against the specific tenets of the liberal school, but they run wholly on the ground afforded by the premises of that school. The ideals of his propaganda are natural-rights ideals, but his theory of the working out of these ideals in the course of history rests on the Hegelian metaphysics of development, and his method of speculation and construction of theory is given by the Hegelian dialectic.

What first and most vividly centered interest on Marx and his speculations was his relation to the revolutionary socialistic movement; and it is those features of his doctrines which bear immediately on the propaganda that still continue to hold the attention of the greater number of his critics. Chief among these doctrines, in the apprehension of his critics, is the theory of value, with its corollaries: (*a*) the doctrines of the exploitation of labor by capital; and (*b*) the laborer's claim to the whole product of his labor. Avowedly, Marx traces his doctrine of labor-value to Ricardo, and through him to the classical economists.[2] The laborer's claim to the whole product of labor, which is pretty constantly implied, though not frequently avowed by Marx, he has in all probability taken from English writers of the early nineteenth century,[3] more particularly from William Thompson. These doctrines are, on their face, nothing but a development of the conceptions of natural rights which then pervaded English speculation and afforded the metaphysical ground of the liberal movement. The more formidable critics of the Marxian socialism have made much of these doctrinal elements that further the propaganda, and have, by laying the stress on these, diverted attention from other elements that are of more vital consequence to the system as a body of theory. Their exclusive interest in this side of " scientific socialism " has even led them to deny the Marxian system all substantial originality, and make it a (doubtfully legitimate) offshoot of English Liberalism and natural rights.[4] But this is one-sided criticism. It may hold as against certain tenets of the so-called " scientific socialism," but it is not altogether to the point as regards the Marxian system of theory. Even the Marxian theory of value, surplus value, and exploitation,

[2] *Cf. Critique of Political Economy,* chap. i, " Notes on the History of the Theory of Commodities," pp. 56–73 (English translation, New York, 1904).

[3] See Menger, *Right to the Whole Produce of Labor,* sections iii–v and viii–ix, and Foxwell's admirable Introduction to Menger.

[4] See Menger and Foxwell, as above, and Schaeffle, *Quintessence of Socialism,* and *The Impossibility of Social Democracy.*

is not simply the doctrine of William Thompson, transcribed and sophisticated in a forbidding terminology, however great the superficial resemblance and however large Marx's unacknowledged debt to Thompson may be on these heads. For many details and for much of his animus Marx may be indebted to the Utilitarians; but, after all, his system of theory, taken as a whole, lies within the frontiers of neo-Hegelianism, and even the details are worked out in accord with the preconceptions of that school of thought and have taken on the complexion that would properly belong to them on that ground. It is, therefore, not by an itemised scrutiny of the details of doctrine and by tracing their pedigree in detail that a fair conception of Marx and his contribution to economics may be reached, but rather by following him from his own point of departure out into the ramifications of his theory, and so overlooking the whole in the prespective which the lapse of time now affords us, but which he could not himself attain, since he was too near to his own work to see why he went about it as he did.

The comprehensive system of Marxism is comprised within the scheme of the Materialistic Conception of History.[5] This materialistic conception is essentially Hegelian,[6] although it belongs with the Hegelian Left, and its immediate affiliation is with Feuerbach, not with the direct line of Hegelian orthodoxy. The chief point of interest here, in identifying the materialistic conception with Hegelianism, is that this identification throws it

[5] See Engels, *The Development of Socialism from Utopia to Science,* especially section ii and the opening paragraphs of section iii; also the preface of *Zur Kritik der politischen Oekonomie.*

[6] See Engels, as above, and also his *Feuerbach: The Roots of Socialist Philosophy* (translation, Chicago, Kerr & Co., 1903).

immediately and uncompromisingly into contrast with Darwinism and the post-Darwinian conceptions of evolution. Even if a plausible English pedigree should be worked out for this Materialistic Conception, or " Scientific Socialism," as has been attempted, it remains none the less true that the conception with which Marx went to his work was a transmuted framework of Hegelian dialectic.[7]

Roughly, Hegelian materialism differs from Hegelian orthodoxy by inverting the main logical sequence, not by discarding the logic or resorting to new tests of truth or finality. One might say, though perhaps with excessive crudity, that, where Hegel pronounces his dictum, *Das Denken ist das Sein,* the materialists, particularly Marx and Engels, would say *Das Sein macht das Denken.* But in both cases some sort of a creative primacy is assigned to one or the other member of the complex, and in neither case is the relation between the two members a causal relation. In the materialistic conception man's spiritual life — what man thinks — is a reflex of what he is in the material respect, very much in the same fashion as the orthodox Hegelian would make the material world a reflex of the spirit. In both, the dominant norm of speculation and formulation of theory is the conception of movement, development, evolution, progress; and in both the movement is conceived necessarily to take place by the method of conflict or struggle. The movement is of the nature of progress,— gradual advance toward a goal, toward the realisation in explicit form of all that is implicit in the substantial activity involved in the movement. The movement is, further, self-conditioned and self-acting: it is an unfolding by inner necessity. The struggle

[7] See *e.g.,* Seligman, *The Economic Interpretation of History,* Part I.

which constitutes the method of movement or evolution is, in the Hegelian system proper, the struggle of the spirit for self-realisation by the process of the well-known three-phase dialectic. In the materialistic conception of history this dialectical movement becomes the class struggle of the Marxian system.

The class struggle is conceived to be " material," but the term " material " is in this connection used in a metaphorical sense. It does not mean mechanical or physical, or even physiological, but economic. It is material in the sense that it is a struggle between classes for the material means of life. " The materialistic conception of history proceeds on the principle that production and, next to production, the exchange of its products is the groundwork of every social order." [8] The social order takes its form through the class struggle, and the character of the class struggle at any given phase of the unfolding development of society is determined by " the prevailing mode of economic production and exchange." The dialectic of the movement of social progress, therefore, moves on the spiritual plane of human desire and passion, not on the (literally) material plane of mechanical and physiological stress, on which the developmental process of brute creation unfolds itself. It is a sublimated materialism, sublimated by the dominating presence of the conscious human spirit; but it is conditioned by the material facts of the production of the means of life.[9] The ultimately active forces involved in the process of unfolding social life are (apparently) the material agencies engaged in the me-

---

[8] Engels, *Development of Socialism,* beginning of section iii.

[9] *Cf.,* on this point, Max Adler, " Kausalität und Teleologie im Streite um die Wissenschaft" (included in *Marx-Studien,* edited by Adler and Hilfendirg, vol. i), particularly section xi; *cf.* also Ludwig Stein, *Die soziale Frage im Lichte der Philosophie,* whom Adler criticises and claims to have refuted.

chanics of production; but the dialectic of the process —
the class struggle — runs its course only among and in
terms of the secondary (epigenetic) forces of human
consciousness engaged in the valuation of the material
products of industry. A consistently materialistic conception, consistently adhering to a materialistic interpretation of the process of development as well as of the
facts involved in the process, could scarcely avoid making
its putative dialectic struggle a mere unconscious and
irrelevant conflict of the brute material forces. This
would have amounted to an interpretation in terms of
opaque cause and effect, without recourse to the concept
of a conscious class struggle, and it might have led to a
concept of evolution similar to the unteleological Darwinian concept of natural selection. It could scarcely have
led to the Marxian notion of a conscious class struggle as
the one necessary method of social progress, though it
might conceivably, by the aid of empirical generalisation,
have led to a scheme of social process in which a class
struggle would be included as an incidental though perhaps highly efficient factor.[10]   It would have led, as Darwinism has, to a concept of a process of cumulative
change in social structure and function; but this process,
being essentially a cumulative sequence of causation,
opaque and unteleological, could not, without an infusion
of pious fancy by the speculator, be asserted to involve
progress as distinct from retrogression or to tend to a
" realisation " or " self-realisation " of the human spirit
or of anything else. Neither could it conceivably be asserted to lead up to a final term, a goal to which all lines
of the process should converge and beyond which the
process would not go, such as the assumed goal of the
Marxian process of class struggle, which is conceived to

[10] *Cf.* Adler, as above.

cease in the classless economic structure of the socialistic final term. In Darwinism there is no such final or perfect term, and no definitive equilibrium.

The disparity between Marxism and Darwinism, as well as the disparity within the Marxian system between the range of material facts that are conceived to be the fundamental forces of the process, on the one hand, and the range of spiritual facts within which the dialectic movement proceeds,— this disparity is shown in the character assigned the class struggle by Marx and Engels. The struggle is asserted to be a conscious one, and proceeds on a recognition by the competing classes of their mutually incompatible interests with regard to the material means of life. The class struggle proceeds on motives of interest, and a recognition of class interest can, of course, be reached only by reflection on the facts of the case. There is, therefore, not even a direct causal connection between the material forces in the case and the choice of a given interested line of conduct. The attitude of the interested party does not result from the material forces so immediately as to place it within the relation of direct cause and effect, nor even with such a degree of intimacy as to admit of its being classed as a tropismatic, or even instinctive, response to the impact of the material force in question. The sequence of reflection, and the consequent choice of sides to a quarrel, run entirely alongside of a range of material facts concerned.

A further characteristic of the doctrine of class struggle requires mention. While the concept is not Darwinian, it is also not legitimately Hegelian, whether of the Right or the Left. It is of a utilitarian origin and of English pedigree, and it belongs to Marx by virtue of his having borrowed its elements from the system of self-interest. It is in fact a piece of hedonism, and is related to Bentham rather than to Hegel. It proceeds on the grounds of the hedonistic calculus, which is equally foreign to the Hegelian notion of an unfolding process and to the post-Darwinian notions of cumulative causation. As regards the tenability of the doctrine, apart from the question of its derivation and its compatibility with the neo-Hegelian postulates, it is to be added that it is quite out of harmony with the later results of psychological inquiry,— just as is true of the use made of the hedonistic calculus by the classical (Austrian) economics.

Within the domain covered by the materialistic conception, that is to say within the domain of unfolding human culture, which is the field of Marxian speculation at large, Marx has more particularly devoted his efforts to an analysis and theoretical formulation of the present situation,— the current phase of the process, the capitalistic system. And, since the prevailing mode of the production of goods determines the institutional, intellectual, and spiritual life of the epoch, by determining the form and method of the current class struggle, the discussion necessarily begins with the theory of " capitalistic production," or production as carried on under the capitalistic system.[11]

[11] It may be noted, by way of caution to readers familiar with the terms only as employed by the classical (English and Austrian) economists, that in Marxian usage " capitalistic production " means production of goods for the market by hired labor under the direction of employers who own (or control) the means of production and are engaged in industry for the sake of a profit. " Capital " is wealth (primarily funds) so employed. In these and other related points of terminological usage Marx is, of course, much more closely in touch with colloquial usage than those economists of the classical line who make capital signify " the products of past industry used as aids to further production." With Marx " Capitalism " implies certain relations of ownership, no less than the " productive use " which is alone insisted on by so many later economists in defining the term.

Under the capitalistic system, that is to say under the system of modern business traffic, production is a production of commodities, merchantable goods, with a view to the price to be obtained for them in the market. The great fact on which all industry under this system hinges is the price of marketable goods. Therefore it is at this point that Marx strikes into the system of capitalistic production, and therefore the theory of value becomes the dominant feature of his economics and the point of departure for the whole analysis, in all its voluminous ramifications.[12]

It is scarcely worth while to question what serves as the beginning of wisdom in the current criticisms of Marx; namely, that he offers no adequate proof of his labor-value theory.[13] It is even safe to go farther, and say that he offers no proof of it. The feint which occupies the opening paragraphs of the *Kapital* and the corresponding passages of *Zur Kritik,* etc., is not to be taken seriously as an attempt to prove his position on this head by the ordinary recourse to argument. It is rather a self-satisfied superior's playful mystification of those readers (critics) whose limited powers do not enable them to see

[12] In the sense that the theory of value affords the point of departure and the fundamental concepts out of which the further theory of the workings of capitalism is constructed,— in this sense, and in this sense only, is the theory of value the central doctrine and the critical tenet of Marxism. It does not follow that the Marxist doctrine of an irresistible drift towards a socialistic consummation hangs on the defensibility of the labor-value theory, nor even that the general structure of the Marxist economics would collapse if translated into other terms than those of this doctrine of labor-value. *Cf.* Böhm-Bawerk, *Karl Marx and the Close of his System;* and, on the other hand, Franz Oppenheimer, *Das Grundgesetz der Marx'schen Gesellschaftslehre;* and Rudolf Goldscheid, *Verelendungs- oder Meliorationstheorie.*

[13] *Cf., e.g.,* Böhm-Bawerk, as above; Georg Adler, *Grundlagen der Karl Marx'schen Kritik.*

that his proposition is self-evident. Taken on the Hegelian (neo-Hegelian) ground, and seen in the light of the general materialistic conception, the proposition that value = labor-cost is self-evident, not to say tautological. Seen in any other light, it has no particular force.

In the Hegelian scheme of things the only substantial reality is the unfolding life of the spirit. In the neo-Hegelian scheme, as embodied in the materialistic conception, this reality is translated into terms of the unfolding (material) life of man in society.[14] In so far as the goods are products of industry, they are the output of this unfolding life of man, a material residue embodying a given fraction of this forceful life-process. In this life-process lies all substantial reality, and all finally valid relations of quantivalence between the products of this life-process must run in its terms. The life-process, which, when it takes the specific form of an expenditure of labor power, goes to produce goods, is a process of material forces, the spiritual or mental features of the life-process and of labor being only its insubstantial reflex. It is consequently only in the material changes wrought by this expenditure of labor power that the metaphysical substance of life — labor power — can be embodied; but in these changes of material fact it cannot but be embodied, since these are the end to which it is directed.

This balance between goods in respect of their magnitude as output of human labor holds good indefeasibly, in

---

[14] In much the same way, and with an analogous effect on their theoretical work, in the preconceptions of the classical (including the Austrian) economists, the balance of pleasure and pain is taken to be the ultimate reality in terms of which all economic theory must be stated and to terms of which all phenomena should finally be reduced in any definitive analysis of economic life. It is not the present purpose to inquire whether the one of these uncritical assumptions is in any degree more meritorious or more serviceable than the other.

point of the metaphysical reality of the life-process, whatever superficial (phenomenal) variations from this norm may occur in men's dealings with the goods under the stress of the strategy of self-interest. Such is the value of the goods in reality; they are equivalents of one another in the proportion in which they partake of this substantial quality, although their true ratio of equivalence may never come to an adequate expression in the transactions involved in the distribution of the goods. This real or true value of the goods is a fact of production, and holds true under all systems and methods of production, whereas the exchange value (the "phenomenal form" of the real value) is a fact of distribution, and expresses the real value more or less adequately according as the scheme of distribution in force at the given time conforms more or less closely to the equities given by production. If the output of industry were distributed to the productive agents strictly in proportion to their shares in production, the exchange value of the goods would be presumed to conform to their real value. But, under the current, capitalistic system, distribution is not in any sensible degree based on the equities of production, and the exchange value of goods under this system can therefore express their real value only with a very rough, and in the main fortuitous, approximation. Under a socialistic régime, where the laborer would get the full product of his labor, or where the whole system of ownership, and consequently the system of distribution, would lapse, values would reach a true expression, if any.

Under the capitalistic system the determination of exchange value is a matter of competitive profit-making, and exchange values therefore depart erratically and incontinently from the proportions that would legitimately be given them by the real values whose only expression they are. Marx's critics commonly identify the concept of " value " with that of " exchange value,"[15] and show that the theory of " value " does not square with the run of the facts of price under the existing system of distribution, piously hoping thereby to have refuted the Marxian doctrine; whereas, of course, they have for the most part not touched it. The misapprehension of the critics may be due to a (possibly intentional) oracular obscurity on the part of Marx. Whether by his fault or their own, their refutations have hitherto been quite inconclusive. Marx's severest stricture on the iniquities of the capitalistic system is that contained by implication in his development of the manner in which actual exchange value of goods systematically diverges from their real (labor-cost) value. Herein, indeed, lies not only the inherent iniquity of the existing system, but also its fateful infirmity, according to Marx.

The theory of value, then, is *contained in* the main postulates of the Marxian system rather than derived from them. Marx identifies this doctrine, in its elements, with the labor-value theory of Ricardo,[16] but the relationship between the two is that of a superficial coincidence in their main propositions rather than a substantial identity of theoretic contents. In Ricardo's theory the source and measure of value is sought in the effort and sacrifice undergone by the producer, consistently, on the whole,

[15] Böhm-Bawerk, *Capital and Interest,* Book VI, chap. iii; also *Karl Marx and the Close of his System,* particularly chap. iv; Adler, *Grundlagen,* chaps. ii. and iii.

[16] *Cf. Kapital,* vol. i, chap. xv, p. 486 (4th ed.). See also notes 9 and 16 to chap. i of the same volume, where Marx discusses the labor-value doctrines of Adam Smith and an earlier (anonymous) English writer, and compares them with his own. Similar comparisons with the early — classical — value theories recur from time to time in the later portions of *Kapital.*

with the Benthamite-utilitarian position to which Ricardo somewhat loosely and uncritically adhered. The decisive fact about labor, that quality by virtue of which it is assumed to be the final term in the theory of production, is its irksomeness. Such is of course not the case in the labor-value theory of Marx, to whom the question of the irksomeness of labor is quite irrelevant, so far as regards the relation between labor and production. The substantial diversity or incompatibility of the two theories shows itself directly when each is employed by its creator in the further analysis of economic phenomena. Since with Ricardo the crucial point is the degree of irksomeness of labor, which serves as a measure both of the labor expended and the value produced, and since in Ricardo's utilitarian philosophy there is no more vital fact underlying this irksomeness, therefore no surplus-value theory follows from the main position. The productiveness of labor is not cumulative, in its own working; and the Ricardian economics goes on to seek the cumulative productiveness of industry in the functioning of the products of labor when employed in further production and in the irksomeness of the capitalist's abstinence. From which duly follows the general position of classical economics on the theory of production.

With Marx, on the other hand, the labor power expended in production being itself a product and having a substantial value corresponding to its own labor-cost, the value of the labor power expended and the value of the product created by its expenditure need not be the same. They are not the same, by supposition, as they would be in any hedonistic interpretation of the facts. Hence a discrepancy arises between the value of the labor power expended in production and the value of the product created, and this discrepancy is covered by the concept of surplus value. Under the capitalistic system, wages being the value (price) of the labor power consumed in industry, it follows that the surplus product of their labor cannot go to the laborers, but becomes the profits of capital and the source of its accumulation and increase. From the fact that wages are measured by the value of labor power rather than by the (greater) value of the product of labor, it follows also that the laborers are unable to buy the whole product of their labor, and so that the capitalists are unable to sell the whole product of industry continuously at its full value, whence arise difficulties of the gravest nature in the capitalistic system, in the way of overproduction and the like.

But the gravest outcome of this systematic discrepancy between the value of labor power and the value of its product is the accumulation of capital out of unpaid labor, and the effect of this accumulation on the laboring population. The law of accumulation, with its corollary, the doctrine of the industrial reserve army, is the final term and the objective point of Marx's theory of capitalist production, just as the theory of labor value is his point of departure.[17]   While the theory of value and surplus value are Marx's explanation of the possibility of existence of the capitalistic system, the law of the accumulation of capital is his exposition of the causes which must lead to the collapse of that system and of the manner in which the collapse will come. And since Marx is, always and

[17] Oppenheimer (*Das Grundgesetz der Marx'schen Gesellschaftslehre*) is right in making the theory of accumulation the central element in the doctrines of Marxist socialism, but it does not follow, as Oppenheimer contends, that this doctrine is the keystone of Marx's economic theories. It follows logically from the theory of surplus value, as indicated above, and rests on that theory in such a way that it would fail (in the form in which it is held by Marx) with the failure of the doctrine of surplus value.

everywhere, a socialist agitator as well as a theoretical economist, it may be said without hesitation that the law of accumulation is the climax of his great work, from whatever point of view it is looked at, whether as an economic theorem or as a tenet of socialistic doctrine.

The law of capitalistic accumulation may be paraphrased as follows: [18] Wages being the (approximately exact) value of the labor power bought in the wage contract; the price of the product being the (similarly approximate) value of the goods produced; and since the value of the product exceeds that of the labor power by a given amount (surplus value), which by force of the wage contract passes into the possession of the capitalist and is by him in part laid by as savings and added to the capital already in hand, it follows (*a*) that, other things equal, the larger the surplus value, the more rapid the increase of capital; and, also (*b*), that the greater the increase of capital relatively to the labor force employed, the more productive the labor employed and the larger the surplus product available for accumulation. The process of accumulation, therefore, is evidently a cumulative one; and, also evidently, the increase added to capital is an unearned increment drawn from the unpaid surplus product of labor.

But with an appreciable increase of the aggregate capital a change takes place in its technological composition, whereby the " constant " capital (equipment and raw materials) increases disproportionately as compared with the " variable " capital (wages fund). " Labor-saving devices " are used to a greater extent than before, and labor is saved. A larger proportion of the expenses of production goes for the purchase of equipment and raw materials, and a smaller proportion — though perhaps an

[18] See *Kapital,* vol. i, chap. xxiii.

absolutely increased amount — goes for the purchase of labor power. Less labor is needed relatively to the aggregate capital employed as well as relatively to the quantity of goods produced. Hence some portion of the increasing labor supply will not be wanted, and an " industrial reserve army," a " surplus labor population," an army of unemployed, comes into existence. This reserve grows relatively larger as the accumulation of capital proceeds and as technological improvements consequently gain ground; so that there result two divergent cumulative changes in the situation,— antagonistic, but due to the same set of forces and, therefore, inseparable: capital increases, and the number of unemployed laborers (relatively) increases also.

This divergence between the amount of capital and output, on the one hand, and the amount received by laborers as wages, on the other hand, has an incidental consequence of some importance. The purchasing power of the laborers, represented by their wages, being the largest part of the demand for consumable goods, and being at the same time, in the nature of the case, progressively less adequate for the purchase of the product, represented by the price of the goods produced, it follows that the market is progressively more subject to glut from overproduction, and hence to commercial crises and depression. It has been argued, as if it were a direct inference from Marx's position, that this maladjustment between production and markets, due to the laborer not getting the full product of his labor, leads directly to the breakdown of the capitalistic system, and so by its own force will bring on the socialistic consummation. Such is not Marx's position, however, although crises and depression play an important part in the course of development that is to lead up to socialism. In Marx's theory, socialism is to come by way of a conscious class movement on the part of the propertyless laborers, who will act advisedly on their own interest and force the revolutionary movement for their own gain. But crises and depression will have a large share in bringing the laborers to a frame of mind suitable for such a move.

Given a growing aggregate capital, as indicated above, and a concomitant reserve of unemployed laborers growing at a still higher rate, as is involved in Marx's position, this body of unemployed labor can be, and will be, used by the capitalists to depress wages, in order to increase profits. Logically, it follows that, the farther and faster capital accumulates, the larger will be the reserve of unemployed, both absolutely and relatively to the work to be done, and the more severe will be the pressure acting to reduce wages and lower the standard of living, and the deeper will be the degradation and misery of the working class and the more precipitately will their condition decline to a still lower depth. Every period of depression, with its increased body of unemployed labor seeking work, will act to hasten and accentuate the depression of wages, until there is no warrant even for holding that wages will, on an average, be kept up to the subsistence minimum.[19] Marx, indeed, is explicit to the effect that such will be the case,— that wages will decline below the subsistence minimum; and he cites English conditions of child labor, misery, and degeneration to substantiate his views.[20] When this has gone far enough, when capital-

[19] The "subsistence minimum" is here taken in the sense used by Marx and the classical economists, as meaning what is necessary to keep up the supply of labor at its current rate of efficiency.
[20] See *Kapital*, vol. i, chap. xxiii, sections 4 and 5.

ist production comes near enough to occupying the whole field of industry and has depressed the condition of its laborers sufficiently to make them an effective majority of the community with nothing to lose, then, having taken advice together, they will move, by legal or extra-legal means, by absorbing the state or by subverting it, to establish the social revolution.

Socialism is to come through class antagonism due to the absence of all property interests from the laboring class, coupled with a generally prevalent misery so profound as to involve some degree of physical degeneration. This misery is to be brought about by the heightened productivity of labor due to an increased accumulation of capital and large improvements in the industrial arts; which in turn is caused by the fact that under a system of private enterprise with hired labor the laborer does not get the whole product of his labor; which, again, is only saying in other words that private ownership of capital goods enables the capitalist to appropriate and accumulate the surplus product of labor. As to what the régime is to be which the social revolution will bring in, Marx has nothing particular to say, beyond the general thesis that there will be no private ownership, at least not of the means of production.

Such are the outlines of the Marxian system of socialism. In all that has been said so far no recourse is had to the second and third volumes of *Kapital*. Nor is it necessary to resort to these two volumes for the general theory of socialism. They add nothing essential, although many of the details of the processes concerned in the working out of the capitalist scheme are treated with greater fullness, and the analysis is carried out with great consistency and with admirable results. For economic theory at large these further two volumes are important enough, but an inquiry into their contents in that connection is not called for here.

Nothing much need be said as to the tenability of this theory. In its essentials, or at least in its characteristic elements, it has for the most part been given up by latter-day socialist writers. The number of those who hold to it without essential deviation is growing gradually smaller. Such is necessarily the case, and for more than one reason. The facts are not bearing it out on certain critical points, such as the doctrine of increasing misery; and the Hegelian philosophical postulates, without which the Marxism of Marx is groundless, are for the most part forgotten by the dogmatists of to-day. Darwinism has largely supplanted Hegelianism in their habits of thought.

The particular point at which the theory is most fragile, considered simply as a theory of social growth, is its implied doctrine of population,— implied in the doctrine of a growing reserve of unemployed workmen. The doctrine of the reserve of unemployed labor involves as a postulate that population will increase anyway, without reference to current or prospective means of life. The empirical facts give at least a very persuasive apparent support to the view expressed by Marx, that misery is, or has hitherto been, no hindrance to the propagation of the race; but they afford no conclusive evidence in support of a thesis to the effect that the number of laborers must increase independently of an increase of the means of life. No one since Darwin would have the hardihood to say that the increase of the human species is not conditioned by the means of living.

But all that does not really touch Marx's position. To Marx, the neo-Hegelian, history, including the economic development, is the life-history of the human species;

and the main fact in this life-history, particularly in the economic aspect of it, is the growing volume of human life. This, in a manner of speaking, is the base-line of the whole analysis of the process of economic life, including the phase of capitalist production with the rest. The growth of population is the first principle, the most substantial, most material factor in this process of economic life, so long as it is a process of growth, of unfolding, of exfoliation, and not a phase of decrepitude and decay. Had Marx found that his analysis led him to a view adverse to this position, he would logically have held that the capitalist system is the mortal agony of the race and the manner of its taking off. Such a conclusion is precluded by his Hegelian point of departure, according to which the goal of the life-history of the race in a large way controls the course of that life-history in all its phases, including the phase of capitalism. This goal or end, which controls the process of human development, is the complete realisation of life in all its fullness, and the realisation is to be reached by a process analogous to the three-phase dialectic, of thesis, antithesis, and synthesis, into which scheme the capitalist system, with its overflowing measure of misery and degradation, fits as the last and most dreadful phase of antithesis. Marx, as a Hegelian,— that is to say, a romantic philosopher,— is necessarily an optimist, and the evil (antithetical element) in life is to him a logically necessary evil, as the antithesis is a necessary phase of the dialectic; and it is a means to the consummation, as the antithesis is a means to the synthesis.

THE SOCIALIST ECONOMICS OF KARL MARX AND HIS FOLLOWERS [1]

II. The Later Marxism

MARX worked out his system of theory in the main during the third quarter of the nineteenth century. He came to the work from the standpoint given him by his early training in German thought, such as the most advanced and aggressive German thinking was through the middle period of the century, and he added to this German standpoint the further premises given him by an exceptionally close contact with and alert observation of the English situation. The result is that he brings to his theoretical work a twofold line of premises, or rather of preconceptions. By early training he is a neo-Hegelian, and from this German source he derives his peculiar formulation of the Materialistic Theory of History. By later experience he acquired the point of view of that Liberal-Utilitarian school which dominated English thought through the greater part of his active life. To this experience he owes (probably) the somewhat pronounced individualistic preconceptions on which the doctrines of the Full Product of Labor and the Exploitation of Labor are based. These two not altogether compatible lines of doctrine found their way together into the tenets of scientific [2] socialism, and gives its characteristic Marxian features to the body of socialist economics.

[1] Reprinted by permission from *The Quarterly Journal of Economics,* Vol. XXI, Feb., 1907.
[2] "Scientific" is here used in the half-technical sense which by

The socialism that inspires hopes and fears to-day is of the school of Marx. No one is seriously apprehensive of any other so-called socialistic movement, and no one is seriously concerned to criticise or refute the doctrines set forth by any other school of "socialists." It may be that the socialists of the Marxist observance are not always or at all points in consonance with the best accepted body of Marxist doctrine. Those who make up the body of the movement may not always be familiar with the details — perhaps not even with the general features — of the Marxian scheme of economics; but with such consistency as may fairly be looked for in any popular movement, the socialists of all countries gravitate toward the theoretical position of the avowed Marxism. In proportion as the movement in any given community grows in mass, maturity, and conscious purpose, it unavoidably takes on a more consistently Marxian complexion. It is not the Marxism of Marx, but the materialism of Darwin, which the socialists of to-day have adopted. The Marxist socialists of Germany have the lead, and the socialists of other countries largely take their cue from the German leaders.

The authentic spokesmen of the current international socialism are avowed Marxists. Exceptions to that rule are very few. On the whole, the substantial truth of the Marxist doctrines is not seriously questioned within the lines of the socialists, though there may be some appreciable divergence as to what the true Marxist position is on one point and another. Much and eager controversy circles about questions of that class.

The keepers of the socialist doctrines are passably agreed as to the main position and the general principles.

usage it often has in this connection, designating the theories of Marx and his followers.

Indeed, so secure is this current agreement on the general principles that a very lively controversy on matters of detail may go on without risk of disturbing the general position. This general position is avowedly Marxism. But it is not precisely the position held by Karl Marx. It has been modernised, adapted, filled out, in response to exigencies of a later date than those which conditioned the original formulation of the theories. It is, of course, not admitted by the followers of Marx that any substantial change or departure from the original position has taken place. They are somewhat jealously orthodox, and are impatient of any suggested " improvements " on the Marxist position, as witness the heat engendered in the " revisionist " controversy of a few years back. But the jealous protests of the followers of Marx do not alter the fact that Marxism has undergone some substantial change since it left the hands of its creator. Now and then a more or less consistent disciple of Marx will avow a need of adapting the received doctrines to circumstances that have arisen later than the formulation of the doctrines; and amendments, qualifications, and extensions, with this need in view, have been offered from time to time. But more pervasive though unavowed changes have come in the teachings of Marxism by way of interpretation and an unintended shifting of the point of view. Virtually, the whole of the younger generation of socialist writers shows such a growth. A citation of personal instances would be quite futile.

It is the testimony of his friends as well as of his writings that the theoretical position of Marx, both as regards his standpoint and as regards his main tenets, fell into a definitive shape relatively early, and that his later work was substantially a working out of what was con-

tained in the position taken at the outset of his career.[3]
By the latter half of the forties, if not by the middle of
the forties, Marx and Engels had found the outlook on
human life which came to serve as the point of departure
and the guide for their subsequent development of theory.
Such is the view of the matter expressed by Engels during
the later years of his life.[4]   The position taken by the
two great leaders, and held by them substantially intact,
was a variant of neo-Hegelianism, as has been indicated
in an earlier section of this paper.[5]   But neo-Hegelianism
was short-lived, particularly considered as a standpoint
for scientific theory.   The whole romantic school of
thought, comprising neo-Hegelianism with the rest, began
to go to pieces very soon after it had reached an approach
to maturity, and its disintegration proceeded with exceptional speed, so that the close of the third quarter of the
century saw the virtual end of it as a vital factor in the

[3] There is, indeed, a remarkable consistency, amounting substantially to an invariability of position, in Marx's writing, from
the *Communist Manifesto* to the last volume of the *Capital*.
The only portion of the great *Manifesto* which became antiquated,
in the apprehension of its creators, is the polemics addressed to
the " Philosophical" socialists of the forties and the illustrative
material taken from contemporary politics.   The main position
and the more important articles of theory — the materialistic conception, the doctrine of class struggle, the theory of value and
surplus value, of increasing distress, of the reserve army, of the
capitalistic collapse — are to be found in the *Critique of Political
Economy* (1859), and much of them in the *Misery of Philosophy*
(1847), together with the masterful method of analysis and construction which he employed throughout his theoretical work.

[4] *Cf.* Engels, *Feuerbach* (English translation, Chicago, 1903),
especially Part IV, and various papers published in the *Neue
Zeit;* also the preface to the *Communist Manifesto* written in
1888; also the preface to volume ii. of *Capital,* where Engels
argues the question of Marx's priority in connection with the
leading theoretical principles of his system.

[5] *Cf. Feuerbach,* as above; *The Development of Socialism from
Utopia to Science,* especially sections ii and iii.

development of human knowledge. In the realm of theory, primarily of course in the material sciences, the new era belongs not to romantic philosophy, but to the evolutionists of the school of Darwin. Some few great figures, of course, stood over from the earlier days, but it turns out in the sequel that they have served mainly to mark the rate and degree in which the method of scientific knowledge has left them behind. Such were Virchow and Max Müller, and such, in economic science, were the great figures of the Historical School, and such, in a degree, were also Marx and Engels. The later generation of socialists, the spokesmen and adherents of Marxism during the closing quarter of the century, belong to the new generation, and see the phenomena of human life under the new light. The materialistic conception in their handling of it takes on the color of the time in which they lived, even while they retain the phraseology of the generation that went before them.[6]

The difference between the romantic school of thought, to which Marx belonged, and the school of the evolution-

[6] Such a socialist as Anton Menger, *e.g.,* comes into the neo-Marxian school from without, from the field of modern scientific inquiry, and shows, at least virtually, no Hegelian color, whether in the scope of his inquiry, in his method, or in the theoretical work which he puts forth. It should be added that his *Neue Staatslehre,* and *Neue Sittenlehre* are the first socialistic constructive work of substantial value as a contribution to knowledge, outside of economic theory proper, that has appeared since Lassalle. The efforts of Engels (*Ursprung der Familie*) and Bebel (*Die Frau*) would scarcely be taken seriously as scientific monographs even by hot-headed socialists if it were not for the lack of anything better. Menger's work is not Marxism, whereas Engels's and Bebel's work of this class is practically without value or originality. The unfitness of the Marxian postulates and methods for the purposes of modern science shows itself in the sweeping barrenness of socialistic literature all along that line of inquiry into the evolution of institutions for the promotion of which the materialistic dialectic was invented.

ists into whose hands the system has fallen,— or perhaps, better, is falling,— is great and pervading, though it may not show a staring superficial difference at any one point, — at least not yet. The discrepancy between the two is likely to appear more palpable and more sweeping when the new method of knowledge has been applied with fuller realisation of its reach and its requirement in that domain of knowledge that once belonged to the neo-Hegelian Marxism. The supplanting of the one by the other has been taking place slowly, gently, in large measure unavowedly, by a sort of precession of the point of view from which men size up the facts and reduce them to intelligible order.

The neo-Hegelian, romantic, Marxian standpoint was wholly personal, whereas the evolutionistic — it may be called Darwinian — standpoint is wholly impersonal. The continuity sought in the facts of observation and imputed to them by the earlier school of theory was a continuity of a personal kind,— a continuity of reason and consequently of logic. The facts were construed to take such a course as could be established by an appeal to reason between intelligent and fair-minded men. They were supposed to fall into a sequence of logical consistency. The romantic (Marxian) sequence of theory is essentially an intellectual sequence, and it is therefore of a teleological character. The logical trend of it can be argued out. That is to say, it tends to a goal. On the other hand, in the Darwinian scheme of thought, the continuity sought in and imputed to the facts is a continuity of cause and effect. It is a scheme of blindly cumulative causation, in which there is no trend, no final term, no consummation. The sequence is controlled by nothing but the *vis a tergo* of brute causation, and is essentially mechanical. The neo-Hegelian (Marxian) scheme of development is drawn in the image of the struggling ambitious human spirit: that of Darwinian evolution is of the nature of a mechanical process.[7]

What difference, now, does it make if the materialistic conception is translated from the romantic concepts of Marx into the mechanical concepts of Darwinism? It distorts every feature of the system in some degree, and throws a shadow of doubt on every conclusion that once seemed secure.[8] The first principle of the Marxian scheme is the concept covered by the term " Materialistic," to the effect that the exigencies of the material means of life control the conduct of men in society throughout, and thereby indefeasibly guide the growth of institutions and shape every shifting trait of human culture. This control of the life of society by the material exigencies takes effect through men's taking thought of material (economic) advantages and disadvantages, and choosing that which will yield the fuller material measure of life. When the materialistic conception passes under the Dar-

---

[7] This contrast holds between the original Marxism of Marx and the scope and method of modern science; but it does not, therefore, hold between the latter-day Marxists — who are largely imbued with post-Darwinian concepts — and the non-Marxian scientists. Even Engels, in his latter-day formulation of Marxism, is strongly affected with the notions of post-Darwinian science, and reads Darwinism into Hegel and Marx with a good deal of *naïveté*. (See his *Feuerbach*, especially pp. 93–98 of the English translation.) So, also, the serious but scarcely quite consistent qualifications of the materialistic conception offered by Engels in the letters printed in the *Sozialistische Akademiker,* 1895.

[8] The fact that the theoretical structures of Marx collapse when their elements are converted into the terms of modern science should of itself be sufficient proof that those structures were not built by their maker out of such elements as modern science habitually makes use of. Marx was neither ignorant, imbecile, nor disingenuous, and his work must be construed from such a point of view and in terms of such elements as will enable his results to stand substantially sound and convincing.

winian norm, of cumulative causation, it happens, first, that this initial principle itself is reduced to the rank of a habit of thought induced in the speculator who depends on its light, by the circumstances of his life, in the way of hereditary bent, occupation, tradition, education, climate, food supply, and the like. But under the Darwinian norm the question of whether and how far material exigencies control human conduct and cultural growth becomes a question of the share which these material exigencies have in shaping men's habits of thought, *i.e.*, their ideals and aspirations, their sense of the true, the beautiful, and the good. Whether and how far these traits of human culture and the institutional structure built out of them are the outgrowth of material (economic) exigencies becomes a question of what kind and degree of efficiency belongs to the economic exigencies among the complex of circumstances that conduce to the formation of habits. It is no longer a question of whether material exigencies rationally should guide men's conduct, but whether, as a matter of brute causation, they do induce such habits of thought in men as the economic interpretation presumes, and whether in the last analysis economic exigencies alone are, directly or indirectly, effective in shaping human habits of thought.

Tentatively and by way of approximation some such formulation as that outlined in the last paragraph is apparently what Bernstein and others of the " revisionists " have been seeking in certain of their speculations,[9] and,

[9] Cf. *Voraussetzungen des Sozialismus,* especially the first two (critical) chapters. Bernstein's reverent attitude toward Marx and Engels, as well as his somewhat old-fashioned conception of the scope and method of science, gives his discussion an air of much greater consonance with the orthodox Marxism than it really has. In his later expressions this consonance and conciliatory animus show up more strongly rather than other-

sitting austere and sufficient on a dry shoal up stream, Kautsky has uncomprehendingly been addressing them advice and admonition which they do not understand.[10] The more intelligent and enterprising among the idealist wing — where intellectual enterprise is not a particularly obvious trait — have been struggling to speak for the view that the forces of the environment may effectually reach men's spiritual life through other avenues than the

wise. (See *Socialism and Science,* including the special preface written for the French edition.) That which was to Marx and Engels the point of departure and the guiding norm — the Hegelian dialectic — is to Bernstein a mistake from which scientific socialism must free itself. He says, *e. g.,* (*Voraussetzungen,* end of ch. iv.), "The great things achieved by Marx and Engels they have achieved not by the help of the Hegelian dialectic, but in spite of it."

The number of the "revisionists" is very considerable, and they are plainly gaining ground as against the Marxists of the older line of orthodoxy. They are by no means agreed among themselves as to details, but they belong together by virtue of their endeavor to so construe (and amend) the Marxian system as to bring it into consonance with the current scientific point of view. One should rather say points of view, since the revisionists' endeavors are not all directed to bringing the received views in under a single point of view. There are two main directions of movement among the revisionists: (*a*) those who, like Bernstein, Conrad Schmidt, Tugan-Baranowski, Labriola, Ferri, aim to bring Marxism abreast of the standpoint of modern science, essentially Darwinists; and (*b*) those who aim to return to some footing on the level of the romantic philosophy. The best type and the strongest of the latter class are the neo-Kantians, embodying that spirit of revulsion to romantic norms of theory that makes up the philosophical side of the reactionary movement fostered by the discipline of German imperialism. (See K. Vorländer, *Die neukantische Bewegung im Sozialismus.*)

Except that he is not officially inscribed in the socialist calendar, Sombart might be cited as a particularly effective revisionist, so far as concerns the point of modernising Marxism and putting the modernised materialistic conception to work.

[10] *Cf.* the files of the *Neue Zeit,* particularly during the controversy with Bernstein, and *Bernstein und das Sozialdemokratische Programm.*

calculus of the main chance, and so may give rise to habitual ideals and aspirations independent of, and possibly alien to, that calculus.[11]

So, again, as to the doctrine of the class struggle. In the Marxian scheme of dialectical evolution the development which is in this way held to be controlled by the material exigencies must, it is held, proceed by the method of the class struggle. This class struggle is held to be inevitable, and is held inevitably to lead at each revolutionary epoch to a more efficient adjustment of human industry to human uses, because, when a large proportion of the community find themselves ill served by the current economic arrangements, they take thought, band together, and enforce a readjustment more equitable and more advantageous to them. So long as differences of economic advantage prevail, there will be a divergence of interests between those more advantageously placed and those less advantageously placed. The members of society will take sides as this line of cleavage indicated by their several economic interests may decide. Class solidarity will arise on the basis of this class interest, and a struggle between the two classes so marked off against each other will set in,— a struggle which, in the logic of the situation, can end only when the previously less fortunate class

---

[11] The "idealist" socialists are even more in evidence outside of Germany. They may fairly be said to be in the ascendant in France, and they are a very strong and free-spoken contingent of the socialist movement in America. They do not commonly speak the language either of science or of philosophy, but, so far as their contentions may be construed from the standpoint of modern science, their drift seems to be something of the kind indicated above. At the same time the spokesmen of this scattering and shifting group stand for a variety of opinions and aspirations that cannot be classified under Marxism, Darwinism, or any other system of theory. At the margin they shade off into theology and the creeds.

gains the ascendancy,— and so must the class struggle proceed until it shall have put an end to that diversity of economic interest on which the class struggle rests. All this is logically consistent and convincing, but it proceeds on the ground of reasoned conduct, calculus of advantage, not on the ground of cause and effect. The class struggle so conceived should always and everywhere tend unremittingly toward the socialistic consummation, and should reach that consummation in the end, whatever obstructions or diversions might retard the sequence of development along the way. Such is the notion of it embodied in the system of Marx. Such, however, is not the showing of history. Not all nations or civilisations have advanced unremittingly toward a socialistic consummation, in which all divergence of economic interest has lapsed or would lapse. Those nations and civilisations which have decayed and failed, as nearly all known nations and civilisations have done, illustrate the point that, however reasonable and logical the advance by means of the class struggle may be, it is by no means inevitable. Under the Darwinian norm it must be held that men's reasoning is largely controlled by other than logical, intellectual forces; that the conclusion reached by public or class opinion is as much, or more, a matter of sentiment than of logical inference; and that the sentiment which animates men, singly or collectively, is as much, or more, an outcome of habit and native propensity as of calculated material interest. There is, for instance, no warrant in the Darwinian scheme of things for asserting *a priori* that the class interest of the working class will bring them to take a stand against the propertied class. It may as well be that their training in subservience to their employers will bring them again to realise the equity and excellence of the established system of subjection and unequal distribution of wealth.   Again, no one, for instance, can tell to-day what will be the outcome of the present situation in Europe and America.   It may be that the working classes will go forward along the line of the socialistic ideals and enforce a new deal, in which there shall be no economic class discrepancies, no international animosity, no dynastic politics.   But then it may also, so far as can be foreseen, equally well happen that the working class, with the rest of the community in Germany, England, or America, will be led by the habit of loyalty and by their sportsmanlike propensities to lend themselves enthusiastically to the game of dynastic politics, which alone their sportsmanlike rulers consider worth while.   It is quite impossible on Darwinian ground to foretell whether the " proletariat " will go on to establish the socialistic revolution or turn aside again, and sink their force in the broad sands of patriotism.   It is a question of habit and native propensity and of the range of stimuli to which the proletariat are exposed and are to be exposed, and what may be the outcome is not a matter of logical consistency, but of response to stimulus.

So, then, since Darwinian concepts have begun to dominate the thinking of the Marxists, doubts have now and again come to assert themselves both as to the inevitableness of the irrepressible class struggle and to its sole efficacy.   Anything like a violent class struggle, a seizure of power by force, is more and more consistently deprecated.   For resort to force, it is felt, brings in its train coercive control with all its apparatus of prerogative, mastery, and subservience.[12]

[12] Throughout the revisionist literature in Germany there is a visible softening of the traits of the doctrine of the class struggle, and the like shows itself in the programmes of the party. Outside of Germany the doctrinaire insistence on this tenet is

So, again, the Marxian doctrine of progressive proletarian distress, the so-called *Verelendungstheorie,* which stands pat on the romantic ground of the original Marxism, has fallen into abeyance, if not into disrepute, since the Darwinian conceptions have come to prevail. As a matter of reasoned procedure, on the ground of enlightened material interest alone, it should be a tenable position that increasing misery, increasing in degree and in volume, should be the outcome of the present system of ownership, and should at the same time result in a well-advised and well-consolidated working-class movement that would replace the present system by a scheme more advantageous to the majority. But so soon as the question is approached on the Darwinian ground of cause and effect, and is analysed in terms of habit and of response to stimulus, the doctrine that progressive misery must effect a socialistic revolution becomes dubious, and very shortly untenable. Experience, the experience of history, teaches that abject misery carries with it deterioration and abject subjection. The theory of progressive distress fits convincingly into the scheme of the Hegelian three-phase dialectic. It stands for the antithesis that is to be merged in the ulterior synthesis; but it has no particular force on the ground of an argument from cause to effect.[13]

It fares not much better with the Marxian theory of value and its corollaries and dependent doctrines when

weakening even more decidedly. The opportunist politicians, with strong aspirations, but with relatively few and ill-defined theoretical preconceptions, are gaining ground.

[13] *Cf.* Bernstein, *Die heutige Sozialdemokratie in Theorie und Praxis,* an answer to Brunhuber, *Die heutige Sozialdemokratie,* which should be consulted in the same connection; Goldscheid, *Verelendungs- oder Meliorationstheorie;* also Sombart, *Sozialismus und soziale Bewegung,* 5th edition, pp. 86–89.

Darwinian concepts are brought in to replace the romantic elements out of which it is built up. Its foundation is the metaphysical equality between the volume of human life force productively spent in the making of goods and the magnitude of these goods considered as human products. The question of such an equality has no meaning in terms of cause and effect, nor does it bear in any intelligible way upon the Darwinian question of the fitness of any given system of production or distribution. In any evolutionary system of economics the central question touching the efficiency and fitness of any given system of production is necessarily the question as to the excess of serviceability in the product over the cost of production.[14]   It is in such an excess of serviceability over cost that the chance of survival lies for any system of production, in so far as the question of survival is a question of production, and this matter comes into the speculation of Marx only indirectly or incidentally, and leads to nothing in his argument.

And, as bearing on the Marxian doctrines of exploitation, there is on Darwinian ground no place for a natural right to the full product of labor.   What can be argued in that connection on the ground of cause and effect simply is the question as to what scheme of distribution will

[14] Accordingly, in later Marxian handling of the questions of exploitation and accumulation, the attention is centered on the "surplus product" rather than on the "surplus value." It is also currently held that the doctrines and practical consequences which Marx derived from the theory of surplus value would remain substantially well founded, even if the theory of surplus value was given up. These secondary doctrines could be saved — at the cost of orthodoxy — by putting a theory of surplus product in the place of the theory of surplus value, as in effect is done by Bernstein (*Socialdemokratie in Theorie und Praxis*, sec. 5. Also various of the essays included in *Zur Geschichte und Theorie des Sozialismus*).

help or hinder the survival of a given people or a given civilisation.[15]

But these questions of abstruse theory need not be pursued, since they count, after all, but relatively little among the working tenets of the movement. Little need be done by the Marxists to work out or to adapt the Marxian system of value theory, since it has but slight bearing on the main question,— the question of the trend towards socialism and of its chances of success. It is conceivable that a competent theory of value dealing with the excess of serviceability over cost, on the one hand, and with the discrepancy between price and serviceability, on the other hand, would have a substantial bearing upon the advisability of the present as against the socialistic régime, and would go far to clear up the notions of both socialists and conservatives as to the nature of the points in dispute between them. But the socialists have not moved in the direction of this problem, and they have the excuse that their critics have suggested neither a question nor a solution to a question along any such line. None of the value theorists have so far offered anything that could be called good, bad, or indifferent in this connection, and the socialists are as innocent as the rest. Economics, indeed, has not at this point yet begun to take on a modern tone, unless the current neglect of value theory by the socialists be taken as a negative symptom of advance, indicating that they at least recognise the futility of the

[15] The "right to the full product of labor" and the Marxian theory of exploitation associated with that principle has fallen into the background, except as a campaign cry designed to stir the emotions of the working class. Even as a campaign cry it has not the prominence, nor apparently the efficacy, which it once had. The tenet is better preserved, in fact, among the "idealists," who draw for their antecedents on the French Revolution and the English philosophy of natural rights, than among the latter-day Marxists.

received problems and solutions, even if they are not ready to make a positive move.

The shifting of the current point of view, from romantic philosophy to matter-of-fact, has affected the attitude of the Marxists towards the several articles of theory more than it has induced an avowed alteration or a substitution of new elements of theory for the old. It is always possible to make one's peace with a new standpoint by new interpretations and a shrewd use of figures of speech, so far as the theoretical formulation is concerned, and something of this kind has taken place in the case of Marxism; but when, as in the case of Marxism, the formulations of theory are drafted into practical use, substantial changes of appreciable magnitude are apt to show themselves in a changed attitude towards practical questions. The Marxists have had to face certain practical problems, especially problems of party tactics, and the substantial changes wrought in their theoretical outlook have come into evidence here. The real gravity of the changes that have overtaken Marxism would scarcely be seen by a scrutiny of the formal professions of the Marxists alone. But the exigencies of a changing situation have provoked readjustments of the received doctrinal position, and the shifting of the philosophical standpoint and postulates has come into evidence as marking the limits of change in their professions which the socialistic doctrinaires could allow themselves.

The changes comprised in the cultural movement that lies between the middle and the close of the nineteenth century are great and grave, at least as seen from so near a standpoint as the present day, and it is safe to say that, in whatever historical perspective they may be seen, they must, in some respects, always assert themselves as unprecedented. So far as concerns the present topic, there are three main lines of change that have converged upon the Marxist system of doctrines, and have led to its latter-day modification and growth. One of these — the change in the postulates of knowledge, in the metaphysical foundations of theory — has been spoken of already, and its bearing on the growth of socialist theory has been indicated in certain of its general features. But, among the circumstances that have conditioned the growth of the system, the most obvious is the fact that since Marx's time his doctrines have come to serve as the platform of a political movement, and so have been exposed to the stress of practical party politics dealing with a new and changing situation. At the same time the industrial (economic) situation to which the doctrines are held to apply — of which they are the theoretical formulation — has also in important respects changed its character from what it was when Marx first formulated his views. These several lines of cultural change affecting the growth of Marxism cannot be held apart in so distinct a manner as to appraise the work of each separately. They belong inextricably together, as do the effects wrought by them in the system.

In practical politics the Social Democrats have had to make up their account with the labor movement, the agricultural population, and the imperialistic policy. On each of these heads the preconceived programme of Marxism has come in conflict with the run of events, and on each head it has been necessary to deal shrewdly and adapt the principles to the facts of the time. The adaptation to circumstances has not been altogether of the nature of compromise, although here and there the spirit of compromise and conciliation is visible enough. A conciliatory party policy may, of course, impose an adaptation of form and color upon the party principles, without thereby seriously affecting the substance of the principles themselves; but the need of a conciliatory policy may, even more, provoke a substantial change of attitude toward practical questions in a case where a shifting of the theoretical point of view makes room for a substantial change.

Apart from all merely tactical expedients, the experience of the past thirty years has led the German Marxists to see the facts of the labor situation in a new light, and has induced them to attach an altered meaning to the accepted formulations of doctrine. The facts have not freely lent themselves to the scheme of the Marxist system, but the scheme has taken on such a new meaning as would be consistent with the facts. The untroubled Marxian economics, such as it finds expression in the *Kapital* and earlier documents of the theory, has no place and no use for a trade-union movement, or, indeed, for any similar non-political organisation among the working class, and the attitude of the Social-Democratic leaders of opinion in the early days of the party's history was accordingly hostile to any such movement,[16]— as much so, indeed, as the loyal adherents of the classical political economy. That was before the modern industrial era

[16] It is, of course, well known that even in the transactions and pronounciamentos of the International a good word is repeatedly said for the trade-unions, and both the Gotha and the Erfurt programmes speak in favor of labor organisations, and put forth demands designed to further the trade-union endeavors. But it is equally well known that these expressions were in good part perfunctory, and that the substantial motive behind them was the politic wish of the socialists to conciliate the unionists, and make use of the unions for the propaganda. The early expressions of sympathy with the unionist cause were made for an ulterior purpose. Later on, in the nineties, there comes a change in the attitude of the socialist leaders toward the unions.

had got under way in Germany, and therefore before the German socialistic doctrinaires had learned by experience what the development of industry was to bring with it. It was also before the modern scientific postulates had begun to disintegrate the neo-Hegelian preconceptions as to the logical sequence in the development of institutions.

In Germany, as elsewhere, the growth of the capitalistic system presently brought on trade-unionism; that is to say, it brought on an organised attempt on the part of the workmen to deal with the questions of capitalistic production and distribution by business methods, to settle the problems of working-class employment and livelihood by a system of non-political, businesslike bargains. But the great point of all socialist aspiration and endeavor is the abolition of all business and all bargaining, and, accordingly, the Social Democrats were heartily out of sympathy with the unions and their endeavors to make business terms with the capitalist system, and make life tolerable for the workmen under that system. But the union movement grew to be so serious a feature of the situation that the socialists found themselves obliged to deal with unions, since they could not deal with the workmen over the heads of the unions. The Social Democrats, and therefore the Marxian theorists, had to deal with a situation which included the union movement, and this movement was bent on improving the workman's conditions of life from day to day. Therefore it was necessary to figure out how the union movement could and must further the socialistic advance; to work into the body of doctrines a theory of how the unions belong in the course of economic development that leads up to socialism, and to reconcile the unionist efforts at improvement with the ends of Social Democracy. Not only were the unions seeking improvement by unsocialistic methods, but the level of comfort among the working classes was in some respects advancing, apparently as a result of these union efforts. Both the huckstering animus of the workmen in their unionist policy and the possible amelioration of working-class conditions had to be incorporated into the socialistic platform and into the Marxist theory of economic development. The Marxist theory of progressive misery and degradation has, accordingly, fallen into the background, and a large proportion of the Marxists have already come to see the whole question of working-class deterioration in some such apologetic light as is shed upon it by Goldscheid in his *Verelendungs-oder Meliorationstheorie.* It is now not an unusual thing for orthodox Marxists to hold that the improvement of the conditions of the working classes is a necessary condition to the advance of the socialistic cause, and that the unionist efforts at amelioration must be furthered as a means toward the socialistic consummation. It is recognised that the socialistic revolution must be carried through not by an anæmic working class under the pressure of abject privation, but by a body of full-blooded workingmen gradually gaining strength from improved conditions of life. Instead of the revolution being worked out by the leverage of desperate misery, every improvement in working-class conditions is to be counted as a gain for the revolutionary forces. This is a good Darwinism, but it does not belong in the neo-Hegelian Marxism.

Perhaps the sorest experience of the Marxist doctrinaires has been with the agricultural population. Notoriously, the people of the open country have not taken kindly to socialism. No propaganda and no changes in the economic situation have won the sympathy of the peasant farmers for the socialistic revolution. Notoriously, too, the large-scale industry has not invaded the agricultural field, or expropriated the small proprietors, in anything like the degree expected by the Marxist doctrinaires of a generation ago.   It is contained in the theoretical system of Marx that, as modern industrial and business methods gain ground, the small proprietor farmers will be reduced to the ranks of the wage-proletariat, and that, as this process of conversion goes on, in the course of time the class interest of the agricultural population will throw them into the movement side by side with the other wage-workmen.[17]   But at this point the facts have hitherto not come out in consonance with the Marxist theory.   And the efforts of the Social Democrats to convert the peasant population to socialism have been practically unrewarded.   So it has come about that the political leaders and the keepers of the doctrines have, tardily and reluctantly, come to see the facts of the agrarian situation in a new light, and to give a new phrasing to the articles of Marxian theory that touch on the fortunes of the peasant farmer.   It is no longer held that either the small properties of the peasant farmer must be absorbed into larger properties, and then taken over by the State, or that they must be taken over by the State directly, when the socialistic revolution is established. On the contrary, it is now coming to be held that the peasant proprietors will not be disturbed in their holdings by the great change.   The great change is to deal with capitalistic enterprise, and the peasant farming is not properly " capitalistic."   It is a system of production in which the producer normally gets only the product of his own labor.   Indeed, under the current régime of markets and credit relations, the small agricultural producer, it is held, gets less than the product of his own labor, since the capitalistic business enterprises with which

---

[17] *Cf. Kapital,* vol. i, ch. xiii, sect. 10.

he has to deal are always able to take advantage of him. So it has become part of the overt doctrine of socialists that as regards the peasant farmer it will be the consistent aim of the movement to secure him in the untroubled enjoyment of his holding, and free him from the vexatious exactions of his creditors and the ruinous business traffic in which he is now perforce involved. According to the revised code, made possible by recourse to Darwinian concepts of evolution instead of the Hegelian three-phase dialectic, therefore, and contrary to the earlier prognostications of Marx, it is no longer held that agricultural industry must go through the capitalistic mill; and it is hoped that under the revised code it may be possible to enlist the interest and sympathy of this obstinately conservative element for the revolutionary cause. The change in the official socialist position on the agricultural question has come about only lately, and is scarcely yet complete, and there is no knowing what degree of success it may meet with either as a matter of party tactics or as a feature of the socialistic theory of economic development. All discussions of party policy, and of theory so far as bears on policy, take up the question; and nearly all authoritative spokesmen of socialism have modified their views in the course of time on this point.

The socialism of Karl Marx is characteristically inclined to peaceable measures and disinclined to a coercive government and belligerent politics. It is, or at least it was, strongly averse to international jealousy and patriotic animosity, and has taken a stand against armaments, wars, and dynastic aggrandisement. At the time of the French-Prussian war the official organisation of Marxism, the International, went so far in its advocacy of peace as to urge the soldiery on both sides to refuse to fight. After the campaign had warmed the blood of the two nations, this advocacy of peace made the International odious in the eyes of both French and Germans. War begets patriotism, and the socialists fell under the reproach of not being sufficiently patriotic. After the conclusion of the war the Socialistic Workingmen's Party of Germany sinned against the German patriotic sentiment in a similar way and with similarly grave results. Since the foundation of the empire and of the Social-Democratic party, the socialists and their doctrines have passed through a further experience of a similar kind, but on a larger scale and more protracted. The government has gradually strengthened its autocratic position at home, increased its warlike equipment, and enlarged its pretensions in international politics, until what would have seemed absurdly impossible a generation ago is now submitted to by the German people, not only with a good grace, but with enthusiasm. During all this time that part of the population that has adhered to the socialist ideals has also grown gradually more patriotic and more loyal, and the leaders and keepers of socialist opinion have shared in the growth of chauvinism with the rest of the German people. But at no time have the socialists been able to keep abreast of the general upward movement in this respect. They have not attained the pitch of reckless loyalty that animates the conservative German patriots, although it is probably safe to say that the Social Democrats of to-day are as good and headlong patriots as the conservative Germans were a generation ago. During all this period of the new era of German political life the socialists have been freely accused of disloyalty to the national ambition, of placing their international aspirations above the ambition of imperial aggrandisement.

The socialist spokesmen have been continually on the

defensive. They set out with a round opposition to any considerable military establishment, and have more and more apologetically continued to oppose any "undue" extension of the warlike establishments and the warlike policy. But with the passage of time and the habituation to warlike politics and military discipline, the infection of jingoism has gradually permeated the body of Social Democrats, until they have now reached such a pitch of enthusiastic loyalty as they would not patiently hear a truthful characterisation of. The spokesmen now are concerned to show that, while they still stand for international socialism, consonant with their ancient position, they stand for national aggrandisement first and for international comity second. The relative importance of the national and the international ideals in German socialist professions has been reversed since the seventies.[18] The leaders are busy with interpretation of their earlier formulations. They have come to excite themselves over nebulous distinctions between patriotism and jingoism. The Social Democrats have come to be German patriots first and socialists second, which comes to saying that they are a political party working for the maintenance of the existing order, with modifications. They are no longer a party of revolution, but of reform, though the measure of reform which they demand greatly exceeds the Hohenzollern limit of tolerance. They are now as much, if not more, in touch with the ideas of English liberalism than with those of revolutionary Marxism.

The material and tactical exigencies that have grown out of changes in the industrial system and in the political situation, then, have brought on far-reaching changes of

---

[18] *Cf.* Kautsky, *Erfurter Programm,* ch. v, sect. 13; Bernstein, *Voraussetzungen,* ch. iv, sect. e.

adaptation in the position of the socialists. The change may not be extremely large at any one point, so far as regards the specific articles of the programme, but, taken as a whole, the resulting modification of the socialistic position is a very substantial one. The process of change is, of course, not yet completed,— whether or not it ever will be,— but it is already evident that what is taking place is not so much a change in amount or degree of conviction on certain given points as a change in kind,— a change in the current socialistic habit of mind.

The factional discrepancies of theory that have occupied the socialists of Germany for some years past are evidence that the conclusion, even a provisional conclusion, of the shifting of their standpoint has not been reached. It is even hazardous to guess which way the drift is setting. It is only evident that the past standpoint, the standpoint of neo-Hegelian Marxism, cannot be regained,— it is a forgotten standpoint. For the immediate present the drift of sentiment, at least among the educated, seems to set toward a position resembling that of the National Socials and the Rev. Mr. Naumann; that is to say, imperialistic liberalism. Should the conditions, political, social, and economic, which to-day are chiefly effective in shaping the habits of thought among the German people, continue substantially unchanged and continue to be the chief determining causes, it need surprise no one to find German socialism gradually changing into a somewhat characterless imperialistic democracy. The imperial policy seems in a fair way to get the better of revolutionary socialism, not by repressing it, but by force of the discipline in imperialistic ways of thinking to which it subjects all classes of the population. How far a similar process of sterilisation is under way, or is likely to overtake the socialist movement in other countries, is an obscure question to which the German object-lesson affords no certain answer.

THE MUTATION THEORY AND THE BLOND RACE [1]

THE theories of racial development by mutation, associated with the name of Mendel, when they come to be freely applied to man, must greatly change the complexion of many currently debated questions of race — as to origins, migrations, dispersion, chronology, cultural derivation and sequence. In some respects the new theories should simplify current problems of ethnology, and they may even dispense with many analyses and speculations that have seemed of great moment in the past.

The main postulate of the Mendelian theories — the stability of type — has already done much service in anthropological science, being commonly assumed as a matter of course in arguments dealing with the derivation and dispersion of races and peoples. It is only by force of this assumption that ethnologists are able to identify any given racial stock over intervals of space or time, and so to trace the racial affinities of any given people. Question has been entertained from time to time as to the racial fixity of given physical traits — as, *e.g.,* stature, the cephalic indices, or hair and eye color — but on the whole these and other standard marks of race are still accepted as secure grounds of identification.[2] Indeed, without some such assumption any ethnological inquiry must degenerate into mere wool-gathering.

[1] Reprinted by permission from *The Journal of Race Development,* Vol. III, No. 4.

[2] *Cf.,* however, W. Ridgeway, "The Application of Zoölogical Laws to Man," *Report, British Association for Advancement of Science* (Dublin), 1908.

But along with this, essentially Mendelian, postulate of the stability of types, ethnologists have at the same time habitually accepted the incompatible Darwinian doctrine that racial types vary incontinently after a progressive fashion, arising through insensible cumulative variations and passing into new specific forms by the same method, under the Darwinian rule of the selective survival of slight and unstable (non-typical) variations. The effect of these two incongruous premises has been to leave discussions of race derivation somewhat at loose ends wherever the two postulates cross one another.

If it be assumed, or granted, that racial types are stable, it follows as a matter of course that these types or races have not arisen by the cumulative acquirement of unstable non-specific traits, but must have originated by mutation or by some analogous method, and this view must then find its way into anthropology as into the other biological sciences. When such a step is taken an extensive revision of questions of race will be unavoidable, and an appreciable divergence may then be looked for among speculations on the mutational affinities of the several races and cultures.

Among matters so awaiting revision are certain broad questions of derivation and ethnography touching the blond race or races of Europe. Much attention, and indeed much sentiment, has been spent on this general topic. The questions involved are many and diverse, and many of them have been subject of animated controversy, without definitive conclusions.

The mutation theories, of course, have immediately to do with the facts of biological derivation alone, but when the facts are reviewed in the light of these theories it will be found that questions of cultural origins and relationship are necessarily drawn into the inquiry. In particular, an inquiry into the derivation and distribution of the blond stock will so intimately involve questions of the Aryan speech and institutions as to be left incomplete without a somewhat detailed attention to this latter range of questions. So much so that an inquiry into the advent and early fortunes of the blond stock in Europe will fall, by convenience, under two distinct but closely related captions: The Origin of the Blond Type, and The Derivation of the Aryan Culture.

(a) It is held, on the one hand, that there is but a single blond race, type or stock (Keane, Lapouge, Sergi), and on the other hand that there are several such races or types, more or less distinct but presumably related (Deniker, Beddoe, and other, especially British, ethnologists). (b) There is no good body of evidence going to establish a great antiquity for the blond stock, and there are indications, though perhaps inconclusive, that the blond strain, including all the blond types, is of relatively late date — unless a Berber (Kabyle) blond race is to be accepted in a more unequivocal manner than hitherto. (c) Neither is there anything like convincing evidence that this blond strain has come from outside of Europe — except, again, for the equivocal Kabyle — or that any blond race has ever been widely or permanently distributed outside of its present European habitat. (d) The blond race is not found unmixed. In point of pedigree all individuals showing the blond traits are hybrids, and the greater number of them show their mixed blood in their physical traits. (e) There is no community, large or small, made up exclusively of blonds, or nearly so, and there is no good evidence available that such an all-blond or virtually all-blond community ever has existed, either in historic or prehistoric times. The race appears never to have lived in isolation. (f) It occurs in several (perhaps hybrid) variants — unless these variants are to be taken (with Deniker) as several distinct races. (g) Counting the dolicho-blond as the original type of the race, its nearest apparent relative among the races of mankind is the Mediterranean (of Sergi), at least in point of physical traits. At the same time the blond race, or at least the dolicho-blond type, has never since neolithic times, so far as known, extensively and permanently lived in contact with the Mediterranean. (h) The various (national) ramifications of the blond stock — or rather the various racial mixtures into which an appreciable blond element enters — are all, and to all appearance have always been, of Aryan (" Indo-European," " Indo-Germanic ") speech — with the equivocal exception of the Kabyle. (i) Yet far the greater number and variety (national and linguistic) of men who use the Aryan speech are not prevailingly blond, or even appreciably mixed with blond. (j) The blond race, or the peoples with an appreciable blond admixture, and particularly the communities in which the dolicho-blond element prevails, show little or none of the peculiarly Aryan institutions — understanding by that phrase not the known institutions of the ancient Germanic peoples, but that range of institutions said by competent philologists to be reflected in the primitive Aryan speech. (k) These considerations raise the presumption that the blond race was not originally of Aryan speech or of Aryan culture, and they also suggest (l) that the Mediterranean, the nearest apparent relative of the dolicho-blond, was likewise not originally Aryan.

Accepting the mutation theory, then, for the purpose in hand, and leaving any questions of Aryanism on one side for the present, a canvass of the situation so outlined may be offered in such bold, crude and summary terms as should be admissible in an analysis which aims to be tentative and provisional only. It may be conceived that the dolichocephalic blond originated as a mutant of the Mediterranean type (which it greatly resembles in its scheme of biometric measurements [3]) probably some time after that race had effected a permanent lodgment on the continent of Europe. The Mediterranean stock may be held (Sergi and Keane) to have come into Europe from Africa,[4] whatever its remoter derivation may have been. It is, of course, not impossible that the mutation which gave rise to the dolicho-blond may have occurred before the parent stock left Africa, or rather before it was shut out of Africa by the submergence of the land connection across Sicily, but the probabilities seem to be against such a view. The conditions would appear to have been less favorable to a mutation of this kind in the African habitat of the parent stock than in Europe, and less favorable in Europe during earlier quaternary time than toward the close of the glacial period.

The causes which give rise to a variation of type have always been sufficiently obscure, whether the origin of species be conceived after the Darwinian or the Mendelian fashion, and the mutation theories have hitherto afforded little light on that question. Yet the Mendelian postulate that the type is stable except for such a mutation as shall establish a new type raises at least the presumption that such a mutation will take place only under exceptional circumstances, that is to stay, under circumstances so substantially different from what the type is best adapted to as to subject it to some degree of physiological strain. It

[3] *Cf.* Sergi, *The Mediterranean Race*, ch. xi, xiii.
[4] Sergi, *Arii e Italici;* Keane, *Man Past and Present,* ch. xii.

is to be presumed that no mutation will supervene so long as the conditions of life do not vary materially from what they have been during the previous uneventful life-history of the type. Such is the presumption apparently involved in the theory and such is also the suggestion afforded by the few experimental cases of observed mutation, as, *e.g.,* those studied by De Vries.

A considerable climatic change, such as would seriously alter the conditions of life either directly or through its effect on the food supply, might be conceived to bring on a mutating state in the race; or the like effect might be induced by a profound cultural change, particularly any such change in the industrial arts as would radically affect the material conditions of life. These considerations, mainly speculative it is true, suggest that the dolicho-blond mutant could presumably have emerged only at a time when the parent stock was exposed to notably novel conditions of life, such as would be presumed (with De Vries) to tend to throw the stock into a specifically unstable (mutating) state; at the same time these novel conditions of life must also have been specifically of such a nature as to favor the survival and multiplication of this particular human type. The climatic tolerance of the dolicho-blond, *e.g.,* is known to be exceptionally narrow. Now, it is not known, indeed there is no reason to presume, that the Mediterranean race was exposed to such variations of climate or of culture before it entered Europe as might be expected to induce a mutating state in the stock, and at the same time a mutant gifted with the peculiar climatic intolerance of the dolicho-blond would scarcely have survived under the conditions offered by northern Africa in late quaternary time. But the required conditions are had later on in Europe, after the Mediterranean was securely at home in that continent.

The whole episode may be conceived to have run off somewhat in the following manner. The Mediterranean race is held to have entered Europe in force during quaternary time, presumably after the quaternary period was well advanced, most likely during the last genial, interglacial period. This race then brought the neolithic culture, but without the domestic animals (or plants?) that are a characteristic feature of the later neolithic age, and it encountered at least the remnants of an older, palaeolithic population. This older European population was made up of several racial stocks, some of which still persist as obscure and minor elements in the later peoples of Europe. The (geologic) date to be assigned this intrusion of the Mediterranean race into Europe is of course not, and can perhaps never be, determined with any degree of nicety or confidence. But there is a probability that it coincides with the recession of the ice-sheet, following one or another of the severer periods of glaciations, that occurred before the submergence of the land connection between Europe and Africa, over Gibraltar, Sicily, and perhaps Crete. How late in quaternary time the final submergence of the Mediterranean basin occurred is still a matter of surmise; the intrusion of the Mediterranean race into Europe appears, on archaeological evidence, to have occurred in late quaternary time, and in the end this archaeological evidence may help to decide the geologic date of the severance of Europe from Africa.

The Mediterranean race seems to have spread easily over the habitable surface of Europe and shortly to have grown numerous and taken rank as the chief racial element in the neolithic population; which argues that no very considerable older population occupied the European continent at the time of the Mediterranean invasion; which in turn implies that the fairly large (Magdalenian) population of the close of the palaeolithic age was in great part destroyed or expelled by the climatic changes that coincided with or immediately preceded the advent of the Mediterranean race. The known characteristics of the Magdalenian culture indicate a technology, a situation and perhaps a race, somewhat closely paralleled by the Eskimo;[5] which argues that the climatic situation before which this Magdalenian race and culture gave way would have been that of a genial interglacial period rather than a period of glaciation.

During this genial (perhaps sub-tropical) inter-glacial period immediately preceding the last great glaciation the Magdalenian stock would presumably find Europe climatically untenable, judging by analogy with the Eskimo; whereas the Mediterranean stock should have found it an eminently favorable habitat, for this race has always succeeded best in a warm-temperate climate. Both the extensive northward range of the early neolithic (Mediterranean) settlements and the total disappearance of the Magdalenian culture from the European continent point to a climatic situation in Europe more favorable to the former race and more unwholesome for the latter than the conditions known to have prevailed at any time since the last interglacial period, especially in the higher latitudes. The indications would seem to be that the whole of Europe, even the Baltic and Arctic seaboards, became climatically so fully impossible for the Magdalenian race during this interglacial period as to result in its extinction or definitive expulsion; for when, in recent times, climatically suitable conditions return, on the Arctic seaboard, the culture which takes the place that should have been occupied by the Magdalenian is the Finnic (Lapp)—

[5] *Cf.* W. J. Sollas, "Palaeolithic Races and their Modern Representatives," *Science Progress,* vol. iv, 1909–1910.

a culture unrelated to the Magdalenian either in race or technology, although of much the same cultural level and dealing with a material environment of much the same character. And this genial interval that was fatal to the Magdalenian was, by just so much, favorable to the Mediterranean race.

But glacial conditions presently returned, though with less severity than the next preceding glacial period; and roughly coincident with the close of the genial interval in Europe the land connection with Africa was cut off by submergence, shutting off retreat to the south. How far communication with Asia may have been interrupted during the subsequent cold period, by the local glaciation of the Caucasus, Elburz and Armenian highlands, is for the present apparently not to be determined, although it is to be presumed that the outlet to the east would at least be seriously obstructed during the glaciation. There would then be left available for occupation, mainly by the Mediterranean race, central and southern Europe together with the islands, notably Sicily and Crete, left over as remnants of the earlier continuous land between Europe and Africa. The southern extensions of the mainland, and more particularly the islands, would still afford a favorable place for the Mediterranean race and its cultural growth. So that the early phases of the great Cretan (Aegean) civilisation are presumably to be assigned to this period that is covered by the last advance of the ice in northern Europe. But the greater portion of the land area so left accessible to the Mediterranean race, in central or even in southern Europe, would have been under glacial or subglacial climatic conditions. For this race, essentially native to a warm climate, this situation on the European mainland would be sufficiently novel and trying, particularly throughout that ice-fringed range of country where they would be exposed to such cold and damp as this race has never easily tolerated.

The situation so outlined would afford such a condition of physiological strain as might be conceived to throw the stock into a specifically unstable state and so bring on a phase of mutation. At the same time this situation, climatic and technological, would be notably favorable to the survival and propagation of a type gifted with all the peculiar capacities and limitations of the dolicho-blond; so that any mutant showing the traits characteristic of that type would then have had an eminently favorable chance of survival. Indeed, it is doubtful, in the present state of the available evidence, whether such a type of man could have survived in Europe from or over any period of quaternary time prior to the last period of glaciation. The last preceding interglacial period appears to have been of a sufficiently genial (perhaps sub-tropical) character throughout Europe to have definitively eliminated the Magdalenian race and culture, and a variation of climate in the genial sense sufficiently pronounced to make Europe absolutely untenable for the Magdalenian — presumed to be something of a counterpart to the Eskimo both in race and culture — should probably have reached the limit of tolerance for the dolicho-blond as well. The latter is doubtless not as intolerant of a genial — warm-temperate — climate as the former, but the dolicho-blond after all stands much nearer to the Eskimo in this matter of climatic tolerance than to either of the two chief European stocks with which it is associated. Apparently no racial stock with a climatic tolerance approximately like that of the Eskimo, the Magdalenian, or the current races of the Arctic seaboard, survived over the last inter-glacial period; and if the dolicho-blond is conceived to have lived through that period it would appear to have been by a precariously narrow margin. So that, on one ground and another, the mutation out of which the dolicho-blond has arisen is presumably to be assigned to the latest period of glaciation in Europe, and with some probability to the time when the latest glaciation was at its maximum, and to the region where glacial and seaboard influences combined to give that racial type a differential advantage over all others.

This dolicho-blond mutation may, of course, have occurred only once, in a single individual, but it should seem more probable, in the light of De Vries' experiments, that the mutation will have been repeated in the same specific form in several individuals in the same general locality and in the same general period of time. Indeed, it would seem highly probably that several typically distinct mutations will have occurred, repeatedly, at roughly the same period and in the same region, giving rise to several new types, some of which, including the dolicho-blond, will have survived. Many, presumably the greater number, of these mutant types will have disappeared, selectively, being unfit to survive under those subglacial seaboard conditions that were eminently favorable to the dolicho-blond; while other mutants arising out of the same mutating period and adapted to climatic conditions of a more continental character, suitable to more of a continental habitat, less humid, at a higher altitude and with a wider seasonal variation of temperature, may have survived in the regions farther inland, particularly eastward of the selectively defined habitat of the dolicho-blond. These latter may have given rise to several blond races, such as are spoken for by Deniker[6] and certain British ethnologists.

[6] *The Races of Mankind;* and "Les six races composant la population de l'Europe," *Journal Anth. Inst.,* 1906.

The same period of mutation may well have given rise also to one or more brunet types, some of which may have survived. But if any new brunet type has come up within a period so recent as this implies, the fact has not been noted or surmised hitherto — unless the brunet races spoken for by Deniker are to be accepted as typically distinct and referred to such an origin. The evidence for the brunet stocks has not been canvassed with a question of this kind in view. These stocks have not been subject of such eager controversy as the dolicho-blond, and the attention given them has been correspondingly less. The case of the blond is unique in respect of the attention spent on questions of its derivation and prehistory, and it is also singular in respect of the facility with which it can be isolated for the purposes of such an inquiry. This large and persistent attention, from all sorts of ethnologists, has brought the evidence bearing on the dolicho-blond into such shape as to permit more confident generalisations regarding that race than any other.

In any case the number of mutant individuals, whether of one or of several specific types, will have been very few as compared with the numbers of the parent stock from which they diverged, even if they may have been somewhat numerous as counted absolutely, and the survivors whose offspring produced a permanent effect on the European peoples will have been fewer still. It results that these surviving mutants will not have been isolated from the parent stock, and so could not breed in isolation, but must forthwith be crossed on the parent stock and could therefore yield none but hybrid offspring. From the outset, therefore, the community or communities in which the blond mutants were propagated would be made up of a mixture of blond and brunet, with the brunet greatly preponderating. It may be added that in all probability there were also present in this community from the start one or more minor brunet elements besides the predominant Mediterranean, and that at least shortly after the close of the glacial period the new brachycephalic brunet (Alpine) race comes into the case; so that the chances favor an early and persistent crossing of the dolicho-blond with more than one brunet type, and hence they favor complications and confusion of types from the start. It follows that, in point of pedigree, according to this view there neither is nor ever has been a pure-bred dolicho-blond individual since the putative original mutant with which the type came in. But under the Mendelian rule of hybrids it is none the less to be expected that, in the course of time and of climatically selective breeding, individuals (perhaps in appreciable numbers) will have come up from time to time showing the type characters unmixed and unweakened, and effectively pure-bred in point of heredity. Indeed, such individuals, effectively pure-bred or tending to the establishment of a pure line, will probably have emerged somewhat frequently under conditions favorable to the pure type. The selective action of the conditions of life in the habitat most favorable to the propagation of the dolicho-blond has worked in a rough and uncertain way toward the establishment, in parts of the Baltic and North Sea region, of communities made up prevailingly of blonds. Yet none of these communities most favorably placed for a selective breeding in the direction of a pure dolicho-blond population have gone far enough in that direction to allow it safely to be said that the composite population of any such given locality is more than half blond.

Placed as it is in a community of nations made up of a hybrid mixture of several racial stocks there is probably no way at present of reaching a convincing demonstration of the typical originality of this dolicho-blond mutant, as contrasted with the other blond types with which it is associated in the European population; but certain general considerations go decidedly, perhaps decisively, to enforce such a view: (a) This type shows such a pervasive resemblance to a single one of the known older and more widely distributed types of man (the Mediterranean) as to suggest descent by mutation from this one rather than derivation by crossing of any two or more known types. The like can not be said of the other blond types, all and several of which may plausibly be explained as hybrids of known types. They have the appearance of blends, or rather of biometrical compromises, between two or more existing varieties of man. Whereas it does not seem feasible to explain the dolicho-blond as such a blend or compromise between any known racial types. (b) The dolicho-blond occurs, in a way, centrally to the other blond types, giving them a suggestive look of being ramifications of the blond stock, by hybridisation, into regions not wholly suited to the typical blond. The like can scarcely be said for any of the other European types or races. The most plausible exception would be Deniker's East-European or Oriental race, Beddoe's Saxon, which stands in a somewhat analogous spacial relation to the other blond types. But this brachycephalic blond is not subject to the same sharp climatic limitations that hedge about the dolicho-blond; it occurs apparently with equally secure viability within the littoral home area of the dolicho-blond and in continental situations where conditions of altitude and genial climate would bar the latter from permanent settlement. The ancient and conventionally accepted center of diffusion of blondness in Europe lies within the seaboard region bordering on the south Baltic, the North Sea and the narrow waters of the Scandinavian peninsulas. Probably, if this broad central area of diffusion were to be narrowed down to a particular spot, the consensus of opinion as to where the narrower area of characteristic blondness is to be looked for, would converge on the lands immediately about the narrow Scandinavian waters. This would seem to hold true for historic and for prehistoric times alike. This region is at the same time, by common consent, the peculiar home of the dolicho-blond, rather than of any other blond type. (c) The well known but little discussed climatic limitation of the blond race applies particularly to the dolicho-blond, and only in a pronouncedly slighter degree to the other blond types. The dolicho-blond is subject to a strict regional limitation, the other blond types to a much less definite and wider limitation of the same kind. Hence these others are distributed somewhat widely, over regions often remote and climatically different from the home area of the dolicho-blond, giving them the appearance of being dispersed outward from this home area as hybrid extensions of the central and typical blond stock. A further and equally characteristic feature of this selective localisation of the dolicho-blond race is the fact that while this race does not succeed permanently outside the seaboard region of the south Baltic and North Sea, there is no similar selective bar against other races intruding into this region. Although the dolicho-blond perhaps succeeds better within its home area than any other competing stock or type, yet several other types of man succeed so well within the same region as to hold it, and apparently always to have held it, in joint tenancy with the dolicho-blond.

A close relationship, amounting to varietal identity, of the Kabyle with the dolicho-blond has been spoken for by Keane and by other ethnologists. But the very different climatic tolerance of the two races should put such an identity out of the question. The Kabyle lives and thrives best, where his permanent home area has always been, in a high and dry country, sufficiently remote from the sea to make it a continental rather than a littoral habitat. The dolicho-blond, according to all available evidence, can live in the long run only in a seaboard habitat, damp and cool, at a high latitude and low altitude. There is no known instance of this race having gone out from its home area on the northern seaboard into such a region as that inhabited by the Kabyle and having survived for an appreciable number of generations. That this type of man should have come from Mauritania, where it could apparently not live under the conditions known to have prevailed there in the recent or the remoter past, would seem to be a biologic impossibility. Hitherto, when the dolicho-blond has migrated into such or a similar habitat it has not adapted itself to the new climatic requirements but has presently disappeared off the face of the land. Indeed, the experiment has been tried in Mauritanian territory. If the Kabyle blond is to be correlated with those of Europe, it will in all probability have to be assigned an independent origin, to be derived from an earlier mutation of the same Mediterranean stock to which the dolicho-blond is to be traced.

Questions of race in Europe are greatly obscured by the prevalence of hybrid types having more or less fixity and being more or less distinctly localised. The existing European peoples are hybrid mixtures of two or more racial stocks. The further fact is sufficiently obvious, though it has received less critical attention than might be, that these several hybrid populations have in the course of time given rise to a number of distinct national and local types, differing characteristically from one another and having acquired a degree of permanence, such as to simulate racial characters and show well marked national and local traits in point of physiognomy and temperament. Presumably, these national and local types of physique and temperament are hybrid types that have been selectively bred into these characteristic forms in adaptation to the peculiar circumstances of environment and culture under which each particular local population is required to live, and that have been so fixed (provisionally) by selective breeding of the hybrid material subject to such locally uniform conditions — except so far as the local characters in question are of the nature of habits and are themselves therefore to be classed as an institutional element rather than as characteristics of race.

It is evident that under the Mendelian law of hybridisation the range of favorable, or viable, variations in any hybrid population must be very large — much larger than the range of fluctuating (non-typical) variations obtainable under any circumstances in a pure-bred race. It also follows from these same laws of hybridisation that by virtue of the mutual exclusiveness of allelomorphic characters or groups of characters it is possible selectively to obtain an effectually " pure line " of hybrids combining characters drawn from each of the two or more parent stocks engaged, and that such a composite pure line may selectively be brought to a provisional fixity [7] in any such hybrid population. And under conditions favorable to a type endowed with any given hybrid combination of characters so worked out the given hybrid type (composite pure line) may function in the racial mixture in which it is so placed very much as an actual racial type would behave under analogous circumstances; so that, *e.g.*, under

[7] Illustrated by the various pure breeds or " races " of domestic animals.

continued intercrossing such a hybrid population would tend cumulatively to breed true to this provisionally stable hybrid type, rather than to the actual racial type represented by any one of the parent stocks of which the hybrid population is ultimately made up, unless the local conditions should selectively favor one or another of these ultimate racial types. Evidently, too, the number of such provisionally stable composite pure lines that may be drawn from any hybrid mixture of two or more parent stocks must be very considerable — indeed virtually unlimited; so that on this ground there should be room for any conceivable number of provisionally stable national or local types of physique and temperament, limited only by the number of characteristically distinguishable local environments or situations that might each selectively act to characterise and establish a locally characteristic composite pure line; each answering to the selective exigencies of the habitat and cultural environment in which it is placed, and each responding to these exigencies in much the same fashion as would an actual racial type — provided only that this provisionally stable composite pure line is not crossed on pure-bred individuals of either of the parent stocks from which it is drawn, pure-bred in respect of the allelomorphic characters which give the hybrid type its typical traits.

When the hybrid type is so crossed back on one or other of its parent stocks it should be expected to break down; but in so slow-breeding a species as man, with so large a complement of unit characters (some 4000 it has been estimated), it will be difficult to decide empirically which of the two lines — the hybrid or the parent stock — proves itself in the offspring effectively to be a racial type; that is to say, which of the two (or more) proves to be an ultimately stable type arisen by a Mendelian mutation, and which is a provisionally stable composite pure line selectively derived from a cross. The inquiry at this point, therefore, will apparently have to content itself with arguments of probability drawn from the varying behavior of the existing hybrid types under diverse conditions of life.

Such general consideration of the behavior of the blond types of Europe, other than the dolicho-blond, and more particularly consideration of their viability under divergent climatic conditions, should apparently incline to the view that they are hybrid types, of the nature of provisionally stable composite pure lines.

So far, therefore, as the evidence has yet been canvassed, it seems probable on the whole that the dolicho-blond is the only survivor from among the several mutants that may have arisen out of this presumed mutating period; that the other existing blond types, as well as certain brunets, are derivatives of the hybrid offspring of the dolicho-blond crossed on the parent Mediterranean stock or on other brunet stocks with which the race has been in contact early or late; and that several of these hybrid lines have in the course of time selectively been established as provisionally stable types (composite pure lines), breakable only by a fresh cross with one or other of the parent types from which the hybrid line sprang, according to the Mendelian rule.[8]

All these considerations may not be convincing, but they are at least suggestive to the effect that if originality is to

[8] Mr. R. B. Bean's discussion of Deniker's " Six Races," *e. g.*, goes far to show that such is probably the standing of the blond types, other than the dolicho-blond, among these six races of Europe; although such is not the conclusion to which Mr. Bean comes. *Philippine Journal of Science,* September, 1909.

be claimed for any one of the blond types or stocks it can best be claimed for the dolicho-blond, while the other blond types may better be accounted for as the outcome of the crossing of this stock on one or another of the brunet stocks of Europe.

THE BLOND RACE AND THE ARYAN CULTURE [1]

IT has been argued in an earlier paper [2] that the blond type or types of man (presumably the dolichocephalic blond) arose by mutation from the Mediterranean stock during the last period of severe glaciation in Europe. This would place the emergence of this racial type roughly coincident with the beginning of the European neolithic; the evidence going presumptively to show that the neolithic technology came into Europe with the Mediterranean race, at or about the same time with that race, and that the mutation which gave rise to the dolicho-blond took place after the Mediterranean race was securely settled in Europe. Since this blond mutant made good its survival under the circumstances into which it so was thrown it should presumably be suited by native endowment to the industrial and climatic conditions that prevailed through the early phases of the neolithic age in Europe; that is to say, it would be a type of man selectively adapted to the technological situation characteristic of the early neolithic but lacking as yet the domestic animals (and crop-plants?) that presently give much of its character to that culture.

Beginning, then, with the period of the last severe glaciation, and starting with this technological equipment, those portions of the European population that contained an

[1] Reprinted by permission from *The University of Missouri Bulletin*, Science Series, vol. ii, No. 3.
[2] "The Mutation Theory and the Blond Race," in *The Journal of Race Development*, April, 1913.

appreciable and increasing admixture of the blond may be conceived to have ranged across the breadth of Europe, particularly in the lowlands, in the belt of damp and cool country that fringed the ice, and to have followed the receding ice-sheet northward when the general climate of Europe began to take on its present character with the returning warmth and dryness. By force of the strict climatic limitation to which this type is subject, the blond element, and more particularly the dolicho-blond, will presently have disappeared by selective elimination from the population of those regions from which the ice-sheet and its fringe of cool and humid climate had receded. The cool and humid belt suited to the propagation of the blond mutant (and its blond hybrids) would shift northward and shorten down to the seaboard as the glacial conditions in which it had originated presently ceased. So that presently, when Europe finally lost its ice-sheet, the blond race and its characteristic hybrids would be found confined nearly within the bounds which have marked its permanent extension in historic times. These limits have, no doubt, fluctuated somewhat in response to secular variations of climate; but on the whole they appear to have been singularly permanent and singularly rigid.

Apparently after the dolicho-blond had come to occupy the restricted habitat which the stock has since continued to hold on the northern seaboard of Europe, toward the close of what is known in Danish chronology as the " older stone age," the early stock of domestic animals appear to have been introduced into Europe from Asia; the like statement will hold more doubtfully for the older staple crop plants, with the reservation that their introduction appears to antedate that of the domestic animals. At least some such date seems indicated by their first appearance in Denmark late in the period of the " kitchen middens." Virtually all of these essential elements of their material civilisation appear to have come to the blond-hybrid communities settled on the narrow Scandinavian waters, as to the rest of Europe, from Turkestan. This holds true at least for the domestic animals as a whole, the possible exceptions among the early introductions being not of great importance. Some of the early crop plants may well have come from what is now Mesopotamian or Persian territory, and may conceivably have reached western Europe appreciably earlier, without affecting the present argument. If the European horse had been domesticated in palæolithic times, as appears at least extremely probable, that technological gain appears to have been lost before the close of the palæolithic age; perhaps along with the extinction of the European horse.

These new elements of technological equipment, the crop plants and animals, greatly affected the character of the neolithic culture in Europe; visibly so as regards the region presumably occupied by the dolicho-blond,— or the blond-hybrid peoples. On the material side of the community's life they would bring change direct and immediate, altering the whole scheme of ways and means and shifting the pursuit of a livelihood to new lines; and on the immaterial side their effect would be scarcely less important, in that the new ways and means and the new manner of life requisite and induced by their use would bring on certain new institutional features suitable to a system of mixed farming. Whatever may have been the manner of their introduction, whether they were transmitted peaceably by insensible diffusion from group to group or were carried in with a high hand by a new intrusive population that overran the country and imposed its own cultural scheme upon the Europeans along with the new ways and means of life,— in any case these new cultural elements will have spread over the face of Europe somewhat gradually and will have reached the blond-hybrid communities in their remote corner of the continent only after an appreciable lapse of time. Yet, it is to be noted, it is after all relatively early in neolithic times that certain of the domestic plants and animals first come into evidence in the Scandinavian region.

The crop plants appear to have come in earlier than the domestic animals, being perhaps brought in by the peoples of the Mediterranean race at their first occupation of Europe in late quaternary time. With tillage necessarily goes a sedentary manner of life. So that at their first introduction the domestic animals were intruded into a system of husbandry carried on by a population living in settled communities, and drawing their livelihood in great part from the tilled ground but also in part from the sea and from the game-bearing forests that covered much of the country at that time. It was into such a situation that the domestic animals were intruded on their first coming into Europe,— particularly into the seaboard region of north Europe.

On the open ranges of western and central Asia, from which these domestic animals came, and even in the hill country of that general region, the peoples that draw their livelihood from cattle and sheep are commonly of a nomadic habit of life, in the sense that the requirements of forage for their herds and flocks hold them to an unremitting round of seasonal migration. It results that, except in the broken hill country, these peoples habitually make use of movable habitations, live in camps rather than in settled, sedentary communities. Certain peculiar institutional arrangements also result from this nomadic manner of life associated with the care of flocks and herds on a large scale. But on their introduction into Europe the domestic animals appear on the whole not to have supplanted tillage and given rise to such a nomadic-pastoral scheme of life, exclusively given to cattle raising, but rather to have fallen into a system of mixed farming which combined tillage with a sedentary or quasi-sedentary grazing industry. Such particularly appears to have been the case in the seaboard region of the north, where there is no evidence of tillage having been displaced by a nomadic grazing industry. Indeed, the small-scale and broken topography of this European region has never admitted a large-scale cattle industry, such as has prevailed on the wide Asiatic ranges. An exception, at least partial and circumscribed, may perhaps be found in the large plains of the extreme Southeast and in the Danube valley; and it appears also that grazing, after the sedentary fashion, took precedence of tillage in prehistoric Ireland as well as here and there in the hilly countries of southern and central Europe.

Such an introduction of tillage and grazing would mean a revolutionary change in the technology of the European stone age, and a technological revolution of this kind will unavoidably bring on something of a radical change in the scheme of institutions under which the community lives; primarily in the institutions governing the details of its economic life, but secondarily also in its domestic and civil relations. When such a change comes about through the intrusion of new material factors the presumption should be that the range of institutions already associated with these material factors in their earlier home will greatly influence the resulting new growth of institutions in the new situation, even if circumstances may not permit these alien institutions to be brought in and put into effect with the scope and force which they may have had in the culture out of which they have come. Some assimilation is to be looked for even if circumstances will not permit the adoption of the full scheme of institutions, and the institutions originally associated with the intrusive technology will be found surviving with least loss or qualification in those portions of the invaded territory where the invaders have settled in force, and particularly where conditions have permitted them to retain something of their earlier manner of life.

The bringers of these new elements of culture, material and immaterial, had acquired what they brought with them on the open sheep and cattle ranges of the central-Asiatic plains and uplands,— as is held to be the unequivocal testimony of the Aryan speech, and as is borne out by the latest explorations in that region. These later explorations indicate west-central Turkestan as the probable center of the domestication and diffusion of the animals, if not also of the crop plants, that have stocked Europe. Of what race these bearers of the new technology and culture may have been, and just what they brought into Europe, is all a matter of inference and surmise. It was once usual to infer, as a ready matter of course, that these immigrant pastoral nomads from the Asiatic uplands were "Aryans," "Indo-Europeans," "Indo-Germans," of a predominantly blond physique. But what has been said above as well as in the earlier paper referred to comes near excluding the possibility of these invaders being blonds, or more specifically the dolicho-blond. It is, of course, conceivable, with Keane (if his speculations on this head are to be taken seriously), that a fragment of the alleged blond race from Mauretania may have wandered off into Turkestan by way of the Levant, and so may there have acquired the habits of a pastoral life, together with the Aryan speech and institutions, and may

then presently have carried these cultural factors into
Europe and imposed them on the European population,
blond and brunet. But such speculations, which once
were allowable though idle, have latterly been put out of
all question, at least for the present, by the recent Pum-
pelly explorations in Turkestan. It is, for climatic reasons, extremely improbable that any blond stock should
have inhabited any region of the central-Asiatic plains or
· uplands long enough to acquire the pastoral habits of life
and the concomitant Aryan speech and institutions, and it
is fairly certain that the dolicho-blond could not have survived for that length of time under the requisite conditions of climate and topography.

It is similarly quite out of the question that the dolicho-
blond, arising as a mutant type late in quaternary time,
should have created the Aryan speech and culture in Europe, since neither the archæological evidence nor the
known facts of climate and topography permit the hypothesis that a pastoral-nomadic culture of home growth
has ever prevailed in Europe on a scale approaching that
required for such a result. And there is but little more
possibility that the bringers of the new (Aryan) culture
should have been of the Mediterranean race; although the
explorations referred to make it nearly certain that the
communities which domesticated the pastoral animals
(and perhaps the crop-plants) in Turkestan were of that
race. The Mediterranean race originally is Hamitic, not
Aryan, it is held by men competent to speak on that matter, and the known (presumably) Mediterranean prehistoric settlements in Turkestan, at Anau, are moreover
obviously the settlements of a notably sedentary people
following a characteristically peaceable mode of life.
The population of these settlements might of course conceivably have presently acquired the nomadic and preda-

tory habits reflected by the Aryan speech and institutions, but there is no evidence of such an episode at Anau, where the finds show an uninterrupted peaceable and sedentary occupation of the sites throughout the period that could come in question. The population of the settlements at Anau could scarcely have made such a cultural innovation, involving the adoption of an alien language, except under the pressure of conquest by an invading people; which would involve the subjection of the peaceable communities of Anau and the incorporation of their inhabitants as slaves or as a servile class in the predatory organisation of their masters. The Mediterranean people of Anau could accordingly have had a hand in carrying this pastoral-predatory (Aryan) culture into the West only as a subsidiary racial element in a migratory community made up primarily of another racial stock.

This leaves the probability that an Asiatic stock, without previous settled sedentary habits of life, acquired the domesticated animals from the sedentary and peaceable communities of Anau, or from some similar village (pueblo) or villages of western Turkestan, and then through a (moderately) long experience of nomadic pastoral life acquired also the predatory habits and institutions that commonly go with a pastoral life on a large scale. These cultural traits they acquired in such a degree of elaboration and maturity as is implied by the primitive Aryan (or, better, proto-Aryan) speech, including a more or less well developed patriarchal system; so that they would presently become a militant and migratory community somewhat after the later-known Tatar fashion, and so made their way westward as a self-sufficient migratory host and carried the new material culture into Europe together with the alien Aryan speech. It is at the same time almost unavoidable that in such an event this migratory host would have carried with them into the West an appreciable servile contingent made up primarily of enslaved captives from the peaceable agricultural settlements of the Mediterranean race, which had originally supplied them with their stock of domestic animals.

Along with these new technological elements and the changes of law and custom which their adoption would bring on, there will also have come in the new language that was designed to describe these new ways and means of life and was adapted to express the habits of thought which the new ways and means bred in the peoples that adopted them. The immigrant pastoral (proto-Aryan) language and the pastoral (patriarchal and predatory) law and custom will in some degree have been bound up with the technological ways and means out of which they arose, and they would be expected to have reached and affected the various communities of Europe in somewhat the same time and the same measure in which these material facts of the pastoral life made their way among these peoples. In the course of the diffusion of these cultural elements, material and immaterial, among the European communities the language and in a less degree the domestic and civil usages and ideals bred by the habits of the pastoral life might of course come to be dissociated from their material or technological basis and might so be adopted by remoter peoples who never acquired any large measure of the material culture of those pastoral nomads whose manner of life had once given rise to these immaterial features of Aryan civilisation.

Certain considerations going to support this far-flung line of conjectural history may be set out more in detail: (a) The Aryan civilisation is of the pastoral type, with such institutions, usages and preconceptions as a large-scale pastoral organisation commonly involves. Such is said by competent philologists to be the evidence of the primitive Aryan speech. It is substantially a servile organisation under patriarchal rule, or, if the expression be preferred, a militant or predatory organisation; these alternative phrases describe the same facts from different points of view. It is characterised by a well-defined system of property rights, a somewhat pronounced subjection of women and children, and a masterful religious system tending strongly to monotheism. A pastoral culture on the broad plains and uplands of a continental region, such as west-central Asia, will necessarily fall into some such shape, because of the necessity of an alert and mobile readiness for offense and defense and the consequent need of soldierly discipline. Insubordination, which is the substance of free institutions, is incompatible with a prosperous pastoral-nomadic mode of life. When worked out with any degree of maturity and consistency the pastoral-nomadic culture that has to do with sheep and cattle appears always to have been a predatory, and therefore a servile culture, particularly when drawn on the large scale imposed by the topography of the central-Asiatic plains, and reënforced with the use of the horse. (The reindeer nomads of the arctic seaboard may appear to be an exception, at least in a degree, but they are a special case, admitting a particular explanation, and their case does not affect the argument for the Aryan civilisation.) The characteristic and pervasive human relation in such a culture is that of master and servant, and the social (domestic and civil) structure is an organisation of graded servitude, in which no one is his own master but the overlord, even nominally. The family is patriarchal, women and children are in strict tutelage, and discretion vests in the male head alone. If the group grows large its civil institutions are of a like coercive character, it commonly shows a rigorous tribal organisation, and in the end, with the help of warlike experience, it almost unavoidably becomes a despotic monarchy.

It has not been unusual to speak of the popular institutions of Germanic paganism — typified, *e.g.*, by the Scandinavian usages of local self-government in pagan times — as being typically Aryan institutions, but that is a misnomer due to uncritical generalisation guided by a chauvinistic bias. These ancient north-European usages are plainly alien to the culture reflected by the primitive Aryan Speech, if we are to accept the consensus of the philological ethnologists to the effect that the people who used the primitive Aryan speech must have been a community of pastoral nomads inhabiting the plains and uplands of a continental region. That many of these philological ethnologists also hold to the view that these Aryans were north-European pagan blonds may raise a personal question of consistency but does not otherwise touch the present argument.

(b) A racial stock that has ever been of first-rate consequence in the ethnology of Europe (the Alpine, brachycephalic brunet, the *homo alpinus* of the Linnean scheme) comes into Europe at this general period, from Asia; and this race is held to have presently made itself at home, if not dominant, throughout middle Europe, where it has in historic times unquestionably been the dominant racial element.

(c) The pastoral-nomadic institutions spoken of above appear to have best made their way in those regions of Europe where this brachycephalic brunet stock has been present in some force if not as a dominant racial factor. The evidence is perhaps not conclusive, but there is at least a strong line of suggestion afforded by the distribution of the patriarchal type of institutions within Europe, including the tribal and gentile organisation. There is a rough concomitance between the distribution of these cultural elements presumably derived from an Aryan source on the one hand, and the distribution past or present of the brachycephalic brunet type on the other hand. The regions where this line of institutions are known to have prevailed in early times are, in the main, regions in which the Alpine racial type is also known to have been present in force, as, *e.g.,* in the classic Greek and Roman republics.

At the same time a gentile organisation seems also to have been associated from the outset with the Mediterranean racial stock and may well have been comprised in the institutional furniture of that race as it stood before the advent of the Alpine stock; but the drift of later inquiry and speculation on this head appears to support the view that this Mediterranean gentile system was of a matrilinear character, such as is found in many extant agricultural communities of the lower barbarian culture, rather than of a patriarchal kind, such as characterises the pastoral nomads. The northern blond communities alone appear, on the available evidence, to have had no gentile or tribal institutions, whether matrilinear or patriarchal. The classic Greek and Roman communities appear originally to have been of the Mediterranean race and to have always retained a broad substratum of the Mediterranean stock as the largest racial element in their population, but the Alpine stock was also largely represented in these communities at the period when their tribal and gentile institutions are known to have counted for much, as, indeed, it has continued ever since.

Apart from these communities of the Mediterranean seaboard, the peoples of the Keltic culture appear to have

had the tribal and gentile system, together with the patriarchal family, in more fully developed form than it is to
be found in Europe at large. The peoples of Keltic
speech are currently believed by ethnologists to have originally been of a blond type, although opinions are not altogether at one on that head,— the tall, perhaps red-haired,
brachycephalic blond, the " Saxon " of Beddoe, the " Oriental " of Deniker. But this blond type is perhaps best
accounted for as a hybrid of the dolicho-blond crossed on
the Alpine brachycephalic brunet. Some such view of its
derivation is fortified by what is known of the prehistory
and the peculiar features of the early Keltic culture.
This culture differs in some respects radically from that
of the dolicho-blond communities, and it bears more of a
resemblance to the culture of such a brunet group of
peoples as the early historic communities of upper and
middle Italy. If the view is to be accepted which is coming into currency latterly, that the Keltic is to be affiliated
with the culture of Hallstatt and La Tène, such affiliation
will greatly increase the probability that it is to be counted
as a culture strongly influenced if not dominated by the
Alpine stock. The Hallstatt culture, lying in the valley
of the Danube and its upper affluents, lay in the presumed
westward path of immigration of the Alpine stock; its
human remains are of a mixed character, showing a
strong admixture of the brachycephalic brunet type; and
it gives evidence of cultural gains due to outside influence
in advance of the adjacent regions of Europe. This Keltic culture, then, as known to history and prehistory, runs
broadly across middle Europe along the belt where blond
and brunet elements meet and blend; and it has some of
the features of that predatory-pastoral culture reflected
by the primitive Aryan speech, in freer development, or
in better preservation, than the adjacent cultural regions

to the north; at the same time the peoples of this Keltic culture show more of affiliation to or admixture with the brachycephalic brunet than the other blond-hybrid peoples do.

On the other hand the communities of dolicho-blond hybrids on the shores of the narrow Scandinavian waters, remote from the centers of the Alpine culture, show little of the institutions peculiar to a pastoral people. These dolicho-blond hybrids of the North come into history at a later date, but with a better preserved and more adequately recorded paganism than the other barbarians of Europe. The late-pagan Germanic-Scandinavian culture affords the best available instance of archaic dolicho-blond institutions, if not the sole instance; and it is to be noted that among these peoples the patriarchal system is weak and vague,— women are not in perpetual tutelage, the discretion of the male head of the household is not despotic nor even unquestioned, children are not held under paternal discretion beyond adult age, the patrimony is held to no clan liabilities and is readily divisible on inheritance, and so forth. Neither is there any serious evidence of a tribal or gentile system among these peoples, early or late, nor are any of them, excepting the late and special instance of the Icelandic colony, known ever to have been wholly or mainly of pastoral habits; indeed, they are known to have been without the pastoral animals until some time in the neolithic period. The only dissenting evidence on these heads is that of the Latin writers, substantially Cæsar and Tacitus, whose testimony is doubtless to be thrown out as incompetent in view of the fact that it is supported neither by circumstantial evidence nor by later and more authentic records. In speaking of " tribes " among the Germanic hordes these Latin writers are plainly construing Germanic facts in Roman terms, very much as the Spanish writers of a later day construed Mexican and Peruvian facts in mediæval-feudalistic terms,— to the lasting confusion of the historians; whereas in enlarging on the pastoral habits of the Germanic communities they go entirely on data taken from bodies of people on the move and organised for raiding, or recently and provisionally settled upon a subject population presumably of Keltic derivation or of other alien origin and inhabiting the broad lands of middle Europe remote from the permanent habitat of the dolicho-blond. Great freedom of assumption has been used and much ingenuity has been spent in imputing a tribal system to the early Germanic peoples, but apart from the sophisticated testimony of these classical writers there is no evidence for it. The nearest approach to a tribal or a gentile organisation within this culture is the " kin," which counts for something in early Germanic law and custom; but the kin is far from being a gens or clan, and it will be found to have more of the force of a clan organisation the farther it has strayed from the Scandinavian center of diffusion of the dolicho-blond and the more protracted the warlike discipline to which the wandering host has been exposed. All these properly Aryan institutions are weakest or most notably wanting where the blond is most indubitably in evidence.

Taking early Europe as a whole, it will appear that among the European peoples at large institutions of the character reflected in the primitive Aryan speech and implied in the pastoral-nomadic life evidenced by the same speech are relatively weak, ill-defined or wanting, arguing that Europe was never fully Aryanised. And the peculiar geographical and ethnic distribution of this Aryanism of institutions argues further that the dolicho-blond culture of the Scandinavian region was less profoundly affected by the Aryan invasion than any other equally well known section of Europe. What is known of this primitive Aryan culture, material, domestic, civil and religious, through the Sanskrit and other early Asiatic sources, may convincingly be contrasted with what is found in early Europe. These Asiatic records, which are our sole dependence for a competent characterisation of the Aryan culture, shows it to have resembled the culture of the early Hebrews or that of the pastoral Turanians more closely than it resembles the early European culture at large, and greatly more than it resembles the known culture of the early communities of dolicho-blond hybrids.

(d) Scarcely more conclusive, but equally suggestive, is the evidence from the religious institutions of the Aryanised Europeans. As would be expected in any predatory civilisation, such as the pastoral-nomadic cultures typically are, the Aryan religious system is said to have leaned strongly toward a despotic monarchical form, a hierarchically graded polytheism, culminating in a despotic monotheism. There is little of all this to be found in early pagan Europe. The nearest well-known approach to anything of the kind is the late-Greek scheme of Olympian divinities with Zeus as a doubtful suzerain,— known through latter-day investigations to have been superimposed on an earlier cult of a very different character. The Keltic (Druidical) system is little known, but it is perhaps not beyond legitimate conjecture, on the scant evidence available, that this system had rather more of the predatory, monarchical-despotic cast than the better known pagan cults of Europe. The Germanic paganism, as indicated by the late Scandinavian — which alone is known in any appreciable degree — was a lax polytheism which imputed little if any coercive power to the highest god, and which was not taken so very seriously anyway by the "worshipers,"— if Snorri's virtually exclusive account is to be accepted without sophistication. The evidence accorded by the religious cults of Europe yields little that is conclusive, beyond throwing the whole loose-jointed, proliferous European paganism out of touch with anything that can reasonably be called Aryan. And this in spite of the fact that all the available evidence is derived from the European cults as they stood after having been exposed to long centuries of Aryanisation. So that it may well be held that such systematisation of myths and observances as these European cults give evidence of, and going in the direction of a despotic monotheism, is to be traced to the influence of the intrusive culture of the Aryan or Aryanised invaders,— as is fairly plain in the instance of the Olympians.

(e) That the languages of early Europe, so far as known, belong almost universally to the Aryan family may seem an insurmountable obstacle to the view here spoken for. But the difficulties of the case are not appreciably lessened by so varying the hypothesis as to impute the Aryan speech to the dolicho-blond, or to any blond stock, as its original bearer. Indeed, the difficulties are increased by such an hypothesis, since the Aryan-speaking peoples of early times, as of later times, have in the main been communities made up of brunets without evidence of a blond admixture, not to speak of an exclusively blond people. (There is no evidence of the existence of an all-blond people anywhere, early or late.)

The early European situation, so far as known, offers no exceptional obstacles to the diffusion of an intrusive language. Certain mass movements of population, or rather mass movements of communities shifting their ground by secular progression, are known to have taken place, as, *e.g.,* in the case of the Hallstatt-La Tène-Keltic culture moving westward on the whole as it gained ground and spread by shifting and ramification outward from its first-known seat in the upper Danube valley. All the while, as this secular movement of growth, ramification and advance was going on, the Hallstatt-La Tène-Keltic peoples continued to maintain extensive trade relations with the Mediterranean seaboard and the Ægean on the one side and reaching the North-Sea littoral on the other side. In all probability it is by trade relations of this kind — chiefly, no doubt, through trade carried on by itinerant merchants — that the new speech made its way among the barbarians of Europe; and it is no far-fetched inference that it made its way, in the North at least, as a trade jargon. All this accords with what is going on at present under analogous circumstances. The superior merit by force of which such a new speech would make its way need be nothing more substantial than a relatively crude syntax and phonetics — such as furthers the dissemination of English to-day in the form of Chinook jargon, Pidgin English, and Beach la Mar. Such traits, which might in some other light seem blemishes, facilitate the mutilation of such a language into a graceless but practicable trade jargon. With jargons as with coins the poorer (simpler) drives out the better (subtler and more complex). A second, and perhaps the chief, point of superiority by virtue of which a given language makes its way as the dominant factor in such a trade jargon, is the fact that it is the native language of the people who carry on the trade for whose behoof the jargon is contrived. The traders, coming in contact with many men, of varied speech, and carrying their varied stock of trade goods, will impose their own names for the articles bartered and so contribute that much to the jargon vocabulary,— and a jargon is at its inception little more than a vocabulary.

The traders at the same time are likely to belong to the people possessed of the more efficient technology, since it is the superior technology that commonly affords them their opportunity for advantageous trade; hence the new or intrusive words, being the names of new or intrusive facts, will in so far find their way unhindered into current speech and further the displacement of the indigenous language by the jargon.

Such a jargon at the outset is little else than a vocabulary comprising names for the most common objects and the most tangible relations. On this simple but practicable framework new varieties of speech will develop, diversified locally according to the kind and quantity of materials and linguistic tradition contributed by the various languages which it supplants or absorbs.

In so putting forward the conjecture that the several forms of Aryan speech have arisen out of trade jargons that have run back to a common source in the language of an intrusive proto-Aryan people, and developing into widely diversified local and ethnic variants according as the mutilated proto-Aryan speech (vocabulary) fell into the hands of one or another of the indigenous barbarian peoples,— in this suggestion there is after all nothing substantially novel beyond giving a collective name to facts already well accepted by the philologists. Working backward analytically step by step from the mature results given in the known Aryan languages they have discovered and divulged — with what prolixity need not be alluded to here — that in their beginnings these several idioms were little else than crude vocabularies covering the commonest objects and most tangible relations, and that by time-long use and wont the uncouth strings of vocables whereby the beginners of these languages sought to express themselves have been worked down through a stupendously elaborate fabric of prefixes, infixes and suffixes, etc., etc., to the tactically and phonetically unexceptionable inflected languages of the Aryan family as they stood at their classical best. And what is true of the European languages should apparently hold with but slight modification for the Asiatic members of the family. These European idioms are commonly said to be, on the whole, less true to the pattern of the inferentially known primitive Aryan than are its best Asiatic representatives; as would be expected in case the latter were an outgrowth of jargons lying nearer the center of diffusion of the proto-Aryan speech and technology.

As regards the special case of the early north-European communities of dolicho-blond hybrids, the trade between the Baltic and Danish waters on the one hand and the Danube valley, Adriatic and Ægean on the other hand is known to have been continued and voluminous during the neolithic and bronze ages,— as counted by the Scandinavian chronology. In the course of this traffic, extending over many centuries and complicated as it seems to have been with a large infiltration of the brachycephalic brunet type, much might come to pass in the way of linguistic substitution and growth.

AN EARLY EXPERIMENT IN TRUSTS [1]

According to Much,[2] following in the main the views of Penka, Wilser, De Lapouge, Sophus Müller, Andreas Hansen, and other spokesmen of the later theories touching Aryan origins, the area of characterisation of the West-European culture, as well as of that dolicho-blond racial stock that bears this culture, is the region bordering on the North Sea and the Baltic, and its center of diffusion is to be sought on the southern shores of the Baltic. This region is in a manner, then, the primary focus of that culture of enterprise that has reshaped the scheme of life for mankind during the Christian era. Its spirit of enterprise and adventure has carried this race to a degree of material success that is without example in history, whether in point of the extent or of the scope of its achievements. Up to the present the culminating achievement of this enterprise is dominion in business, and its most finished instrument is the quasi-voluntary coalition of forces known as a Trust.

In its method and outward form this enterprise of the Indo-germanic racial stock has varied with the passage of time and the change of circumstances; but in its spirit and objective end it has maintained a singularly consistent character through all the mutations of name and external circumstance that have passed over it in the course of history.

In its earlier, more elemental expression this enterprise

[1] Reprinted by permission from *The Journal of Political Economy*, Vol. XII, March, 1904.

[2] Matthaeus Much, *Die Heimat der Indogermanen.*

takes the form of raiding, by land and sea. A shrewd interpretation might, without particular violence to the facts, find a coalition of forces of the kind which is later known as a Trust in the Barbarian raids spoken of as the *Völkerwanderung*. Such an interpretation would seem remote, however, and not particularly apt. The beginnings of a *bona fide* trust enterprise are of a more businesslike character and have left a record more amenable to the tests of accountancy. A trust, as that term is colloquially understood, is a business organisation.

Now, the line of enterprise, of indigenous growth in the north-European cultural region, which first falls into settled shape as an orderly, organised business is the traffic of those seafaring men of the North known to fame as the Vikings. And it is in this traffic, so far as the records show, that a trust, with all essential features, is first organised. The term "viking" covers, somewhat euphemistically, two main facts: piracy and slave-trade. Without both of these lines of business the traffic could not be maintained in the long run; and both, but more particularly the latter, presume, as an indispensable condition to their successful prosecution, a regular market and an assured demand for the output. It is a traffic in which, in order to get the best results, a relatively large initial investment must be sunk, and the period of turn-over — the "period of production" — is necessarily of some duration; the risk is also considerable. Further, certain technological prerequisites must be met, in the way particularly of shipbuilding, navigation, and the manufacture of weapons; an adequate accumulation of capital goods must be had, coupled with a sagacious spirit of adventure; there must also be an available supply of labor. There appears to have been a concurrence of all these circumstances, together with favorable market conditions, in the south-Baltic region from about the sixth century onward; the circumstances apparently growing gradually more favorable through the succeeding four centuries.

The viking trade appears to have grown up gradually on the Baltic seaboard, as well as in the Sound country and throughout the fjord region of Norway, as a by-occupation of the farming population. Its beginnings are earlier than any records, so that the earliest traditions speak of it as an institution well understood and fully legitimate. The well-to-do freehold farmers, including some who laid claim to the rank of *jarl,* seem to have found it an agreeable and honorable diversion, as well as a lucrative employment for their surplus wealth and labor supply. From such sporadic and occasional beginnings it passed presently into an independently organised and self-sustaining line of business enterprise, and in the course of time it attained a settled business routine and a defined code of professional ethics. Syndication, of a loose form, had begun as early as the oldest accounts extant, but it is evident from the way in which the matter is spoken of that combination had not at that date — say, about the beginning of the ninth century — long been the common practice. It was not then a matter of course. The early combinations were relatively small and transient. They took the form of " gentlemen's agreements," pools, working arrangements, division of territory, etc., rather than hard and fast syndicates. In those early days a combine would be formed for a season between two or more capitalist-undertakers, for the most part employing their own capital only, without recourse to credit; although credit arrangements occur quite early, but are not very common in the earlier recorded phases of the trade. Such a loose combine, say about the middle of the ninth century, might comprise from two to a dozen boats. What may be called the normal unit in the trade at that time was a boat of perhaps thirty tons' burden, with an effective crew of some eighty men. Boats and crews gradually increase both in size and efficiency for a century and a half after that time.

Syndication, of an increasingly close texture and increasingly permanent effect, appears to have rapidly grown in favor through the ninth and tenth centuries. The reasons for this movement of coalition are plain. The volume of the trade, as well as its territorial extension, increased uninterruptedly. The technique of the trade was gradually improved, and the equipment and management were improved and reduced to standard forms. The tonnage employed at any given time can, of course, not be ascertained with anything like a confident approximation; but its steady increase is unmistakable. Year by year the boats and crews increase in average size as well as in number, until by the middle of the tenth century the number of men and ships engaged, as well as the volume of capital invested in the trade, are probably larger than the corresponding figures for any other form of lucrative enterprise at that time. It is, at that time, altogether the best-organised line of enterprise in the West-European region in respect of its business management, and the most efficient and progressive in respect of its equipment and technology. At a conservative guess, the aggregate number of ships engaged about the middle of the tenth century must have appreciably exceeded six hundred, and may have reached one thousand; with crews which had also grown gradually larger until they may by this time have averaged 150 or 200 men. There was consequently what would in

modern phrase be called an " overproduction " of pirat-
ical craft — overinvestment in the viking trade and consequent cut-throat competition. The various coalitions
came into violent conflict, and many of them went under,
with great resultant loss of capital, impoverishment of
well-to-do families, hardship and demoralisation of the
entire trade.

Added to these untoward conditions within the trade
was the open disfavor of the crown, in each of the three
Scandinavian kingdoms. The traffic had long passed out
of the stage at which it had offered a lucrative opening
for farmers' sons who were tired of the farm and eager
to find excitement, reputation, and creature comforts in
that wider human contact and busier life for which the
tedium of the farm had sharpened their appetites. The
larger capitalists alone could succeed as organisers or
directors of a viking concern under the changed conditions. The common run of well-to-do farmers had
neither the tangible assets nor the " good-will " requisite
to the successful promotion of a new company of freebooters. At the best, their sons could enter the business
only as employees and with but a very uncertain outlook
to speedy promotion to an executive position. On the
other hand, as the trade became better organised in
stronger hands, with a larger equipment, and as the competition within the trade grew more severe, the black-
mail from which much of the profits of the trade was
drawn grew more excessive and more uncertain, both as
to its amount and as to the manner and incidents with
which it was levied. As competition grew severe and
the small vikings practically disappeared, and as the
demoralisation that goes with cut-throat competition set
in, the livelihood of the common people, at whose expense
the vikings lived, grew progressively more precarious, and

even their domestic peace and household industry grew insecure. Popular sentiment was running strongly against the whole traffic. So much so, indeed, as to threaten the tenure of courts and sovereigns if the popular hardship incident to the continuance of the trade were not abated.

The politicians, therefore, made a strenuous show of effort to regulate, or even to repress, the viking organisations. Outright and indiscriminate repression was scarcely a feasible remedy, certainly not an agreeable one. The viking companies were a source of strength to the country, both in that they might be drawn on for support in case of war and in that they brought funds into the country. The remedy to which the politicians turned, by preference, therefore, was a regulation of the companies in such a manner as to let " the foreigners pay the tax," to adapt a modern phrase. If the freebooters of a given state could be induced, by stringent regulations, to prey upon the people of the neighboring states, and particularly if they worked at cross-purposes with similar companies of freebooters domiciled in such neighboring states, it was then plain to the sagacious politicians of those days that the companies might be more of a blessing than a curse. On trial it was found that this policy of control gave at the best but very dubious results, and consequently the repressive hand of the authorities perforce fell with increasingly rigorous pressure on the viking organisations, particularly on the smaller ones which were scarcely of national importance. The competition in the trade was too severe to admit of a consistent avoidance of excesses and irregularities on the part of the vikings, and these irregularities obliged the authorities to interfere.

Under these circumstances it is plain that no viking

combine could hope to prosper in the long run unless it were strong enough to take an international position and to maintain a practical monopoly of the trade. " International " in these premises means within the Scandinavian countries. In the days of its finest development the viking trade was domiciled in the Scandinavian countries, almost exclusively. This means the two Scandinavian peninsulas, with Iceland, the Faroes, Orkneys, Hebrides, and the Scandinavian portions of Scotland. To this, for completeness of statement, is to be added a stretch of Wendish seaboard on the south of the Baltic and a negligible patch of German territory. The trade, so far as regards its home offices, to use a modern phrase, gathered in the main about two chief centers: the Orkneys and the south end of the Baltic. Outlying regions, such as the Norwegian fjord country and the Hebrides, are by no means negligible, but the two regions named above are after all the chief seats of the traffic; and of these two centers the Baltic — chiefly Danish — region is in many respects the more notable. Its viking traffic is better, more regularly organised, is carried on with a more evident sense of a solidarity of interests and a more consistent view to a long-term prosperity. As one might say, looking at the matter from the modern standpoint, it has more of a look of stability and conservative management, such as belongs to an investment business, and has less of a speculative air, than the trade that centers in the western isles.

Perhaps it is just on this account, because of its greater stability of interests and more conservative animus, that the traffic of this region responds with greater alacrity to the pressure of excessive competition and political interference, and so enters on a policy of larger and closer coalition. It may be added that many of the great

captains of adventure in this region are men of good
family and substantial standing in the community. As
may often happen in a like conjuncture, when the irksomeness of this competitive situation in the Baltic was
fast becoming intolerable, there arose a man of far-seeing
sagacity and settled principles, of executive ability and
businesslike integrity, who saw the needs of the hour and
the available remedy, and who saw at the same glance
his own opportunity of gain. This man was Pálnatoki,
the descendant of an honorable line of country gentlemen in the island of Funen, whose family had from
time immemorial borne an active and prudent part in the
trade, and had been well seen at court and in society.
He was a man of mature experience, with a large investment in the traffic, and with a body of " good-will " that
gave him perhaps his most decisive advantage.

During the reign of Harald Gormsson, about the middle of the tenth century, Pálnatoki seems to have cast
about for a basis on which to promote an international
coalition of vikings, such as would put an end to headlong competition in the trade and would at the same time
be placed above the accidents of national politics. To
this end it was necessary to find a neutral ground on
which to establish the home office of the concern. Such
a mediæval-Scandinavian New Jersey was the Wendish
kingdom at the south of the Baltic.

Jómsborg (on the island of Wollin, at the mouth of
the Oder) seems to have been a resort of vikings before
Pálnatoki organised his company there and strengthened
the harbor, which may have been fortified by those who
held it before him. Here the new company was incorporated under a special franchise from the Wendish
crown, with the stipulation that it was to do business only
outside the Wendish territories. The tangible assets of

the corporation were the harbor and fortified town of Jómsborg, together with the ships and other equipment of such vikings as were admitted to fellowship; its intangible assets were its franchise and the good-will of the promoter and the underlying companies. Its by-laws were very strict, both as to the discipline of the personnel and as to the distribution of earnings. The promoter, who was the first president of the corporation, was given extreme powers for the enforcement of the by-laws, and throughout his long incumbency of office he exercised his powers with the greatest discretion and with a most salutary effect.

This neutral, international corporation of piracy rapidly won a great prestige. In modern phrase, its intangible assets grew rapidly larger. Backed by the competitive pressure which the new corporation was able to bring upon the smaller companies and syndicates, this prestige of the Jómsvikings brought a steady run of applications for admission into the trust. The trust's policy was substantially the same as has since become familiar in other lines of enterprise, with the difference that in those early days the competitive struggle took a less sophisticated form. Outstanding syndicates and private firms were given the alternative of submission to the trust's terms or retirement from the traffic. There was great hardship among the outstanding concerns, especially among that large proportion of them that were unable to meet the scale of requirements imposed on applicants for admission into the trust. The qualifications both as to equipment and personnel were extremely strict, so that a large percentage of the applicants were excluded; and the unfortunates who failed of admission found themselves in a doubtful position that grew more precarious with every year that passed. Practically, such concerns were either frozen out of the business or forced into a liquidation which permanently wound up their affairs and terminated their corporate existence.

The accounts extant are of course not reliable in minute details, being not strictly contemporary, nor are they cast in such modern terms as would give an easy comparison with present-day facts. The chief documents in the case are *Jómsvikingasaga, Saxo Grammaticus, Heimskringla,* and *Olafssaga Tryggvasonar;* but nearly the whole of the saga literature bears on the development of the viking trade, and characteristic references to the Jómsviking trust occur throughout. The evidence afforded by these accounts converges to the conclusion that toward the close of the tenth century the trust stood in a high state of prosperity and was in a position virtually to dictate the course of the traffic for all that portion of the viking trade that centered in the Baltic. Its prestige and influence were strong wherever the traffic extended, even in the region of the western isles and in the fjord country of Norway. It had even come to be a factor of first-rate consequence in international politics, and its power was feared and courted by those two sovereigns who established the Danish rule in England, as well as by their Swedish, Norwegian, and Russian contemporaries. It is probably not an overstatement to say that the Danish conquest of England would not have been practicable except for the alliance of the trust with Svend, which enabled him to turn his attention from the complications of Scandinavian politics to his English interests.

The extent of the trust's material equipment at the height of its prosperity is a matter of surmise rather than of statistical information. Some notion of its strength may be gathered from the statement that the fortified harbor of Jómsborg included within its castellated sea-wall an inclosed basin capable of floating three hundred ships at anchor. In the great raid against the kingdom of Norway, whose failure inaugurated the disintegration of the trust, the number of ships sent out is variously given by different authorities. The *Jómsvikingasaga* says that they numbered one long hundred. This fleet, however, was made up of craft selected from among the ships that were under the immediate command of four of the great captains of adventure. The fleet, as it lay in the Sound before the final selection, is said to have numbered 185, but the context shows that this fleet was but a fraction of the aggregate Jómsviking tonnage. Of this disastrous expedition but a fraction returned; yet various later expeditions of the Jómsvikings are mentioned in which some scores of their ships took part.

The trust having become an international power, it undertook to shape the destiny of nations and dynasties, and it broke under the strain. It, or its directors, took a contract to bring Norway into subjection to the Danish crown. Partly through untoward accidents, partly through miscalculation and hurried preparations, it failed in this undertaking, which brought the affairs of the trust to a spectacular crisis. From this disaster it never recovered. With the opening of the eleventh century the viking trust fell into abeyance, and in a few years it disappeared from the field. There are several good reasons for its failure. On the death of its founder the management had passed into the hands of Sigvaldi, a man of less sagacity and less integrity as well as of more unprincipled personal ambition, and somewhat given to flighty ventures in the field of politics. It was Sigvaldi's overweening personal ambition that committed the corporation to the ill-advised expedition against Norway. The trust, moreover, being supreme within its field, the discipline grew lax and its exactions grew arbitrary, sometimes going to unprovoked excesses. As one might say, too little thought was given to "economies of production," and the charges were pushed beyond "what the traffic would bear." But for all that, in spite of its meddling in politics, and in spite of jobbery and corruption in its management, the trust still had a fair outlook for continued success, except that the bottom dropped out of the trade. For better or worse, the slave-trade in the north of Europe collapsed on the introduction of Christianity, at least so far as regards the trade in Christians; and without a slave market the viking enterprise had no chance of reasonable earnings. At the same time, the risk and hardships of the traffic — the "cost of production"— grew heavier as the countries to the south became better able to defend their shores. The passenger traffic failed almost entirely, and the goods traffic was in a disorganised and unprofitable state. The costs were fast becoming prohibitive, even to men so enterprising and necessitous as the Norwegian freebooters. The situation changed in such a way as to leave the trust out.

Some show of corporate existence was still maintained for a short period after the trust's great crisis, but there was an end of discipline and authoritative control. The minor concerns and private establishments that had once formed part of the trust continued in the trade on an independent footing, but with decreasing regularity and with diminishing strength. As the equipment wore out it was not replaced, and the trade lapsed. The great captains of the industry, like Sigvaldi, Thorkel Haraldson, Sigurd Kápa, and Vagn Akason, turned their holdings to the service of the dynastic politics which were then engaging the attention of the northern countries. Much of this body of enterprise and wealth was exhausted in working out the imperialistic schemes of expansion of Svend and Knut the Great; and what was left over shared the fortunes of the other available forces of the Scandinavian countries, being dissipated in political dissensions, extortionate government organisations, and the establishment of a church and a nobility.

THE END